The Theory of Quantum Liquids

Normal Fermi Liquids

VOLUME I

The Theory of Quantum Liquids

Normal Fermi Liquids

DAVID PINES
University of Illinois

PHILIPPE NOZIÈRES
Institut Max Von Laue-Paul Langevin

Advanced Book Program

CRC Press
Taylor & Francis Group
Boca Raton London New York

CRC Press is an imprint of the
Taylor & Francis Group, an **informa** business

Originally published in 1966 by W.A. Benjamin, Inc.

Published 1989 by Westview Press

Published 2018 by CRC Press
Taylor & Francis Group
6000 Broken Sound Parkway NW, Suite 300
Boca Raton, FL 33487-2742

CRC Press is an imprint of the Taylor & Francis Group, an informa business

Visit the Taylor & Francis Web site at
http://www.taylorandfrancis.com

and the CRC Press Web site at
http://www.crcpress.com

Library of Congress Cataloging-in-Publication Data

Nozières, P. (Philippe)
Theory of quantum liquids / Philippe Nozières and David Pines
p. cm. -- (Advanced book classics series)
Pines' name appears first on the earlier edition.
Includes bibliographies and index.
Contents: v. 1. Normal fermi liquids.
1. Quantum liquids. I. Pines, David. II. Title. III. Series.
QC174.4.N69 1988 530.1'33--dc19 88-24104
ISBN 0-201-09429-0 (H) ISBN 0-201-40774-4 (P)

ISBN 13: 978-0-201-40774-7 (pbk)

Publisher's Foreword

"Advanced Book Classics" is a reprint series which has come into being as a direct result of public demand for the individual volumes in this program. That was our initial criterion for launching the series. Additional criteria for selection of a book's inclusion in the series include:

- Its intrinsic value for the current scholarly buyer. It is not enough for the book to have some historic significance, but rather it must have a timeless quality attached to its content, as well. In a word, "uniqueness."
- The book's global appeal. A survey of our international markets revealed that readers of these volumes comprise a boundaryless, worldwide audience.
- The copyright date and imprint status of the book. Titles in the program are frequently fifteen to twenty years old. Many have gone out of print, some are about to go out of print. Our aim is to sustain the lifespan of these very special volumes.

We have devised an attractive design and trim-size for the "ABC" titles, giving the series a striking appearance, while lending the individual titles unifying identity as part of the "Advanced Book Classics" program. Since "classic" books demand a long-lasting binding, we have made them available in hardcover at an affordable price. We envision them being purchased by individuals for reference and research use, and for personal and public libraries. We also foresee their use as primary and recommended course materials for university level courses in the appropriate subject area.

The "Advanced Book Classics" program is not static. Titles will continue to be added to the series in ensuing years as works meet the criteria for inclusion which we've imposed. As the series grows, we naturally anticipate our book buying audience to grow with it. We welcome your support and your suggestions concerning future volumes in the program and invite you to communicate directly with us.

Advanced Book Classics

V.I. Arnold and A. Avez, *Ergodic Problems of Classical Mechanics*

E. Artin and J. Tate, *Class Field Theory*

Michael F. Atiyah, *K-Theory*

David Bohm, *The Special Theory of Relativity*

P.C. Clemmow and J. P. Dougherty, *Electrodynamics of Particles and Plasmas*

Ronald C. Davidson, *Theory of Nonneutral Plasmas*

P.G. deGennes, *Superconductivity of Metals and Alloys*

Bernard d'Espagnat, *Conceptual Foundations of Quantum Mechanics, 2nd Edition*

Richard Feynman, *Photon-Hadron Interactions*

Dieter Forster, *Hydrodynamic Fluctuations, Broken Symmetry, and Correlation Functions*

William Fulton, *Algebraic Curves: An Introduction to Algebraic Geometry*

Kurt Gottfried, *Quantum Mechanics*

Leo Kadanoff and Gordon Baym, *Quantum Statistical Mechanics*

I.M. Khalatnikov, *An Introduction to the Theory of Superfluidity*

George W. Mackey, *Unitary Group Representations in Physics, Probability and Number Theory*

A. B. Migdal, *Qualitative Methods in Quantum Theory*

Philippe Nozières and David Pines, *The Theory of Quantum Liquids, Volume II* - new material, 1990 copyright

David Pines and Philippe Nozières, *The Theory of Quantum Liquids, Volume I: Normal Fermi Liquids*

F. Rohrlich, *Classical Charged Particles - Foundations of Their Theory*

David Ruelle, *Statistical Mechanics: Rigorous Results*

Julian Schwinger, *Particles, Source and Fields, Volume I*

Julian Schwinger, *Particles, Sources and Fields, Volume II*

Julian Schwinger, *Particles, Sources and Fields, Volume III* - new material, 1989 copyright

Jean-Pierre Serre, *Abelian ℓ-Adic Representations and Elliptic Curves*

R.F. Streater and A.S. Wightman, *PCT Spin and Statistics and All That*

René Thom, *Structural Stability and Morphogenesis*

Vita

David Pines

Professor of Physics at the University of Illinois at Urbana-Champagne, he is affiliated with the Center for Advanced Study. He has made pioneering contributions to an understanding of many-body problems in condensed matter and nuclear physics, and to theoretical astrophysics. Editor of Addison-Wesley's *Frontiers in Physics* series and the American Physical Society's *Reviews of Modern Physics*, Dr. Pines is a member of the National Academy of Sciences and is a Fellow of the American Academy of Arts and Sciences and the American Association for the Advancement of Science. He is a past Board Chairman of the Santa Fe Institute and currently serves as Co-Chairman of the Institute's Science Board. Dr. Pines received the Eugene Feenberg Memorial Medal for Contributions to Many-Body Theory in 1985, the P.A.M. Dirac Silver Medal for the Advancement of Theoretical Physics in 1984, and the Friemann Prize in Condensed Matter Physics in 1983.

Philippe Nozières

Professor of Physics at the Collegé de France, Paris, he studied at the Ecole Normale Superieure in Paris and conducted research at Princeton University. Dr. Nozières has served as a Professor at the University of Paris and at the University of Grenoble. His research is currently based at the Laue Langevin Institute in Grenoble. A member of the Académie des Sciences, he has been awarded the Holweck Award of the French Physical Society and the Institute of Physics. Dr. Nozières' work has been concerned with various facets of the many-body problem, and his work currently focuses on crystal growth and surface physics.

Special Preface

We began writing this book at a time when field theoretical methods in statistical mechanics were expanding rapidly. Our aim was to focus on the physics which lies behind such sophisticated techniques, to describe simple physical facts in a simple language. Hence our deliberate choice of "elementary" methods in explaining such fundamental concepts as elementary excitations, their interactions and collisions, etc.... Rather than elaborating on calculations, we tried to explain *qualitative* and *unifying* aspects of an extremely broad and diversified field. Such a limited scope—albeit ambitious—probably explains why our book has retained popularity throughout the years. It is a comforting thought to evolve from a "frontier" level to a "classic" status. We hope it is not only a matter of age!

The book was originally organized in two volumes. Volume I dealt with "normal" Fermi fluids, *i.e.*, those which display no order of any type. Typical examples are liquid ^{3}He or electron liquids at temperatures above a possible superfluid transition. We discussed at length the nature of elementary excitations, the central concept of response functions, the new features brought about by the long range of Coulomb interactions in charged systems. Volume II was supposed to deal with superfluid systems, both bosons (^{4}He) and fermions (metallic superconductors); it was never completed. The main reason was a matter of timing. The year, 1965, marked an explosive growth of the work on superconductors, with such new concepts as phase coherence, the Josephson effect, etc. Things were moving fast, while our ambition was to provide a carefully thought out picture, in which concepts and methods were put in perspective. It was definitely not the appropriate time, and consequently Volume II fell into oblivion. We nevertheless had completed a long chapter on Bose condensation and liquid ^{4}He, which has been widely circulated in the community. After some hesitation, we have decided to take the opportunity of this "classic" series to publish as Volume II our text, written in 1964, as it stands. We do this partly because it contains physical concepts that have perhaps not been pursued in the detail they deserve (e.g., the interaction of elementary excitations), partly because we hope our early work will provide a perspective on the field of ^{4}He which will help the reader appreciate the subsequent evolution of ideas.

Altogether, the present volume is centered around a mean field approach, appropriately generalized in order to cope with strong coupling situations. Subsequent developments involved in a number of fluctuation dominated problems, such as critical phenomena, or the Kondo problem in magnetic alloys; these were in a process of development in 1965, and we did not consider them. We did, however, emphasize the importance of interactions between elementary excitations, "ancestor" of mode-mode coupling.

During the past twenty years, the subject of Fermi liquids has developed significantly. New Fermi liquids have been discovered (superfluid ^{3}He, dilute mixtures of ^{3}He in ^{4}He, spin polarized ^{3}He, superfluid neutron matter in neutron stars) while understanding of ^{3}He, the "archetypical" normal Fermi liquid, has been increased substantially by careful measurements of its "Landau" properties, such as transport coefficients, specific heat, and zero sound, and by neutron scattering experiments which probe the density and spin excitation spectra in the "non-Landau" regime.

Review articles which deal with these topics at a level comparable to that of this book include:

P.W. Anderson and W.F. Brinkman, Theory of Anistropic Superfluidity in ^{3}He, in the Physics of Liquid and Solid Helium, Part II, ed. by K. H. Bennemann and J. B. Ketterson, J. Wiley Pub., pp. 177-286 (1978).

A. J. Leggett, A Theoretical Description of the New Phases of Liquid ^{3}He, in the Rev. Mod. Phys. *47*, pp. 331 (1975).

G. Baym and C. J. Pethick, Landau Fermi Liquid Theory and Low Temperature Properties of Normal Liquid ^{3}He, and Low Temperature Properties of Dilute Solutions of ^{3}He in Superfluid ^{4}He, in The Physics of Liquid and Solid Helium, Part II, ed. K. H. Bennemann and J. B. Ketterson, J. Wiley Pub., pp. 1-175 (1978).

D. Pines, Excitations and Transport in Quantum Liquids, in Highlights of Condensed Matter Theory, Soc. Italiano di Fisica, pp. 580 (1985); and Can. Jour. Phys. *65*, pp. 1357 (1987).

The subject of broken symmetry in condensed matter physics represents another significant concept which has been developed considerably since the appearance of our book. A theoretical discussion of magnetic instabilities, Stoner's theory of ferromagnetism, and spin density waves may be found in the review articles in "Magnetism," ed. G. Rado and H. Suhl, Academic Press, 1963, with a general discussion of broken symmetry may be found inter alia, in "Basic Notions of Condensed Matter Physics," P. W. Anderson, Addison Wesley, 1983, as well as in the review articles on superfluid ^{3}He listed above.

Looking over the past twenty-five years since we began work on this book, the scene has changed in the field of many-body physics. What was new has become standard wisdom, while new phenomena have taken center stage. Nevertheless, the feeling of excitement remains the same—as witnessed by the upsurge of work on localizations, narrow band materials, the quantum Hall effect, heavy electron systems, high T_c superconductors, etc. All of that must be anchored on solid ground—a ground which we tried to lay down in the original edition of this book. We hope that our 1965 baby has aged well, and it gives us pleasure to "launch her" again.

David Pines
Philippe Nozières

Preface

Our aim in writing this book has been to provide a unified, yet elementary, account of the theory of quantum liquids. Strictly speaking, a quantum liquid is a spatially homogeneous system of strongly interacting particles at temperatures sufficiently low that the effects of quantum statistics are important. In this category fall liquid ^{3}He and ^{4}He. In practice, the term is used more broadly, to include those aspects of the behavior of conduction electrons in metals and degenerate semiconductors which are not sensitive to the periodic nature of the ionic potential. The conduction electrons in a metal may thus be regarded as a normal Fermi liquid, or a superfluid Fermi liquid, depending on whether the metal in question is normal or superconducting.

While the theory of quantum liquids may be said to have had its origin some twenty-five years ago in the classic work of Landau on ^{4}He, it is only within the past decade that it has emerged as a well-defined subfield of physics. Thanks to the work of many people, we possess today a unified point of view and a language appropriate for the description of many-particle systems. We understand where elementary excitations afford an apt description and where they do not; we appreciate the relationship between quasiparticle excitations and collective modes, and how both derive from the basic interactions of the system particle. There now exists a number of model solutions for the many-body problem, solutions which can be shown to be valid for a given class of particle interactions and system densities: examples are an electron system at high densities and low temperatures, and a dilute boson system at low temperatures. In addition, there is a semiphenomenological theory, due to Landau, which describes the macroscopic behavior of an arbitrary normal Fermi liquid at low temperatures. Finally, and most important, has been the development of a successful microscopic theory of superconductivity by Bardeen, Cooper, and Schrieffer.

These developments have profoundly altered the main lines of research in quantum statistical mechanics: it has changed from the study of dilute, weakly interacting gases to an investigation of quantum liquids in which the interaction between particles plays an essential role. The resulting body of theory has developed to the point that it should be possible to present a coherent account of quantum liquids for the non-specialist, and such is our aim.

In writing this book, we have had three sorts of readers in mind:

(i) Students who have completed the equivalent of an undergraduate physics major, and have taken one year of a graduate course in quantum mechanics.

(ii) Experimental physicists working in the fields of low-temperature or solid-state physics.

(iii) Theoretical physicists or chemists who have not specialized in many-particle problems.

Our book is intended both as a text for a graduate course in quantum statistical mechanics or low temperature theory and as a monograph for reference and self-study. The reader may be surprised by its designation as a text for a course on statistical mechanics, since a perusal of the table of contents shows few topics that are presently included in most such courses. In fact, we believe it is time for extension of our teaching of statistical mechanics, to take into account all that we have learned in the past decade. We hope that this book may prove helpful in that regard and that it may also prove useful as a supplementary reference for an advanced course in solid-state physics.

We have attempted to introduce the essential physical concepts with a minimum of mathematical complexity; therefore, we have not made use of either Green's functions or Feynman diagrams. We hope that their absence is compensated for by our book being more accessible to the experimentalist and the nonspecialist. Accounts of field-theoretic methods in many-particle problems may be found in early books by the authors [D. Pines, *The Many-Body Problem*, Benjamin, New York (1962), P. Nozières, *The Theory of Interacting Fermi Systems*, Benjamin, New York (1963)] and in L. P. Kadanoff and G. Baym, *Quantum Statistical Mechanics*, Benjamin, New York (1962), and by A. A. Abrikosov, L. P. Gor'kov, and I. E. Dzyaloshinski, *Methods of Quantum Field Theory in Statistical Physics*, Prentice-Hall, New York (1964), to mention but a few reference works.

The decision to publish the book in two volumes stems, in part, from its length, and in part, from the natural division of quantum liquids into two classes, normal and superfluid. A third and perhaps controlling factor has been that a single volume would have meant a delay in publication of the present material of well over a year.

Although our book is a large one, we have not found it possible to describe all quantum liquids, or every aspect of the behavior of a given liquid. For example, we have not included a description of nuclear matter, of phase transitions, or of variational calculations of the ground state of various many-particle systems. On the other hand, we have compared theory with experiment in a number of places and, where appropriate, have compared and contrasted the behavior of different quantum liquids.

We have chosen to begin the book with the Landau theory of a neutral Fermi liquid in order to illustrate, in comparatively elementary fashion, the way both quantum statistics and particle interaction determine system behavior. We next consider the description of an arbitrary quantum liquid; we discuss the mathematical theory of linear response and correlations, which establishes the language appropriate for that description. We then go on to discuss, in Volume I, charged Fermi liquids, and in Volume II, the superfluid Bose liquid and superconductors.

The authors began work on this book in Paris, at the Laboratoire de Physique of the Ecole Normale Superiéure, in the fall of 1962, when one of us was on leave from the University of Illinois. Since then we worked on the book both in Paris and at the University of Illinois. We should like to thank Professor Yves Rocard, of the Université de Paris, and Professor G. M. Almy, head of the Physics Department at the University of Illinois, for their support and encouragement. One of us (D.P.) would also like to thank the John Simon Guggenheim Memorial Foundation for their support during 1962 and 1963, and the Army Research Office (Durham) for their support during 1963 and 1964.

We should like to express our gratitude to the many friends and colleagues to whom we have turned for advice and discussion, and particularly to Professor John Bardeen for his advice and encouragement. We are deeply indebted to Dr. Conyers Herring for his careful reading of a preliminary version of Chapter 1, and to Professor Gordon Baym, who read carefully the entire manuscript and whose comments have improved both the accuracy and the clarity of our presentation. We owe an especial debt of gratitude to Dr. Odile Betbeder-Matibet, who has been of substantial assistance in the correction of the proof, and to Mme. M. Audouin, who has helped in the preparation of the index.

David Pines
Philippe Nozières

December 1966

Contents

The Theory of Quantum Liquids

Normal Fermi Liquids

INTRODUCTION

Let us consider a gas of neutral atoms interacting through a short-range binary potential. At high enough temperatures and low enough pressures, the gas is dilute. Each atom moves as if it were essentially free, apart from infrequent collisions with other atoms or with the container walls. The system is well described by the elementary kinetic theory of gases. It displays the usual properties of a classical gas; the specific heat C_v is temperature independent; in the case of fermions (particles with spin) the spin susceptibility varies inversely with the temperature, according to Curie's law.

As the pressure is increased and the temperature is lowered, the above picture tends to break down for two distinct reasons. On the one hand, because of the increase in density, the interaction between particles becomes far more efficient. On the other hand, the decrease in the temperature weakens the kinetic energy compared to the particle "interaction" energy. At some stage, the gas undergoes a first-order transition to a liquid state. This state is characterized by strong particle correlations, which insure the cohesion of the liquid. The transition is essentially dynamic in character, since it arises from the particle interaction. It corresponds to a purely classical effect, giving rise to a *classical liquid*.

As the temperature is lowered still further, the kinetic energy of the liquid is further decreased, while the interaction between the particles plays a correspondingly more important role. As a result, in almost all cases one observes a first-order phase transition from the liquid state to a solid state. The only exceptions are the isotopes of helium, ^{3}He and ^{4}He, which remain liquid down to the lowest attainable temperatures. Helium is anomalous because the forces between the atoms are relatively weak, while, because of its low mass, the zero-point oscillations of the individual atoms are large.

Helium thus remains liquid through a temperature regime in which quantum effects must be taken into account. These become important when the thermal de Broglie wavelength of a particle, $(\hbar^2/2M\kappa T)^{1/2}$, becomes comparable to the average spacing between particles; this occurs at about 3–4°K for helium. Quantum effects may be viewed as deriving from the symmetry properties of the many-body wave function, and are essentially *statistical* in their nature. One expects (and finds) that at sufficiently low temperatures, where the quantum nature of the liquid has become manifest, ^{3}He, which obeys Fermi–Dirac statistics, and ^{4}He, which obeys Bose–Einstein statistics, will behave quite differently. We are thus led to consider the theory of *quantum liquids*, in which an important role is played by both the degeneracy characteristic of a quantum many-particle system, and the interaction between the particles.

^{3}He (a Fermi liquid) and ^{4}He (a Bose liquid) are the only "real" quantum liquids found in nature. However, one can also regard the conduction electrons in metals, semimetals, and degenerate semiconductors as a quantum liquid; this electron "liquid" is not homogeneous, since the electrons in a solid move in the periodic field of the ion cores. For many purposes, however, one can neglect the influence of this periodic potential. For conduction electrons in metals the degeneracy temperature (at which quantum effects become of importance) is of the order of 50,000°K; it is ~100°K for a semimetal, and ~3°K for a typical semiconductor with a conduction electron density of 10^{16}.

Strangely enough, despite the often quite sizable particle interaction and despite the fact that one is dealing with a quantum-mechanical many-particle system, quantum liquids at sufficiently low temperatures are better understood than their classical counterparts. The explanation lies in the concept of *elementary excitations*, which under suitable circumstances provide a complete description of the low-lying excited states of the quantum liquid. At very low temperatures, only a few such excitations are present; the excitations are long-lived and interact weakly with one another; most properties of the system can be explained in terms of them.

In Volume I of this book we shall be concerned with a single group of quantum liquids, *normal Fermi liquids*. A normal Fermi liquid may be roughly defined as a degenerate Fermi liquid in which the properties of the system are *not* drastically modified by the particle interactions, no matter how strong they might be. In other words, the liquid retains the essential properties of the noninteracting fermion system. (It has a well-defined Fermi surface, its specific heat varies linearly with the temperature, etc.) Examples of normal Fermi liquids are ^{3}He above 4 millidegrees and conduction electrons in metals which are not superconducting. In 1956 Landau constructed an elegant, semiphenomenological theory of the macroscopic behavior of normal Fermi liquids in the low-temperature limit. We present the Landau theory for neutral Fermi liquids in Chapter 1, and apply it to ^{3}He.

There are shortcomings to the Landau theory. It is not applicable to microscopic phenomena, those which involve distances of the order of the interparticle spacing, or energies comparable to that of a particle on the Fermi surface. Moreover, it is, in a certain sense, too complete in that it provides far more information than any experiment will ever sample. It is therefore of interest to develop a direct description of experimental measurements on many-particle systems. An exact formalism can be developed so long as the system responds *linearly* to the measuring apparatus. The general theory of linear response, applicable to both microscopic and macroscopic phenomena, is presented in Chapter 2. It establishes the connections between response and correlation functions and the extent to which these may be related to the spectrum of elementary excitations. The theory provides a number of exact results of great practical importance. More important, it establishes the *language* one should use in discussing the properties of quantum liquids, in both the microscopic and macroscopic regimes, and thus enables one to appreciate the physical features which are common to all quantum liquids; these unifying aspects are too often obscured by a diversity of mathematical descriptions.

Chapters 3 and 4 are devoted to charged Fermi liquids. Because of the long range of the Coulomb interaction, a charged Fermi liquid differs appreciably from its neutral counterpart. In Chapter 3 the new physical features introduced by the Coulomb interaction, screening and plasma oscillation, are introduced, and described in detail. The Landau theory is then generalized and applied to the description of a number of macroscopic phenomena encountered in electron liquids. Chapter 3 is analogous to Chapter 1 in that important physical concepts are introduced and described in the macroscopic limit. Chapter 4 resembles Chapter 2; it contains a formal description of measurements on charged particle systems and is concerned with microscopic as well as macroscopic phenomena; the various dielectric functions of interest are defined, and applied to a number of problems of physical interest.

In Chapters 1 through 4, we deal with certain exact relationships between various physical quantities, or with macroscopic theories, whose validity is limited to phenomena in the macroscopic regime. Detailed microscopic theories are, by contrast, subject to certain limitations: either they represent solutions to model problems (the description of a physical system in a limited range of densities or interaction strength, one which rarely corresponds to physical reality); or they provide an approximate account of real physical systems, to an extent which is difficult to estimate with precision. Both kinds of microscopic theories are considered in Chapter 5. The random phase approximation is developed and applied to the high-density electron gas, a model problem for which it provides an accurate description. The structure of the generalized random phase approximation for both neutral and charged particle systems is described, and connection is made, in the macroscopic limit, to the Landau theory. Approximate microscopic theories, which are intended for an electron liquid at metallic densities, are developed and applied to the description of simple metals.

Volume II of this book is devoted to the theory of superfluid quantum liquids. Mathematically, a superfluid is characterized by macroscopic occupation of a single quantum state, the "condensate." Physically, the most dramatic manifestation of superfluid behavior is the resistance-free motion associated with that single state. Bose liquids, such as ^{4}He, may be regarded as the simplest superfluid systems, since the existence of a condensate is already evident in the lowest-order perturbation-theoretic treatment. Superfluidity in Fermi liquids is, on the other hand, a far more subtle matter. As a result, many years passed between the experimental discovery of superconductivity by Kammerlingh Onnes in 1911, and the successful microscopic theory of Bardeen, Cooper, and Schrieffer in 1957.

The theory of Bose liquids is developed and applied to ^{4}He in Chapter 6, while the theory of superfluid Fermi liquids is developed and applied to superconductivity in Chapter 7. The essential features of superfluid behavior are described in Chapters 8 to 14. We attempt to present the treatments in parallel as much as possible, in order to emphasize the similarities in behavior of the two kinds of superfluids. There is, however, one important difference. For the superfluid Fermi liquid, there exists an excellent microscopic theory, that of Bardeen, Cooper, and Schrieffer, which is in close agreement with essentially all experiments on superconductivity. On the other hand, no satisfactory microscopic theory exists for the only Bose liquid found in nature, ^{4}He. As a result, our primary emphasis in developing the Bose liquid theory is on phenomenological considerations, and on the macroscopic theory to which Landau and his collaborators have made so many important contributions.

One final comment: although the title of our book indicates that we shall confine our attention to quantum liquids, this is not quite the case in practice. Since the interacting fermion system of principal physical interest is that formed by conduction electrons in a metal, we have attempted to indicate, at the appropriate places, the generalizations required to take into account the effects of the ionic periodic potential. These "solid-state" effects are discussed in Sections 1.3, 3.8, 4.5, 5.6, and 7.6.

CHAPTER 1

NEUTRAL FERMI LIQUIDS

Let us consider a noninteracting Fermi gas in equilibrium at a temperature T. The probability that a single particle has energy ϵ is given by the well-known expression

$$f(\epsilon) = \frac{1}{1 + \exp\left[(\epsilon - \mu)/\kappa T\right]}$$

where κ is Boltzmann's constant. The constant μ, known as the chemical potential, is adjusted in such a way as to give the correct total number of particles. At high temperatures μ is negative and very much smaller than $-\kappa T$; $f(\epsilon)$ reduces to the usual Maxwell–Boltzmann expression; the gas is "classical." In the opposite limit, $T \to 0$, $f(\epsilon)$ becomes a Fermi–Dirac step function, which jumps from 1 to zero at the positive chemical potential μ_0: the gas is said to be *fully degenerate*. The transition from one regime to the other occurs around the "degeneracy temperature,"

$$T_F = \frac{\mu_0}{\kappa}.$$

In the degenerate region, the number of excited states available to the system is very much reduced by the exclusion principle, which acts to "freeze" the distribution; at a temperature T, only those particles whose energy is within κT of the Fermi energy are affected by a change in temperature. This reduction has striking physical consequences: the specific heat becomes proportional to T, instead of being constant; the spin susceptibility becomes temperature independent, instead of varying as $1/T$.

For real fermion systems, the particle interaction and the exclusion principle act simultaneously; we are thus led to study *degenerate Fermi liquids*, in which both effects are important. In some systems, the nature of the degenerate gas is drastically modified by the particle interactions. Such is the case, for instance, in a superconducting electron gas. Frequently, the interacting liquid retains many properties of the gas: it is then said to be *normal* (a definition which will be made more explicit in the course of this chapter). A normal Fermi liquid at $T = 0$ has a sharply defined Fermi surface S_F; its elementary excitations may be pictured as *quasiparticles* outside S_F and *quasiholes* inside S_F, in close analogy with the single-particle excitations of a noninteracting Fermi gas. Such a resemblance explains why so many properties of the liquid can be interpreted in terms of a "one-particle approximation." To consider another example, the one-electron theory of solids provides a correct account of a large number of "sophisticated" phenomena in metals (de Haas–van Alphen effect, transport properties, etc.) even though it ignores the not-inappreciable particle interaction. Again, the explanation of this success is found in the concept of quasiparticle excitations. Let us emphasize that such single-particle theories are not complete; there exist "many-body" effects which arise as a consequence of particle interaction, and which are characteristic of the liquid state.

At the present time, we do not possess a theory that completely describes the properties of an interacting Fermi liquid at an arbitrary temperature. The problem may be formulated by means of sophisticated field-theoretic techniques. However, the general solutions that have been obtained (see, for example, Balian and de Dominicis) are of a somewhat formal character, and have not led, as yet, to explicit results which may be compared with experiment. Fortunately, one may obtain a number of simple results in the limit of *low temperatures* ($T \ll T_F$), for phenomena occurring on a *macroscopic* scale. The relevant theory was constructed by Landau (1956) on a semiphenomenological basis; Landau's assumptions have since been substantiated by detailed microscopic analysis [Pitaevskii (1959), Luttinger and Nozières (1962)]. In this chapter, we shall adopt Landau's semiphenomenological point of view, and explore the applications of the theory to neutral systems such as degenerate liquid ^{3}He.

As we have pointed out in the introduction, ^{3}He is the only degenerate Fermi liquid found in nature. With minor modifications the theory of Fermi liquids may also be applied to electrons in metals or semimetals (and to nuclear matter). The extension of Landau's theory to charged systems (such as electrons in metals) involves certain difficulties which arise from the long-range character of the Coulomb interaction. For that reason we postpone a detailed discussion of charged Fermi liquids until Chapter 3.

In Section 1.1, we shall introduce the notion of a quasiparticle by means of a careful study of the relation between interacting and noninteracting systems. In a real Fermi liquid, the quasiparticles are not quite independent; there remains a certain *interaction energy* between excited quasiparticles. This new physical feature is the key to the Landau theory, the fundamentals of which are presented in Section 1.2. The theory is applied to various equilibrium properties in Section 1.3; its extension to electrons in metals is also briefly discussed. We then proceed to study nonequilibrium properties. The transport equation for quasiparticles is set up in Section 1.4; it is applied to a calculation of the current density in Section 1.5. Section 1.6 is devoted to an extensive discussion of "localized" quasiparticle excitations, and of their interaction with the surrounding medium; we are led to develop a formal solution of the transport equation, which proves useful in other respects. The important concept of collective modes is introduced in Section 1.7; the damping of the collective modes is discussed in some detail, and the stability of the ground state against collective excitations is studied from both a static and a dynamic point of view. Section 1.8 is concerned with the consequences of real collisions between quasiparticles; physical phenomena that are discussed include the lifetime of quasiparticles, the usual transport coefficients, and the collision-induced damping of collective modes. A comparison between zero sound and first sound is carried out in the following section. Finally, the theory is applied to degenerate ^{3}He in Section 1.10.

This whole chapter is based on the Landau theory of Fermi liquids, which shows clearly all the important features brought about by the interaction. This theory is rigorous only in well-defined limits. We shall stress the conditions of applicability throughout our discussion of the theory.

1.1. THE QUASIPARTICLE CONCEPT

ELEMENTARY EXCITATIONS OF A NONINTERACTING FERMI GAS

Let us first consider a system of N noninteracting free fermions, each of mass m, enclosed in a volume Ω. The eigenstates of the total system are antisymmetrized combinations of N different single-particle states. Each single particle is characterized by two quantum numbers, its momentum $\mathbf{p}$ and its spin, $\sigma = \pm\frac{1}{2}$; its normalized wave function in configuration space is a simple plane wave:

$$\psi_{\mathbf{p}}(\mathbf{r}) = \frac{1}{\sqrt{\Omega}} e^{i\mathbf{p}\cdot\mathbf{r}}. \tag{1.1}$$

The total wave function is a Slater determinant made up of N such plane waves. All the eigenstates of the system can be characterized by the distribution function $n_{\mathbf{p}\sigma}$, which is equal to 1 if the state $\mathbf{p}, \sigma$ is occupied, to zero otherwise. (In what follows, we shall omit the spin index σ, and include it in $\mathbf{p}$, unless specified otherwise.)

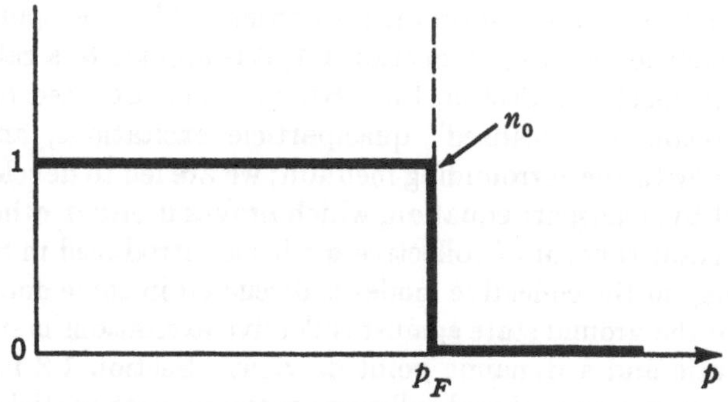

FIGURE 1.1. *The ground state distribution of quasiparticles.*

A particle with momentum $\mathbf{p}$ possesses a kinetic energy $p^2/2m$. In the absence of interaction, the energies of the particles are simply additive: the total energy E of the system is given by

$$E = \sum_{\mathbf{p}} n_{\mathbf{p}} \frac{p^2}{2m}. \tag{1.2}$$

The ground state is obtained by filling the N plane wave states of lowest energy. The ground state distribution is shown in Fig. 1.1, the Fermi momentum p_F being given by

$$\frac{N}{\Omega} = \frac{1}{3\pi^2}\left(\frac{p_F}{\hbar}\right)^3. \tag{1.3}$$

All the plane wave states lying inside the *Fermi surface* S_F (here the sphere of radius p_F) are filled in the ground state; those lying outside S_F are empty.

Let us add a single particle to the system. The ground state of the $(N + 1)$-particle system is obtained if the additional particle is added in the lowest available momentum state, one on the Fermi surface. The chemical potential μ, defined as

$$\mu = E_0(N + 1) - E_0(N) = \frac{\partial E_0}{\partial N}, \tag{1.4a}$$

is thus given by

$$\mu = p_F{}^2/2m. \tag{1.4b}$$

The chemical potential is equal to the energy of a particle *on the Fermi surface*. This result will be seen to apply equally well in the presence of particle interaction.

Excited states of the system are best specified with reference to the ground state. A given excited state is obtained by "exciting" a certain number of particles across the Fermi surface. Such a procedure is equivalent to creating an equal number of *particles* outside S_F and of *holes* inside S_F. Particles and holes thus appear as "elementary excitations," whose configurations give rise to all excited states. The amount of "excitation" is characterized by the departure of the distribution function from its value in the ground state

$$\delta n_{\mathbf{p}} = n_{\mathbf{p}} - n_{\mathbf{p}}{}^0. \tag{1.5}$$

A particle excitation of momentum $\mathbf{p}' > p_F$ corresponds to $\delta n_{\mathbf{p}} = \delta_{\mathbf{pp}'}$, while a hole excitation of momentum $\mathbf{p}' < p_F$ corresponds to $\delta n_{\mathbf{p}} = -\delta_{\mathbf{pp}'}$. For the noninteracting system, the excitation energy is simply

$$E - E_0 = \sum_{\mathbf{p}} \frac{p^2}{2m} \delta n_{\mathbf{p}}. \tag{1.6}$$

At low temperatures, particles and holes will only be excited near the Fermi surface; $\delta n_{\mathbf{p}}$ will typically be of order 1 in a small region surrounding S_F, and will otherwise be negligible.

In an isolated system, the total number of particles is conserved: the number of excited particles must therefore be equal to that of excited holes. This restriction is sometimes inconvenient. It is then preferable to work with what is equivalent to the grand canonical ensemble of statistical mechanics, a system which is characterized by its chemical potential μ rather than by its number of particles N. Such a situation may be realized by imagining the system to be in contact with a reservoir of particles. In such cases, the quantity of interest is not the energy E, but rather the *free energy*, given by $F = E - \mu N$ at zero temperature. It follows from Eq. (1.6) that the excitation free energy associated with the distribution $\delta n_{\mathbf{p}}$ is given by

$$F - F_0 = \sum_{\mathbf{p}} \left(\frac{p^2}{2m} - \mu\right) \delta n_{\mathbf{p}}. \tag{1.7}$$

Equation (1.7) obviously reduces to (1.6) when the number of particles is conserved, i.e., when $\Sigma_{\mathbf{p}}\delta n_{\mathbf{p}} = 0$.

According to (1.7), the free energy of a particle with momentum $\mathbf{p}$ is $(p^2/2m - \mu)$; it corresponds to the free energy of an elementary excitation *outside* the Fermi surface. Inside S_F, the excitations are *holes*, for which $\delta n_{\mathbf{p}} = -1$. The free energy associated with these excitations is then $(\mu - p^2/2m)$, rather than $(p^2/2m - \mu)$. Since μ is equal to $p^2/2m$ on the Fermi surface, the free energy of an *elementary excitation* of momentum $\mathbf{p}$ (not to be confused with that of a particle) may be written as $|p^2/2m - \mu|$, a result valid both inside and outside S_F. Note that the excitation free energy is always positive: this is necessary to ensure the stability of the ground state.

DEFINITION OF QUASIPARTICLES AND QUASIHOLES

Let us now turn to the case of an interacting Fermi liquid. We are interested in the nature of its elementary excitations. A "frontal" attack on the problem involves the introduction of Green's functions, and the mathematical apparatus of many-body perturbation theory, an approach which lies far beyond the scope of this book. We therefore adopt an alternative approach, which consists in comparing the interacting "real" liquid with the noninteracting "ideal" gas; we establish a one-to-one correspondence between the eigenstates of the two systems. Such an approach will provide us with a qualitative understanding of the excitation spectrum of an interacting system.

Consider an eigenstate of the ideal system, characterized by a distribution function $n_{\mathbf{p}}$. In order to establish a connection with the real system, we imagine that the interaction between the particles is switched on infinitely slowly. Under such "adiabatic" conditions, the ideal eigenstates will progressively transform into certain eigenstates of the real interacting system. However, there is no *a priori* reason why such a procedure should generate *all* real eigenstates. For instance, it may well happen that the real ground state may not be obtained in that way; superconductors furnish us with a specific example of such a failure. We therefore must *assume* that the real ground state may be adiabatically generated starting from some ideal eigenstate with a distribution $n_{\mathbf{p}}^{\circ}$. This statement may be considered as the definition of a *normal* fermion system.

For reasons of symmetry, the distribution $n_{\mathbf{p}}^{\circ}$ of an isotropic system is spherical. As a result, the spherical Fermi surface is not changed when the interaction between particles is switched on: the real ground state is generated adiabatically from the ideal ground state. Matters are otherwise when the Fermi surface for the noninteracting system is anisotropic (as is the case for metals). Under these circumstances, the Fermi surface will certainly be deformed when the interaction is switched on. In such cases, the real ground state may be shown to follow adiabatically from some *excited* state of the noninteracting system. This situation will not cause any major difficulty with the theory.

Let us now add a particle with momentum **p** to the ideal distribution $n_{\mathbf{p}}^{0}$ and, again, turn on the interaction between the particles adiabatically. We generate an excited state of the real liquid, which likewise has momentum **p**, since momentum is conserved in particle collisions. As the interaction is increased we may picture the bare particle as slowly perturbing the particles in its vicinity; if the change in interaction proceeds sufficiently slowly, the entire system of $N + 1$ particles will remain in equilibrium. Once the interaction is completely turned on, we find that our particle moves together with the surrounding particle distortion brought about by the interaction. In the language of field theory, we would say that the particle is "dressed" with a self-energy cloud. We shall consider the "dressed" particle as an independent entity, which we call a *quasiparticle.* The above excited state corresponds to the real ground state plus a quasiparticle of momentum **p**.

Let S_F be the Fermi surface characterizing the unperturbed distribution $n_{\mathbf{p}}^{0}$ from which the real ground state is built up. Because of the exclusion principle, quasiparticle excitations can be generated only if their momentum **p** lies *outside* S_F. The quasiparticle distribution in **p** space is thus sharply bounded by the Fermi surface S_F.

Using the same adiabatic switching procedure, we can define a *quasihole,* with a momentum **p** lying *inside* the Fermi surface S_F; we may do likewise for higher configurations involving several excited quasiparticles and quasiholes. The quasiparticles and quasiholes thus appear as *elementary excitations* of the real system which, when combined, give rise to a large class of excited states. We have established our desired one-to-one correspondence between ideal and real eigenstates.

Actually, our adiabatic switching method is likely to run into difficulties when the real state under study is damped as a consequence of particle interaction. If the time over which the interaction is turned on is longer than the lifetime of the state that we wish to generate, the switching on of the interaction is no longer *reversible,* since the state has decayed long before the physical value of the interaction is reached. On the other hand, if the interaction is turned on too fast, the process is no longer *adiabatic,* and we do not generate eigenstates of the interacting system. This limitation is not due to mathematical clumsiness: it follows from the physical uncertainty arising from the finite lifetime of the state under consideration.

Such difficulties do not arise for the ground state, which, being stable, can be precisely specified. On the other hand, since quasiparticles and quasiholes undergo real collisions, which lead to damping, any definition of the elementary excitations is somewhat imprecise. Fortunately, the quasiparticle lifetime becomes sufficiently long in the immediate vicinity of the Fermi surface, that the quasiparticle concept makes sense in that region. In pure systems at zero temperature, the lifetime varies as the inverse square of the energy separation from S_F. Since the quasiparticle is better and better defined as one gets closer and closer to the Fermi surface, S_F remains sharply defined. It should be kept in mind that a quasiparticle is only strictly defined if it is right on the Fermi surface. In order for our theory to make sense, we must be careful to introduce quasiparticles only in the neighborhood of S_F.

Let us consider an eigenstate of the noninteracting system, characterized by some distribution function $n_{\mathbf{p}}$ for the usual "bare" particles. By switching on the interaction adiabatically, we obtain an eigenstate of the real system, which may be labeled by the same function $n_{\mathbf{p}}$. In the interacting system, $n_{\mathbf{p}}$ describes the *distribution of quasiparticles.* The "excitation" of the system is measured by the departure $\delta n_{\mathbf{p}}$ from the ground state distribution

$$\delta n_{\mathbf{p}} = n_{\mathbf{p}} - n_{\mathbf{p}}^{\circ}. \tag{1.8}$$

At low temperatures, one only samples low-lying excited states, for which $\delta n_{\mathbf{p}}$ is restricted to the immediate vicinity of the Fermi surface. Under such conditions, quasiparticle damping is negligible, and our overall picture becomes meaningful.

Let us emphasize that the physically meaningful quantity is $\delta n_{\mathbf{p}}$, rather than $n_{\mathbf{p}}$. (It does not make much sense to define an equilibrium distribution $n_{\mathbf{p}}^{\circ}$ in a range where quasiparticles are unstable.) We shall thus be careful to formulate our theory in terms of the departure from equilibrium, $\delta n_{\mathbf{p}}$; $n_{\mathbf{p}}^{\circ}$ will be used only as an intermediate step. In fact, our results will always involve the gradient $\nabla n_{\mathbf{p}}^{\circ}$, a quantity which is localized on the Fermi surface.

ENERGY OF QUASIPARTICLES

For the ideal system, there exists a simple linear relation between the energy of a given state and the corresponding distribution function. When particle interaction is taken into account, the relation between the state energy, E, and the quasiparticle distribution function, $n_{\mathbf{p}}$, becomes much more complicated. It may be expressed in a functional form, $E[n_{\mathbf{p}}]$, which one cannot in general specify explicitly. If, however, $n_{\mathbf{p}}$ is sufficiently close to the ground state distribution $n_{\mathbf{p}}^{\circ}$, we can carry out a Taylor expansion of this functional. On writing $n_\mathbf{p}$ in the form (1.8), and taking $\delta n_\mathbf{p}$ to be small, or to extend over a small region in momentum space, we obtain

$$E[n_\mathbf{p}] = E_0 + \sum_p \epsilon_\mathbf{p} \delta n_\mathbf{p} + O(\delta n^2), \tag{1.9}$$

where $\epsilon_\mathbf{p}$ is the first functional derivative of E. Since each summation over $\mathbf{p}$ carries a factor Ω, $\epsilon_\mathbf{p}$ is of order $\Omega^0 = 1$.

If $\delta n_\mathbf{p}$ describes a state with one extra quasiparticle $\mathbf{p}$, the energy of the state is $(E_0 + \epsilon_\mathbf{p})$: $\epsilon_\mathbf{p}$ is the energy of the quasiparticle. According to Eq. (1.9), the energies of several excited quasiparticles are simply additive, within corrections of order $(\delta n)^2$. We shall assume that $\epsilon_\mathbf{p}$ is continuous when $\mathbf{p}$ crosses the Fermi surface. This statement is not obvious and, again, should be considered as a characteristic of "normal" systems.

On the Fermi surface, $\epsilon_\mathbf{p}$ must equal a constant ϵ_F (the Fermi energy): otherwise, we could lower the ground state energy by transferring a particle from a state inside the Fermi surface to one of lower energy outside S_F. Since the ground state for $(N + 1)$ particles is obtained by adding a quasiparticle on the Fermi surface, ϵ_F is simply the *chemical potential*, $\mu = \partial E_0/\partial N$, at zero temperature. This very important property was first established by Hugenholtz and Van Hove (1958).

In practice, we need only values of $\epsilon_\mathbf{p}$ in the vicinity of the Fermi surface, where we can use a series expansion. The gradient of $\epsilon_\mathbf{p}$,

$$\mathbf{v}_\mathbf{p} = \nabla_\mathbf{p} \epsilon_\mathbf{p}, \tag{1.10}$$

plays the role of a "group velocity" of the quasiparticle. We shall see later that $\mathbf{v}_\mathbf{p}$ is the velocity of a quasiparticle wave packet. In the absence of a magnetic field, and for a system which is reflection invariant, $\epsilon_\mathbf{p}$ and $\mathbf{v}_\mathbf{p}$ do not depend on the spin σ. For an isotropic system, $\epsilon_\mathbf{p}$ depends only on p, and may be denoted as ϵ_p. The velocity $\mathbf{v}_\mathbf{p}$ is then parallel to $\mathbf{p}$, so that we can write (on the Fermi surface)

$$v_{p_F} = p_F/m^*, \tag{1.11}$$

where m^* is called the *effective mass* of the quasiparticle. We remark that the present definition of the effective mass differs from that adopted to describe the motion of independent particles in a periodic lattice; in the latter case, $1/m^*$ is defined as the second derivative of ϵ_p with respect to p.

In an anisotropic system, the velocity $|\mathbf{v}_p|$ varies over the Fermi surface; the notion of an effective mass is somewhat artificial. It is then convenient to introduce the density $\nu(\epsilon)$ of quasiparticle states having an energy $(\mu + \epsilon)$:

$$\nu(\epsilon) = \sum_{\mathbf{p}} \delta(\epsilon_{\mathbf{p}} - \mu - \epsilon). \tag{1.12}$$

At low temperatures, all physical properties will depend on the density of states on the Fermi surface, $\nu(0)$.

This discussion shows the similarity between bare particles in a Fermi gas and quasiparticles in a Fermi liquid. They have the same distribution in momentum space, bounded at $T = 0$ by a sharp Fermi surface. Both follow Fermi statistics. The main difference that we have pointed out is the change of energy and velocity brought about by the interaction with the surrounding medium. Actually, quasiparticles have physical features that do not appear in noninteracting systems: these are discussed in the next section.

1.2. INTERACTION BETWEEN QUASIPARTICLES: LANDAU'S THEORY OF FERMI LIQUIDS

EXPANSION OF THE FREE ENERGY

As we pointed out earlier, the quantity of physical interest is the free energy $F = E - \mu N$, rather than the energy E. The "excitation" free energy, as measured from the ground state value F_o, is given by

$$F - F_o = E - E_o - \mu(N - N_o), \tag{1.13}$$

where N_o is the number of particles in the ground state. In order to generalize the expansion (1.9), we need to calculate $(N - N_o)$. For that purpose, we note that by adding one quasiparticle to the ground state, we add *exactly one* bare particle to the system as a whole. This follows at once from our "adiabatic" definition of quasiparticles: the state with one extra quasiparticle is derived from an ideal state containing $(N + 1)$ particles, and the total number of particles is conserved when the interaction is switched on adiabatically. The difference $(N - N_o)$ may thus be written as

$$N - N_o = \sum_{\mathbf{p}} \delta n_{\mathbf{p}}. \tag{1.14}$$

Using Eqs. (1.9) and (1.14), we obtain, within corrections of order $(\delta n)^2$,

$$F - F_o = \sum_{\mathbf{p}} (\epsilon_p - \mu)\delta n_{\mathbf{p}}. \tag{1.15}$$

If the number of particles of the system is kept constant, the difference

(1.14) vanishes: $(E - E_0)$ is then equal to $(F - F_0)$, and may be written in the form (1.15), as well as (1.9).

Most of the properties that we shall consider will involve a displacement of the Fermi surface by a small amount δ. The corresponding value of $\delta n_{\mathbf{p}}$ is equal to ± 1 in a thin sheet of width δ centered on the Fermi surface; it vanishes outside this sheet. Where $\delta n_{\mathbf{p}}$ is nonzero, $(\epsilon_p - \mu)$ is of order δ. The difference $(F - F_0)$, given by Eq. (1.15), is thus of order δ^2. The expansion (1.15), which looks like a first-order expansion, is actually a second-order one. Our approach is therefore consistent only if we push the Taylor expansion of the functional $(F - F_0)$ one step further, to include all terms of second order in the displacement of the Fermi surface. We are thus led to write

$$F - F_0 = \sum_{\mathbf{p}} (\epsilon_p - \mu)\delta n_{\mathbf{p}} + \frac{1}{2}\sum_{\mathbf{pp}'} f_{\mathbf{pp}'}\delta n_{\mathbf{p}}\delta n_{\mathbf{p}'} + O(\delta n^3). \quad (1.16)$$

Equation (1.16) is the heart of the phenomenological theory of Fermi liquids proposed by Landau (1956). Its most important feature is the new quadratic term, which describes the interaction between quasiparticles. Such a term is not present in the approximate theory of Sommerfeld; we shall see later that it considerably modifies a number of physical properties of the system.

Equation (1.16) furnishes the leading terms of an expansion of $(F - F_0)$ in powers of the relative number of excited quasiparticles. The latter is measured by the ratio

$$\alpha = \frac{\Sigma_{\mathbf{p}}|\delta n_{\mathbf{p}}|}{N}. \quad (1.17)$$

Landau's approximation is valid whenever α is small. We must be careful to maintain consistent approximations throughout any calculation, and to keep in our results only the leading terms with respect to the parameter α.

The coefficient $f_{\mathbf{pp}'}$ is the second variational derivative of E (or F) with respect to $n_{\mathbf{p}}$. It is accordingly invariant under permutation of $\mathbf{p}$ and $\mathbf{p}'$. Since each summation over a momentum index carries a factor Ω, $f_{\mathbf{pp}'}$ is of order $1/\Omega$. This is easily understood if we realize that $f_{\mathbf{pp}'}$ is the *interaction energy* of the excited quasiparticles $\mathbf{p}$ and $\mathbf{p}'$. Each of the latter is spread out over the whole volume Ω; the probability that they interact with one another is of order a^3/Ω, where a is the range of the interaction.

We shall assume that $f_{\mathbf{pp}'}$ is continuous when $\mathbf{p}$ or $\mathbf{p}'$ crosses the Fermi surface. Once again, this may be considered as characteristic of a "normal system." In practice, we shall only need values of f *on* the

Fermi surface, at points such that $\epsilon_p = \epsilon_{p'} = \mu$. Then $f_{\mathbf{pp}'}$ depends only on the direction of $\mathbf{p}$ and $\mathbf{p}'$, and on the spins σ and σ'.

If there is no applied magnetic field, the system is invariant under time reversal, which implies

$$f_{\mathbf{p}\sigma,\mathbf{p}'\sigma'} = f_{-\mathbf{p}-\sigma,-\mathbf{p}'-\sigma'} \tag{1.18}$$

(we have explicitly introduced the spin indices). If furthermore the Fermi surface is invariant under reflection $\mathbf{p} \to -\mathbf{p}$, Eq. (1.18) becomes

$$f_{\mathbf{p}\sigma,\mathbf{p}'\sigma'} = f_{\mathbf{p}-\sigma,\mathbf{p}'-\sigma'}. \tag{1.19}$$

In that case, $f_{\mathbf{p}\sigma,\mathbf{p}'\sigma'}$ depends only on the relative orientation of the spins σ and σ'; there are only two independent components, corresponding respectively to parallel spins and antiparallel spins. It is convenient to write these in the form

$$\begin{aligned} f^{\uparrow\uparrow}_{\mathbf{pp}'} &= f^{s}_{\mathbf{pp}'} + f^{a}_{\mathbf{pp}'}, \\ f^{\uparrow\downarrow}_{\mathbf{pp}'} &= f^{s}_{\mathbf{pp}'} - f^{a}_{\mathbf{pp}'}, \end{aligned} \tag{1.20}$$

where $f^{s}_{\mathbf{pp}'}$ and $f^{a}_{\mathbf{pp}'}$ are the spin symmetric and spin antisymmetric parts of the quasiparticle interaction. The antisymmetric term $f^{a}_{\mathbf{pp}'}$ may be interpreted as due to an *exchange* interaction energy $2f^{a}_{\mathbf{pp}'}$, which appears only when the spins are parallel. In the Russian literature, Eq. (1.20) is usually replaced by

$$f_{\mathbf{p}\sigma,\mathbf{p}'\sigma'} = \varphi_{\mathbf{pp}'} + \boldsymbol{\sigma} \cdot \boldsymbol{\sigma}' \psi_{\mathbf{pp}'}, \tag{1.21}$$

where $\boldsymbol{\sigma}$ and $\boldsymbol{\sigma}'$ are the spin matrices. The coefficient ψ is four times bigger than our f^a. We prefer to use the more symmetric decomposition (1.20).

If the system is isotropic, (1.20) may be further simplified. In that case, for $\mathbf{p}$ and $\mathbf{p}'$ on the Fermi surface, $f^{a}_{\mathbf{pp}'}$ and $f^{s}_{\mathbf{pp}'}$ depend only on the angle ξ between the directions of $\mathbf{p}$ and $\mathbf{p}'$. They may be expanded in a series of Legendre polynomials

$$f^{s(a)}_{\mathbf{pp}'} = \sum_{l=0}^{\infty} f^{s(a)}_{l} P_l(\cos \xi). \tag{1.22}$$

f is completely determined by the set of coefficients f_l^s and f_l^a. It is convenient to express the latter in "reduced" units by setting

$$\nu(0) f_l^{s(a)} = \frac{\Omega m^* p_F}{\pi^2 \hbar^3} f_l^{s(a)} = F_l^{s(a)}. \tag{1.23}$$

The dimensionless quantities F_l^s and F_l^a measure the strength of the interaction as compared to the kinetic energy.

LOCAL ENERGY OF A QUASIPARTICLE

Let us consider a state of the system with a certain distribution of excited quasiparticles $\delta n_{\mathbf{p}'}$. To this system we now add an extra quasiparticle, $\mathbf{p}$. According to Eq. (1.16), the free energy of the additional quasiparticle is equal to

$$\tilde{\epsilon}_{\mathbf{p}} - \mu = (\epsilon_p - \mu) + \sum_{\mathbf{p}'} f_{\mathbf{p}\mathbf{p}'}\delta n_{\mathbf{p}'}. \tag{1.24}$$

If $\mathbf{p}$ is close enough to the Fermi surface, both terms of Eq. (1.24) have the same order of magnitude. The free energy of a quasiparticle thus depends on the state of the system through its interaction energy with other excited quasiparticles. This physical effect, which does not exist in noninteracting systems, arises from the quadratic term of (1.16).

The quantity $\tilde{\epsilon}_{\mathbf{p}}$ plays a central role in the development of the Landau theory. In order to clarify its physical meaning, let us suppose that the system is slightly inhomogeneous. The departure from equilibrium may be characterized by a position-dependent distribution of excited quasiparticles, $\delta n_{\mathbf{p}'}(\mathbf{r})$ (see Section 1.4). As a result, $\tilde{\epsilon}_{\mathbf{p}}$ depends on $\mathbf{r}$: it may be regarded as a *local energy*, perturbed by the surrounding distortion of the medium. We shall see that the gradient of $\tilde{\epsilon}_{\mathbf{p}}$ in ordinary space

$$\nabla_{\mathbf{r}}\tilde{\epsilon}_{\mathbf{p}} = \nabla_{\mathbf{r}}\left\{\sum_{\mathbf{p}'} f_{\mathbf{p}\mathbf{p}'}\delta n_{\mathbf{p}'}\right\} \tag{1.25}$$

may be interpreted as an average *force* exerted by the surrounding medium on the quasiparticle $\mathbf{p}$.

It is mathematically convenient to introduce the distribution function

$$\tilde{n}_{\mathbf{p}}^{\circ} = n^{\circ}(\tilde{\epsilon}_{\mathbf{p}} - \mu), \tag{1.26}$$

where n° is the usual Fermi–Dirac step function defined by

$$n^{\circ}(x) = \begin{cases} 1 & \text{if } x < 0, \\ 0 & \text{if } x > 0. \end{cases} \tag{1.27}$$

It will be seen that for the slightly inhomogeneous system, the distribution $\tilde{n}_{\mathbf{p}}^{\circ}$ corresponds to a "local equilibrium" of quasiparticles, in much the same way as $n_{\mathbf{p}}^{\circ} = n^{\circ}(\epsilon_p - \mu)$ describes the true equilibrium.

The departure from this state of local equilibrium is measured by the difference

$$\delta\tilde{n}_{\mathbf{p}} = n_{\mathbf{p}} - \tilde{n}_{\mathbf{p}}^{\circ}. \tag{1.28}$$

On comparing the definitions of $\delta n_{\mathbf{p}}$ and $\delta\tilde{n}_{\mathbf{p}}$, we see that

$$\delta n_{\mathbf{p}} = \delta\tilde{n}_{\mathbf{p}} + \frac{\partial n_{\mathbf{p}}^{\circ}(\epsilon_p - \mu)}{\partial \epsilon_p}(\tilde{\epsilon}_{\mathbf{p}} - \epsilon_p). \tag{1.29}$$

Using Eq. (1.24), we obtain

$$\delta\bar{n}_{\mathbf{p}} = \delta n_{\mathbf{p}} - \frac{\partial n^0}{\partial \epsilon_p} \sum_{\mathbf{p}'} f_{\mathbf{pp}'} \delta n_{\mathbf{p}'}. \tag{1.30}$$

In practice, $\delta n_{\mathbf{p}}$ will always contain a factor $\partial n^0/\partial \epsilon_p$. At zero temperature, we have

$$\frac{\partial n^0}{\partial \epsilon_p} = -\delta(\epsilon_p - \mu); \tag{1.31}$$

it follows that both δn and $\delta\bar{n}$ are restricted to the Fermi surface S_F.

For an isotropic system at zero temperature, the relation between $\delta n_{\mathbf{p}}$ and $\delta\bar{n}_{\mathbf{p}}$ may be greatly simplified. We split $\delta n_{\mathbf{p}}$ and $\delta\bar{n}_{\mathbf{p}}$ into a spin symmetric and a spin antisymmetric part, by setting

$$\delta n_{\mathbf{p},\pm} = \delta n_{\mathbf{p}}{}^s \pm \delta n_{\mathbf{p}}{}^a \tag{1.32}$$

(with a similar expression for $\delta\bar{n}_{\mathbf{p}}$). We further expand these quantities in a series of normalized spherical harmonics, such as

$$\delta n_{\mathbf{p}}{}^s = \sum_{\ell m} \delta(\epsilon_p - \mu) \delta n_{\ell m}^s Y_{\ell m}(\theta, \varphi), \tag{1.33}$$

where (θ, φ) are the polar angles of $\mathbf{p}$. By putting these expanded forms of $\delta n_{\mathbf{p}}$ and $\delta\bar{n}_{\mathbf{p}}$ back into Eq. (1.30), together with the expansion (1.22) of $f_{\mathbf{pp}'}$, and by making use of the addition theorem for spherical harmonics, we find

$$\begin{aligned} \delta\bar{n}_{\ell m}^s &= \left(1 + \frac{F_\ell{}^s}{2\ell + 1}\right) \delta n_{\ell m}^s, \\ \delta\bar{n}_{\ell m}^a &= \left(1 + \frac{F_\ell{}^a}{2\ell + 1}\right) \delta n_{\ell m}^a, \end{aligned} \tag{1.34}$$

where the coefficients $F_\ell{}^s$ and $F_\ell{}^a$ are given by (1.23). The passage from $\delta n_{\mathbf{p}}$ to $\delta\bar{n}_{\mathbf{p}}$ is thus straightforward.

The quantity $\delta\bar{n}_{\mathbf{p}}$ will prove to be extremely useful in the following sections. Many properties are worked out much more easily in terms of $\delta\bar{n}_{\mathbf{p}}$ than of $\delta n_{\mathbf{p}}$. The interplay of the two quantities is important and Eqs. (1.30) and (1.34) will be used very often.

EQUILIBRIUM DISTRIBUTION OF QUASIPARTICLES AT A FINITE TEMPERATURE

Let us consider a system at temperature T, with a chemical potential μ: the latter may be defined as the adiabatic derivative

$$\mu = (\partial E/\partial N)_S. \tag{1.35}$$

Using elementary thermodynamics, we may show that

$$\mu = (\partial E/\partial N)_T - T(\partial S/\partial N)_T. \tag{1.36}$$

At very low temperatures, the second term of Eq. (1.36) is negligible compared to the first; the chemical potential is then equal to the isothermal derivative of E with respect to N, which is easily calculated.

We obtain the equilibrium distribution of quasiparticles, $n_{\mathbf{p}}^{\circ}(T, \mu)$, by a well-known procedure: given a distribution $n_{\mathbf{p}}$, one counts the number of ways W in which the quasiparticles may be distributed among the various occupied states; one then chooses $n_{\mathbf{p}}$ in such a way as to maximize W, while keeping the total free energy constant. We find

$$n_{\mathbf{p}}^{\circ}(T, \mu) = \frac{1}{1 + \exp[(\bar{\epsilon}_{\mathbf{p}} - \mu)/\kappa T]}, \tag{1.37}$$

where $\bar{\epsilon}_{\mathbf{p}}$ is the quasiparticle *local* energy appropriate to the distribution $n_{\mathbf{p}}^{\circ}(T, \mu)$. From the maximum value $W_{\max}$, we may obtain the entropy $S = \kappa \log W_{\max}$.

The "local" energy $\bar{\epsilon}_{\mathbf{p}}$ is given by Eq. (1.24), $\delta n_{\mathbf{p}}$ being here equal to

$$\delta n_{\mathbf{p}} = n_{\mathbf{p}}^{\circ}(T, \mu) - n_{\mathbf{p}}^{\circ}(0, \mu). \tag{1.38}$$

At very low temperatures, the integral

$$\int \delta n_{\mathbf{p}} p^2 \, dp$$

is of order T^2 in any direction of momentum space. The interaction energy between excited quasiparticles vanishes to the same order, and may therefore be neglected in comparison to $(\epsilon_p - \mu)$, which is of order T. Hence in the low temperature limit, we may replace $\bar{\epsilon}_{\mathbf{p}}$ by ϵ_p in the equilibrium distribution (1.37). By the same token, we prove that the number of particles at constant μ (or conversely the chemical potential at constant N) remains constant up to order T^2.

The thermal motion will only excite quasiparticles within a distance κT from the Fermi surface. The percentage of excited quasiparticles is $\kappa T/\mu$, which is essentially the expansion parameter of (1.16). In neglecting the terms of order δn^3, we make an error in the energy of the order of $(\kappa T/\mu)^3$. In view of this uncertainty, it would be meaningless to evaluate to this order the temperature dependence of any other factor entering the theory. To keep in the spirit of Landau's theory, we should retain only the leading terms with respect to T (usually of order 0 or 1). There is no need to consider higher-order corrections, which are of the same order as the terms of (1.16) that are regarded as negligible.

1.3. EQUILIBRIUM PROPERTIES

We shall now apply the Landau theory to the study of a number of macroscopic properties, characteristic of the system at equilibrium. We shall find that some of these properties are affected by the interaction between quasiparticles, while others are not.

SPECIFIC HEAT

Let us first consider the specific heat, defined as

$$C_v = (\partial E/\partial T)_N. \tag{1.39}$$

It is left as a problem to the reader to show that at low temperature C_v may be put in the more convenient form

$$C_v = (\partial F/\partial T)_\mu. \tag{1.40}$$

In order to calculate C_v, we write the free energy in the form (1.16), δn_p being given by (1.38). We have seen that the integral of δn_p in any direction of momentum space vanishes up to order T^2. The interaction energy in Eq. (1.16) is therefore of order T^4, and is thus negligible with respect to the main term of order T^2. The "thermal" free energy $[F(T) - F_0]$ is given by an expression of the same form as for a noninteracting system. A straightforward calculation leads to

$$F(T) - F_0 = \frac{\pi^2}{6}\nu(0)(\kappa T)^2, \tag{1.41}$$

where the density of states $\nu(0)$ is defined by (1.12). For an isotropic system, $\nu(0)$ is given by

$$\nu(0) = \frac{\Omega m^* p_F}{\pi^2 \hbar^3}, \tag{1.42}$$

from which we obtain the specific heat

$$C_v = \frac{m^* p_F}{3\hbar^3}\kappa^2 T. \tag{1.43}$$

A measurement of the slope of the linear specific heat therefore yields the *effective mass* of the quasiparticles. We remark that C_v is not affected by the interaction between quasiparticles.

COMPRESSIBILITY AND SOUND VELOCITY

Let $E_0(\Omega)$ be the ground state energy of the system, regarded as a function of the volume Ω. The pressure P may be defined as

$$P = -\partial E_0/\partial \Omega. \tag{1.44}$$

(At an arbitrary temperature, Eq. (1.44) should be an adiabatic derivative: we have seen that near $T = 0$ there is no difference between adiabatic and isothermal processes). The compressibility κ is then given by

$$\frac{1}{\kappa} = -\Omega \frac{\partial P}{\partial \Omega}. \tag{1.45}$$

For a large system, E_o is an "extensive" quantity, proportional to the volume when the density $\rho = N/\Omega$ is kept constant. We may thus write

$$E_o = \Omega f(\rho). \tag{1.46}$$

One finds directly from Eqs. (1.44) and (1.46) that

$$\frac{1}{\kappa} = \rho^2 f''(\rho). \tag{1.47}$$

These quantities may be related to the chemical potential $\mu = \partial E_o/\partial N$. Indeed, it is straightforward to establish that

$$\frac{1}{\kappa} = N\rho(\partial\mu/\partial N)_\Omega. \tag{1.48}$$

The compressibility is related to the velocity of sound, s, in the usual way:

$$s^2 = \frac{1}{\kappa m \rho} = \frac{N}{m}\frac{\partial \mu}{\partial N}. \tag{1.49}$$

We now calculate $\partial\mu/\partial N$ or, rather, its inverse $\partial N/\partial\mu$. Let us increase μ by an infinitesimal amount $d\mu$. The Fermi surface swells slightly. (We assume reflection invariance, in order to have the same Fermi surface for both spin directions.) An arbitrary point A of the original Fermi surface undergoes a normal displacement

$$dp_F = (\partial p_F/\partial\mu)\, d\mu \tag{1.50}$$

which brings it to the position B (Fig. 1.2). On the new Fermi surface, the quasiparticle energy must be equal to $(\mu + d\mu)$: this fixes the displacement dp_F. More precisely, B must be such that

$$\epsilon_B(\mu + d\mu) - \epsilon_A(\mu) = d\mu. \tag{1.51}$$

The left-hand side of Eq. (1.51) contains two terms:

(i) When passing from A to B, the momentum $\mathbf{p}$ changes, giving rise to an energy shift $v_\mathbf{p}\, dp_F$;

(ii) Increasing μ by an amount $d\mu$ means adding quasiparticles in the shaded area of Fig. 1.2. ϵ_p is then increased by the interaction energy with these extra quasiparticles.

Let $\delta n_{\mathbf{p}}$ be the change of the distribution function resulting from the displacement of the Fermi surface. $\delta n_{\mathbf{p}}$ is equal to $+1$ in the shaded area of Fig. 1.2 and to 0 outside. Collecting the two contributions to Eq. (1.51), we obtain

$$v_{\mathbf{p}}\,dp_F + \sum_{\mathbf{p}'} f_{\mathbf{pp}'}\delta n_{\mathbf{p}'} = d\mu. \tag{1.52}$$

Since $f_{\mathbf{pp}'}$ is a smooth function of $\mathbf{p}$ and $\mathbf{p}'$, we may replace the above $\delta n_{\mathbf{p}}$ by a suitably normalized δ-function. The correct choice is easily verified to be (at zero temperature)

$$\delta n_{\mathbf{p}} = -\frac{dn^\circ}{d\epsilon_p}\frac{\partial \epsilon_p}{\partial p}\,dp_F = \delta(\epsilon_p - \mu)v_{\mathbf{p}}\,dp_F. \tag{1.53}$$

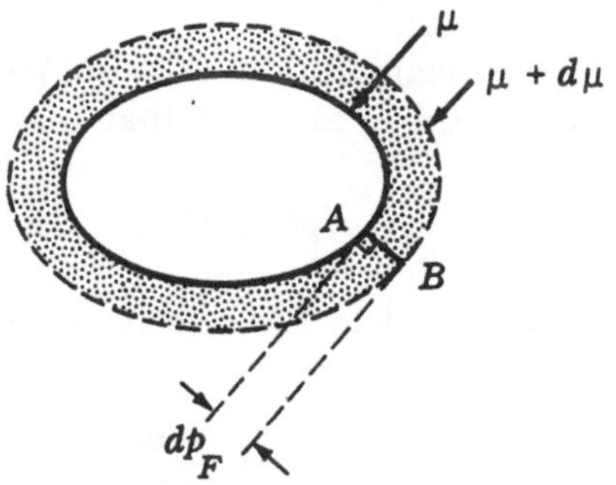

FIGURE 1.2. *Deformation of the Fermi surface when the chemical potential is increased from μ to $(\mu + d\mu)$.*

Inserting this into Eq. (1.52), and dividing by $d\mu$, we obtain

$$v_{\mathbf{p}}(\partial p_F/\partial\mu) + \sum_{\mathbf{p}'} f_{\mathbf{pp}'}v_{\mathbf{p}'}(\partial p'_F/\partial\mu)\delta(\epsilon_{p'} - \mu) = 1. \tag{1.54}$$

This integral equation gives $\partial p_F/\partial\mu$ in every direction: we have thus derived the change of the Fermi surface as μ varies. In order to get $\partial N/\partial\mu$, we remark that

$$dN = \sum_{\mathbf{p}} \delta n_{\mathbf{p}} = \sum_{\mathbf{p}} \delta(\epsilon_p - \mu)v_{\mathbf{p}}\,dp_F. \tag{1.55}$$

The sound velocity is then given by

$$N/ms^2 = \partial N/\partial\mu = \sum_{\mathbf{p}} \delta(\epsilon_p - \mu)v_{\mathbf{p}}(\partial p_F/\partial\mu). \tag{1.56}$$

For an isotropic system, $\partial p_F/\partial\mu$ is independent of direction. In that case, only the $\ell = 0$, spin-independent term of $f_{\mathbf{pp}'}$ contributes to Eq. (1.54). Using the definition (1.23), we find that

$$v_{\mathbf{p}}(\partial p_F/\partial\mu) = \frac{1}{1 + F_0^s}. \tag{1.57}$$

Inserting this into Eq. (1.56), we finally obtain

$$\frac{N}{ms^2} = \frac{\partial N}{\partial\mu} = \frac{\nu(0)}{1 + F_0^s}. \tag{1.58}$$

In Eq. (1.58) we have the usual result for independent particles (of mass m^*) modified by the factor $(1 + F_0^s)^{-1}$; the latter term is a direct consequence of the interaction between the quasiparticles.

We have gone at length through this demonstration to show clearly how the new features brought about by the particle interaction came into the problem. Actually, the result (1.58) may be readily obtained if we note that the *local energy* of a quasiparticle on the Fermi surface is always equal to the chemical potential, μ. When μ increases by an amount $d\mu$, the local Fermi energy is increased by the same amount. The new distribution may thus be written as $n^\circ(\bar{\epsilon}_{\mathbf{p}} - \mu - d\mu)$. The departure from the local equilibrium distribution $n^\circ(\bar{\epsilon}_{\mathbf{p}} - \mu)$ is equal to

$$\delta\bar{n}_{\mathbf{p}} = -d\mu(\partial n^\circ/\partial\epsilon_p). \tag{1.59}$$

For an isotropic system, $\delta\bar{n}_{\mathbf{p}}$ is isotropic and spin independent. Making use of (1.34), we obtain

$$\delta n_{\mathbf{p}} = \frac{\delta\bar{n}_{\mathbf{p}}}{1 + F_0^s} = -\frac{\partial n^\circ/\partial\epsilon_p}{1 + F_0^s}\,d\mu, \tag{1.60}$$

from which Eqs. (1.57) and (1.58) follow at once. This derivation stresses the usefulness of $\delta\bar{n}_{\mathbf{p}}$ for practical calculations.

Let us now return to the result (1.58). Replacing $\nu(0)$ by its expression (1.42), we obtain the sound velocity

$$s^2 = (p_F^2/3mm^*)(1 + F_0^s). \tag{1.61}$$

Since m^* is known from the specific heat, a measurement of s yields F_0^s. In the limit of weak interactions, $m^* \to m$, while $F_0^s \to 0$. The velocity of sound then tends toward $v_F/\sqrt{3}$. We note that Eq. (1.61) only makes sense if $(1 + F_0^s) > 0$. Otherwise, the system is unstable: an imaginary value of sound velocity signifies an exponential buildup of density fluctuations.

SPIN SUSCEPTIBILITY

Let us apply to our Fermi liquid a uniform dc magnetic field $\mathcal{H}$. As usual, the system will develop a paramagnetic spin moment and a diamagnetic orbital moment. We shall postpone until Sections 4.7 and 5.4 consideration of the orbital effects which, although important, are much more difficult to treat.

In the field $\mathcal{H}$, a particle of spin $\sigma = \pm\frac{1}{2}$ acquires a magnetic energy $-g\beta\sigma\mathcal{H}$, where $g = 2$ is the Landé g-factor and β the Bohr magneton, $e\hbar/2mc$. The two spin orientations are no longer in equilibrium, having different chemical potentials $(\mu \pm g\beta\mathcal{H}/2)$. In order to restore equilibrium, the Fermi surfaces for spins $\pm\frac{1}{2}$ must split, in such a way as to have the same chemical potential. In weak fields, the two Fermi surfaces are displaced by opposite amounts, the final chemical potential remaining equal to μ (within corrections of order $\mathcal{H}^2$).

The displacement of the Fermi surface with spin σ must be such as to bring the corresponding chemical potential from $(\mu - g\beta\sigma\mathcal{H})$ back to μ. Since the chemical potential is equal to the *local* energy of a quasiparticle on the Fermi surface, the equilibrium distribution in the presence of the field may be written as

$$n_{\mathbf{p}\sigma}(\mathcal{H}) = n^{\circ}(\bar{\epsilon}_{\mathbf{p}} - \mu - g\beta\sigma\mathcal{H}). \tag{1.62}$$

By repeating the argument which led to Eq. (1.59), we obtain the departure from local equilibrium

$$\delta\bar{n}_{\mathbf{p}\sigma} = -g\beta\sigma\mathcal{H}\frac{\partial n^{\circ}}{\partial\epsilon_p}. \tag{1.63}$$

In order to calculate the total magnetization

$$\mathfrak{M} = \sum_{\mathbf{p}} g\beta\sigma\,\delta n_{\mathbf{p}\sigma}, \tag{1.64}$$

we must pass from $\delta\bar{n}_{\mathbf{p}\sigma}$ to $\delta n_{\mathbf{p}\sigma}$. This is easily done if we assume the system to be isotropic. $\delta\bar{n}_{\mathbf{p}\sigma}$ is then isotropic and spin *antisymmetric*. Making use of Eq. (1.34), we find

$$\delta n_{\mathbf{p}\sigma} = \frac{\delta\bar{n}_{\mathbf{p}\sigma}}{1 + F_0{}^a} = -\frac{g\beta\sigma\mathcal{H}}{1 + F_0{}^a}\frac{\partial n^{\circ}}{\partial\epsilon_p}. \tag{1.65}$$

The calculation of $\mathfrak{M}$ is then straightforward. One finds (putting $g = 2$)

$$\mathfrak{M} = \beta^2\frac{\nu(0)}{1 + F_0{}^a}\mathcal{H}. \tag{1.66}$$

The spin susceptibility χ_P is given by

$$\chi_P = \frac{\mathfrak{M}}{\Omega \mathcal{H}} = \frac{m^* p_F}{\pi^2 \hbar^3} \frac{\beta^2}{1 + F_0{}^a}. \tag{1.67}$$

We see in (1.67) that the spin susceptibility is modified by the exchange interaction $F_0{}^a$. We cannot derive χ_P from the simple knowledge of the density of states, as is the case for a noninteracting system. Indeed, by comparing experimental values of the specific heat (i.e., of m^*) and of the susceptibility, one can "measure" the coefficient $F_0{}^a$. Again, Eq. (1.67) is meaningful only when

$$1 + F_0{}^a > 0. \tag{1.68}$$

Otherwise, the long wavelength fluctuations of the magnetic moment become unstable: the system becomes ferromagnetic.

EXTENSION TO ELECTRONS IN METALS

To what extent can the Landau theory be applied to electrons in metals? In the present section, we shall answer this question in a rather general way, putting aside until Chapter 3 the new features brought about by the long range of the Coulomb interaction.

Let us first summarize the main properties of a system of *noninteracting* electrons which move in the periodic potential of the crystalline lattice of a solid. The single-particle eigenstates are Bloch waves, with a wave function

$$\psi_{\mathbf{p}n}(\mathbf{r}) = e^{i\mathbf{p}\cdot\mathbf{r}} u_{\mathbf{p}n}(\mathbf{r}), \tag{1.69}$$

where $u_{\mathbf{p}n}$ has the periodicity of the lattice. The eigenstates are characterized by two quantum numbers, a "band index" n, and a wave vector $\mathbf{p}$, the latter lying in the first Brillouin zone of the crystal. The Bloch wave (1.69) may be considered as a mixture of plane waves, each having a wave vector $(\mathbf{p} + \mathbf{K})$ where $\mathbf{K}$ is a vector of the reciprocal lattice. If, as above, the wave vectors are restricted to the first Brillouin zone, we may conclude that the wave vector (or the momentum) is still a good quantum number, despite the presence of a periodic potential.

The ground state of an N-electron system is obtained by filling the N lowest Bloch wave states. The solid is a metal whenever there remain unfilled bands. The ground state distribution then levels off on some surface in momentum space, lying in the first Brillouin zone, which is called the Fermi surface S_F. We note that several bands may remain unfilled: the Fermi surface then possesses several sheets, one per unfilled band.

The Fermi surface has the symmetry of the crystalline lattice; it will certainly be anisotropic. Due to invariance under time reversal, the Fermi surface is invariant under combined spin inversion and spatial reflection with respect to the origin. If the crystal does not have a center of symmetry, the Fermi surfaces corresponding to the two directions of spin will be different.

Having surveyed very quickly the fundamentals of the one-electron approximation in metals, we now ask how Landau's theory can be used to take account of the Coulomb interaction between electrons. By using the adiabatic switching procedure introduced in Section 1.2, we can again establish a one-to-one correspondence between the eigenstates of the real system and those of the noninteracting system. For a *normal* metal, the ground state will be adiabatically generated starting from some "ideal" eigenstate characterized by a Fermi surface S_F (as we have mentioned S_F will most likely correspond to an excited state of the noninteracting system). In the same way, we can define a quasiparticle by adding one particle with quantum numbers $(n, \mathbf{p})$ to the noninteracting system, and then switching on the interaction very slowly. Since the total momentum is conserved in the interaction, the final quasiparticle will possess the same quantum numbers as the original Bloch wave, namely a momentum $\mathbf{p}$ in the first Brillouin zone and a band index. The quasiparticle thus acquires the same characteristics as a Bloch wave, and we may define a Fermi surface, etc.

Because of the damping due to particle collisions, the concept of a quasiparticle is only valid in the immediate vicinity of the Fermi surface, where damping effects are negligible. This represents a major departure from the one-electron approximation, in which Bloch waves were defined over the entire Brillouin zone. However, for most practical purposes this restriction is unimportant. Most physical effects, such as transport properties, cyclotron resonance, etc., involve thermally excited electrons, lying within a distance κT from the Fermi surface. κT is typically 10^{-4} to 10^{-2} eV, far smaller than the Fermi energy, of the order of 10 eV. The percentage of excited quasiparticles will always remain very small, so that we can neglect their damping.

In the absence of interaction, a Bloch wave possesses an energy $\epsilon_{n\mathbf{p}}^0$, which depends on $\mathbf{p}$ in a rather complicated way. Everywhere on the Fermi surface, we may define a velocity $\mathbf{v_p}^0$, equal to the gradient of $\epsilon_{n\mathbf{p}}^0$. As a result of the periodic potential $\mathbf{v_p}^0$ will be different from the velocity $\hbar\mathbf{p}/m$ of a free particle. Let us now turn on the interaction. The quasiparticles have an energy $\epsilon_{n\mathbf{p}}$, equal to the first derivative of the energy functional. $\epsilon_{n\mathbf{p}}$ is constant over the Fermi surface, and is equal to the chemical potential μ. We can again define at every point of S_F a velocity $\mathbf{v_p}$, which will be different from $\mathbf{v_p}^\circ$. The difference between $\mathbf{v_p}$ and $\hbar\mathbf{p}/m$ is now due to two distinct effects:

(a) The influence of the periodic lattice on each electron;
(b) The "many-body" effect arising from the Coulomb interaction.

In the case of a nearly isotropic Fermi surface, such as one finds in the alkali metals, we can write

$$v_{p_F} = \frac{\hbar p_F}{m^*}, \tag{1.69a}$$

where m^*, the effective mass, represents the combined influence of the periodic field and electron interactions. Usually these two "influences" are so deeply intermingled that it is not possible to disentangle them. We return to this question in Sections 3.8 and 5.6.

Just as for the Fermi liquid, by considering the second variational derivative of the energy with respect to $n_\mathbf{p}$, we can define an interaction energy between quasiparticles, $f_{\mathbf{pp'}}$. However, since the Fermi surface of a metal is not, in general, isotropic, $f_{\mathbf{pp'}}$ depends on the directions of both $\mathbf{p}$ and $\mathbf{p'}$. As a result, the calculations of quasiparticle properties are very much more complicated. Consider, for example, the preceding calculation of the paramagnetic susceptibility, which is, in principle, still valid for metals. Where there is anisotropy, the displacement of the Fermi surface, $dp_F/d\mathcal{H}$, varies over S_F, and satisfies an integral equation which must be solved to obtain χ_P. Gone, then, is the simplicity of a result like Eq. (1.67); indeed, there may appear so many parameters that a comparison between theory and experiment is not very fruitful. We conclude that comparison between theory and experiment for equilibrium properties is likely to be fruitful only in cases in which the Fermi surface is essentially isotropic.

Another complication in dealing with electrons in metals arises from the long range of the Coulomb interaction. We shall discuss ways of dealing with this in Chapter 3; here we merely remark that the above considerations go through formally provided one regards $f_{\mathbf{pp'}}$ as the *screened* quasiparticle interaction. Thus when due account is taken of screening it is possible to define an electronic compressibility which, for a metal with an isotropic Fermi surface, is related to the state density and $F_0{}^s$ through Eqs. (1.48) and (1.58),

$$N\rho\kappa = \frac{\nu(0)}{1 + F_0{}^s}. \tag{1.69b}$$

Again, let us emphasize that a simple formula like (1.69b) applies only to the case of an essentially isotropic electronic Fermi surface.

Because of the difficulties with anisotropy, one of the most interesting applications of the Landau theory to metal physics is the determination of those effects which are *not* influenced by particle interaction. Phenomena in which the interaction, $f_{\mathbf{pp}'}$, plays no role may then be safely used to determine the Fermi surface and other properties. We return to this question in our consideration of electron transport phenomena in Chapter 3.

We mention one final complication for electrons in metals. Not only do the electrons interact with each other, they interact as well with the lattice vibrations or *phonons* of the crystal. One must therefore inquire in detail as to which phenomena are influenced by this further electron-phonon interaction. Of the equilibrium properties considered here, it has recently been shown [Simkin (1963), Kadanoff and Prange (1964)] that the spin susceptibility and compressibility are *not* influenced by the electron-phonon interaction, while the specific heat is altered. Thus Eqs. (1.67) and (1.69b) may be applied directly to the case of an isotropic electron system, while the m^* one measures in a specific heat experiment will differ from that one would estimate via Eq. (1.69a) on the basis of electron-electron interactions alone. A comparison between theory and experiment for the equilibrium properties of some simple. isotropic metals is given in Section 5.6.

1.4. TRANSPORT EQUATION FOR QUASIPARTICLES

DEFINITION OF NONHOMOGENEOUS DISTRIBUTION FUNCTIONS

Until now, we have considered stable, homogeneous distributions, for which the function $n_{\mathbf{p}}$ was independent of both position and time. We shall now consider a more general problem, in which a weak time-dependent inhomogeneous perturbation is applied to the ground state. Such a perturbation may arise either from an external field, or as a consequence of an internal fluctuation of the system. We assume the perturbation is sufficiently weak that the response of the system is linear. (The formal theory of linear response functions is developed in Chapter 2.)

Let us consider a distribution function $n_{\mathbf{p}}(\mathbf{r}, t)$, which depends explicitly on both the position $\mathbf{r}$ and the time t. Such a description obviously violates the uncertainty principle: it is not possible to specify simultaneously both the momentum $\mathbf{p}$ of a quasiparticle and its position $\mathbf{r}$, or its energy ϵ_p and the time t. This is a major difficulty, which fortunately disappears if we restrict ourselves to *macroscopic* phenomena, such that the typical wave vectors and frequencies remain much smaller than the corresponding atomic parameters.

Let us consider the Fourier transform of $n_{\mathbf{p}}(\mathbf{r}, t)$ with respect to space and time. Within our assumption of linear response, each Fourier component may be treated independently. Thus, it suffices to consider a particular plane wave perturbation, of wave vector $\mathbf{q}$ and frequency ω. We shall write the distribution function as

$$n_{\mathbf{p}}(\mathbf{r}, t) = n_{\mathbf{p}}^{\circ} + \delta n_{\mathbf{p}}(\mathbf{q}, \omega) e^{i(\mathbf{q}\cdot\mathbf{r} - \omega t)}. \tag{1.70}$$

The uncertainty principle gives rise to an uncertainty $\hbar\mathbf{q}$ in the momentum $\mathbf{p}$, and $\hbar\omega$ in the energy ϵ_p. This will be relatively unimportant if the distribution function $n_{\mathbf{p}}$ is smooth enough over that range of $\mathbf{p}$ and ϵ_p. At a temperature T, the characteristic "width" of the Fermi surface is κT for the energy, $\kappa T/v_F$ for the momentum. We might thus expect the distribution function $\delta n_{\mathbf{p}}(\mathbf{q}, \omega)$ to make sense if the following conditions are met:

$$\begin{aligned} \hbar q v_F &\ll \kappa T, \\ \hbar\omega &\ll \kappa T. \end{aligned} \tag{1.71}$$

Under such conditions, the concept of a distribution function is meaningful: we are essentially in a *classical* regime, at least as far as the uncertainty principle is concerned.

The conditions (1.71) are rather restrictive. Actually, the Landau theory applies in the much broader range

$$\begin{aligned} \hbar q v_F &\ll \mu, \\ \hbar\omega &\ll \mu, \end{aligned} \tag{1.72}$$

with, however, a different interpretation of the distribution function $\delta n_{\mathbf{p}}(\mathbf{q}, \omega)$. Thus far, we have defined $\delta n_{\mathbf{p}}$ as the *probability* for finding a quasiparticle with wave vector $\mathbf{p}$. In order to extend the Landau theory to the range (1.72), we must consider $\delta n_{\mathbf{p}}$ as the *probability amplitude* for finding a "pair," consisting in a quasiparticle of momentum $(\mathbf{p} + \hbar\mathbf{q}/2)$ and a quasihole with momentum $(\mathbf{p} - \hbar\mathbf{q}/2)$. (Obviously, this no longer violates the uncertainty principle, since the momentum is not exactly equal to $\mathbf{p}$.) Such a description is equivalent to Wigner's semiclassical approach to statistical mechanics. We shall discuss this point further in Chapter 5. Here we simply assume that the uncertainty principle can be ignored when the conditions (1.72) are met.

According to (1.72), Landau's theory may be applied to a *macroscopic* perturbation, the scale of which is very large compared to the atomic scale. Thus the wavelength of the perturbation must be much larger than the interparticle distance, the frequency much smaller than typical atomic frequencies. Put another way, the Landau result represents the leading term of an expansion of the system response in powers of q/k_F and ω/μ. This limitation of the theory must be kept in mind.

EXPANSION OF THE ENERGY

Let us consider a state characterized by a distribution function

$$n_{\mathbf{p}}(\mathbf{r}, t) = n_{\mathbf{p}}^{\circ} + \delta n_{\mathbf{p}}(\mathbf{r}, t). \tag{1.73}$$

The departure $\delta n_{\mathbf{p}}$ from the ground state is supposed to be small. We assume $\delta n_{\mathbf{p}}$ to contain only long wavelength fluctuations, and to be restricted to the vicinity of the Fermi surface, where quasiparticles are well defined. Note that $n_{\mathbf{p}}$ gives the distribution in a *unit* volume centered at point $\mathbf{r}$: the momentum $\mathbf{p}$ is accordingly quantized for $\Omega = 1$.

As before, the total energy is a functional $E[n_{\mathbf{p}}(\mathbf{r}, t)]$ of the distribution function. If $\delta n_{\mathbf{p}}$ is small, we may perform a Taylor expansion of this functional:

$$E = E_{\circ} + \sum_{\mathbf{p}} \int d^3\mathbf{r}\, \epsilon(\mathbf{p}, \mathbf{r})\delta n_{\mathbf{p}}(\mathbf{r}) + \frac{1}{2}\sum_{\mathbf{p}\mathbf{p}'} \int\!\!\int d^3\mathbf{r}\, d^3\mathbf{r}'\, f(\mathbf{p}\mathbf{r}, \mathbf{p}'\mathbf{r}')\delta n_{\mathbf{p}}(\mathbf{r})\delta n_{\mathbf{p}'}(\mathbf{r}') + \cdots. \tag{1.74}$$

Let us assume our system is invariant under spatial translation. It follows that $\epsilon(\mathbf{p}, \mathbf{r})$ does not depend on $\mathbf{r}$, and is thus equal to ϵ_p. Furthermore $f(\mathbf{p}\mathbf{r}, \mathbf{p}'\mathbf{r}')$ then depends only on the difference $(\mathbf{r} - \mathbf{r}')$.

At this stage, we must take account of the nature of the interaction. In this chapter we are only interested in *short range* forces, which are substantial only over distances of atomic size: this is the case for ^{3}He. If the perturbation corresponds to a macroscopic variation in space and time, $\delta n_{\mathbf{p}}$ may be considered as constant over the range of the interaction: in (1.74) we may replace $\delta n_{\mathbf{p}'}(\mathbf{r}')$ by $\delta n_{\mathbf{p}'}(\mathbf{r})$. The energy can then be written as

$$\begin{aligned} E &= E_{\circ} + \int d^3\mathbf{r}\, \delta E(\mathbf{r}), \\ \delta E(\mathbf{r}) &= \sum_{\mathbf{p}} \epsilon_p \delta n_{\mathbf{p}}(\mathbf{r}) + \frac{1}{2}\sum_{\mathbf{p}\mathbf{p}'} f_{\mathbf{p}\mathbf{p}'}\delta n_{\mathbf{p}}(\mathbf{r})\delta n_{\mathbf{p}'}(\mathbf{r}), \end{aligned} \tag{1.75}$$

where the interaction energy $f_{\mathbf{p}\mathbf{p}'}$ is defined by

$$f_{\mathbf{p}\mathbf{p}'} = \int d^3\mathbf{r}'\, f(\mathbf{p}\mathbf{r}, \mathbf{p}'\mathbf{r}'). \tag{1.76}$$

The energy is thus a *local* function of the distribution $\delta n_{\mathbf{p}}$: at every point, we find the same relation as for an homogeneous system. The $f_{\mathbf{pp'}}$ defined in (1.76) is the same as that of the preceding sections, calculated for $\Omega = 1$. (Let us emphasize again that this conclusion holds only for *macroscopic* perturbations.) It is physically clear that the *local* character of the energy is a direct consequence of the *short range* of the interaction.

TRANSPORT EQUATION FOR QUASIPARTICLES

According to Eq. (1.75), the local excitation energy of a quasiparticle $\mathbf{p}$ is equal to

$$\tilde{\epsilon}_{\mathbf{p}}(\mathbf{r}) = \epsilon_p + \sum_{\mathbf{p'}} f_{\mathbf{pp'}} \delta n_{\mathbf{p'}}(\mathbf{r}). \tag{1.77}$$

Note that this energy depends on $\mathbf{p}$ *and* $\mathbf{r}$. The gradient $\nabla_{\mathbf{p}}\tilde{\epsilon}$ gives the velocity of the quasiparticle, as usual. On the other hand, $(-\nabla_{\mathbf{r}}\tilde{\epsilon})$ is equivalent to a kind of "diffusion" force, which tends to push quasiparticles toward regions of minimum energy.

In order to describe the transport properties, Landau considered the quasiparticles as independent, described by a classical Hamiltonian $\tilde{\epsilon}_{\mathbf{p}}(\mathbf{r})$. The problem is thus reduced to the development of the appropriate kinetic theory for a gas of quasiparticles. We shall discuss the physical meaning of this assumption in a moment: we first establish the transport equation.

We shall use a well-known procedure of kinetic theory: we consider a small volume element in phase space $d^3\mathbf{p}\, d^3\mathbf{r}$, and we calculate the flow of quasiparticles through each side of this element. By establishing the balance of the flow inward and outward, we obtain the equation

$$\frac{\partial n_{\mathbf{p}}}{\partial t} + \nabla_{\mathbf{r}} n_{\mathbf{p}} \cdot \nabla_{\mathbf{p}} \tilde{\epsilon}_{\mathbf{p}} - \nabla_{\mathbf{p}} n_{\mathbf{p}} \cdot \nabla_{\mathbf{r}} \tilde{\epsilon}_{\mathbf{p}} = 0. \tag{1.78}$$

This well-known result governs the flow of quasiparticles in phase space, in the absence of collisions and external forces.

Equation (1.78) refers to the total distribution function for all quasiparticles. But we know that $n_{\mathbf{p}}$ is only defined in the vicinity of the Fermi surface. Furthermore, the concept of independent quasiparticles assumed by Landau cannot be true for all values of the momentum $\mathbf{p}$: it is physically obvious that only *excited* quasiparticles in the immediate vicinity of the Fermi surface can be considered as independent. We must therefore extract from (1.78) a transport equation describing the flow of only excited quasiparticles. This is easily done if we write

$n_{\mathbf{p}}$ in the form (1.73). If we keep only terms of first order in $\delta n_{\mathbf{p}}$, and use (1.77), we obtain

$$\frac{\partial \delta n_{\mathbf{p}}(\mathbf{r}t)}{\partial t} + \nabla_{\mathbf{r}} \delta n_{\mathbf{p}}(\mathbf{r}t) \cdot \mathbf{v}_{\mathbf{p}} - \nabla_{\mathbf{p}} n_{\mathbf{p}}{}^{\circ} \cdot \sum_{\mathbf{p}'} f_{\mathbf{p}\mathbf{p}'} \nabla_{\mathbf{r}} \delta n_{\mathbf{p}'}(\mathbf{r}t) = 0. \qquad (1.79)$$

This linearized transport equation involves only values of $\mathbf{p}$ close to the Fermi surface, thanks to the factor $\nabla_{\mathbf{p}} n_{\mathbf{p}}{}^{\circ} = -\mathbf{v}_{\mathbf{p}} \delta(\epsilon_p - \mu)$.

Let us analyze in some detail the physical meaning of Eq. (1.79). The first two terms (which provide the usual transport equation for a noninteracting system), describe a flow of totally independent excited quasiparticles. The last term, which arises from their interaction, may be interpreted as a flow of the ground state particles dragged by the inhomogeneities of the excitation distribution. We are thus led to the following picture: the elementary excitations, few in number, are completely independent. In homogeneous systems, they do not interact with the "ground state" particles. However, if $\delta n_{\mathbf{p}}$ is not homogeneous, the excited quasiparticles create a force field which acts on the ground state distribution, and distorts it; this effect is of first order in $\delta n_{\mathbf{p}}$, and thus quite important.

The neglect of the interaction between two *excited* quasiparticles is quite reasonable at low temperatures, where their density is small, except for very low-frequency phenomena (see Section 1.9). On the other hand, reducing the interaction between the excited particles and the ground state particles to an *average* macroscopic force is a rather bold assumption. One might expect microscopic correlations to be of importance. According to Landau, they do not play any role: this assumption is indeed true, and may be proved within the framework of perturbation theory. One may think of this result as arising from the exclusion principle, which renders the ground state distribution essentially rigid.

The linearized transport equation may be written very simply in terms of the departure from local equilibrium $\delta \bar{n}_{\mathbf{p}}$. On comparing the definition (1.30) with (1.79), we may write the latter in the form

$$\frac{\partial \delta n_{\mathbf{p}}(\mathbf{r}, t)}{\partial t} + \mathbf{v}_{\mathbf{p}} \cdot \nabla_{\mathbf{r}} \delta \bar{n}_{\mathbf{p}}(\mathbf{r}, t) = 0. \qquad (1.80)$$

This compact result may also be obtained directly from Eq. (1.78), by writing $n_{\mathbf{p}}$ in the form (1.28), i.e.,

$$n_{\mathbf{p}}(\mathbf{r}, t) = n^{\circ}(\bar{\epsilon}_{\mathbf{p}}) + \delta \bar{n}_{\mathbf{p}}(\mathbf{r}, t). \qquad (1.81)$$

The function $n^\circ(\bar{\epsilon}_\mathbf{p})$ gives a contribution to the gradient terms which may be written as

$$\left(\frac{\partial n^\circ}{\partial \epsilon_p} \nabla_\mathbf{r}\bar{\epsilon}_\mathbf{p}\right) \cdot \nabla_\mathbf{p}\bar{\epsilon}_\mathbf{p} - \left(\frac{\partial n^\circ}{\partial \epsilon_p} \nabla_\mathbf{p}\bar{\epsilon}_\mathbf{p}\right) \cdot \nabla_\mathbf{r}\bar{\epsilon}_\mathbf{p}$$

and which therefore vanishes. The only contribution to these gradients arises from $\delta\bar{n}_\mathbf{p}$; if we keep only the first-order terms, we arrive at Eq. (1.80).

According to (1.80), the transport equation involves the time derivative of $\delta n_\mathbf{p}$, and the spatial derivative of $\delta\bar{n}_\mathbf{p}$. This difference may be understood by noting that the gradient terms describe the *diffusion* of quasiparticles, which is certainly governed by the local energy. In this respect, the departure from equilibrium is measured by $\delta\bar{n}_\mathbf{p}$, not by $\delta n_\mathbf{p}$.

In its form (1.80), our transport equation misses two important effects: the influence of external forces and the collisions between excited quasiparticles. Let us first consider collisions, which represent the "dissipative," irreversible part of the interaction between excited quasiparticles (the interaction energy $f_{\mathbf{pp}'}$ being, by contrast, a reactive, reversible effect of that interaction). The collisions are similar in nature to those between molecules in the usual kinetic theory of gases. Their importance is qualitatively measured by some collision frequency ν.

At the low temperatures in which we are interested, collisions are inhibited by the exclusion principle, and are thus comparatively infrequent: ν is small. Collisions will therefore play a role only in very low-frequency phenomena ($\omega \lesssim \nu$) (viscosity, thermal conduction, etc.) where they limit the response to an external force. We shall discuss these problems in some detail in Section 1.8, where we shall see that one can take account of collisions by adding to the right-hand side of the transport equation a "collision integral" $I(\delta n_\mathbf{p})$, which measures the rate of change of $\delta n_\mathbf{p}$ due to collisions. If we work at a frequency ω, much larger than the collision frequency ν, we expect the collisions to play no role: we can then drop the collision integral.

Finally, let us apply a force $\mathfrak{F}_\mathbf{p}$ to the quasiparticle $\mathbf{p}$: the distribution drifts in momentum space. This gives rise to an additional term on the left-hand side of Eq. (1.80), equal to

$$\mathfrak{F}_\mathbf{p} \cdot \nabla_\mathbf{p} n_\mathbf{p}(\mathbf{r}, t). \tag{1.82}$$

Usually, the excitation $\delta n_\mathbf{p}$ of the system is proportional to the applied force $\mathfrak{F}$; in expression (1.82), we may then replace $n_\mathbf{p}$ by the equilibrium distribution $n_\mathbf{p}^\circ$, since we are concerned only with first-order changes in n.

In most problems, we know from first principles the force exerted on a *bare* particle. In order to deduce the force $\mathfrak{F}_\mathbf{p}$ felt by a *quasi*particle, we generally need a detailed knowledge of the structure of the self-energy cloud: this is well beyond the scope of the Landau theory. Actually, in dealing with neutral systems one is primarily interested in the response to a *scalar* field, which produces a force proportional to the density of the system. Since the number of quasiparticles is equal to that of bare particles, the force $\mathfrak{F}_{\mathbf{p}}$ felt by the quasiparticle is then equal to that felt by a bare particle. We can thus obtain an explicit expression for the transport equation in this case. (We shall see in Chapter 3 that a similar conclusion can be reached if we apply a vector field to a charged system.)

By collecting all the new terms, we write the transport equation (1.80) in its final form

$$\frac{\partial \delta n_{\mathbf{p}}}{\partial t} + \mathbf{v}_{\mathbf{p}} \cdot \nabla_{\mathbf{r}} \delta \bar{n}_{\mathbf{p}} + \mathfrak{F}_{\mathbf{p}} \cdot \mathbf{v}_{\mathbf{p}} \frac{\partial n^{\circ}}{\partial \epsilon_{p}} = I(\delta n_{\mathbf{p}}). \tag{1.83}$$

Combined with the relation (1.30) between $\delta n_{\mathbf{p}}$ and $\delta \bar{n}_{\mathbf{p}}$, the transport equation appears as an integral equation with respect to $\delta n_{\mathbf{p}}$, even in the absence of collisions. It is thus markedly more complicated than the corresponding equation for a noninteracting system (in which one has $\delta n_{\mathbf{p}} = \delta \bar{n}_{\mathbf{p}}$).

The transport equation, (1.83), is extremely important. From it we shall derive all the transport properties of a Fermi liquid, as well as the fluctuation spectrum of the system on a macroscopic scale. The remainder of Chapter 1 is devoted to an exposition of the consequences of Eq. (1.83).

1.5. CALCULATION OF THE CURRENT DENSITY

DERIVATION VIA THE CONSERVATION LAW

Let $\mathbf{J}(\mathbf{r}, t)$ be the particle current density at point $\mathbf{r}$, time t. It is tempting to argue that a given quasiparticle carries a current $\mathbf{v}_{\mathbf{p}}$, and that the total particle current is

$$\mathbf{J} = \sum_{\mathbf{p}} \mathbf{v}_{\mathbf{p}} \delta n_{\mathbf{p}} \quad (?) \tag{1.84}$$

This result is *false*, as we shall now demonstrate.

In order to calculate $\mathbf{J}$, we note that it is related to the particle density ρ by the conservation law

$$\frac{\partial \rho}{\partial t} + \operatorname{div} \mathbf{J} = 0, \tag{1.85}$$

which insures that the number of particles is everywhere conserved. The density fluctuation $\delta\rho$ may be written as

$$\delta\rho(\mathbf{r}, t) = \sum_{\mathbf{p}} \delta n_{\mathbf{p}}(\mathbf{r}, t) \tag{1.86}$$

(since the numbers of particles and quasiparticles are equal). We wish to find an expression for $\mathbf{J}(\mathbf{r}, t)$ in terms of the distribution $\delta n_{\mathbf{p}}(\mathbf{r}, t)$ such that (1.85) is satisfied whatever the spatial variation of $\delta n_{\mathbf{p}}$.

For that purpose, we write the transport equation, in the absence of an external force

$$\frac{\partial \delta n_{\mathbf{p}}}{\partial t} + \mathbf{v}_{\mathbf{p}} \cdot \nabla_{\mathbf{r}} \delta \bar{n}_{\mathbf{p}} = I(\delta n_{\mathbf{p}}). \tag{1.87}$$

Let us now sum Eq. (1.87) over the variable $\mathbf{p}$: the contribution of the collision integral is zero, since collisions conserve the number of particles. We thus get

$$\frac{\partial}{\partial t} \delta\rho + \nabla_{\mathbf{r}} \cdot \sum_{\mathbf{p}} \delta \bar{n}_{\mathbf{p}} \mathbf{v}_{\mathbf{p}} = 0. \tag{1.88}$$

Comparing Eqs. (1.85) and (1.88), we obtain our desired expression for the current density

$$\mathbf{J} = \sum_{\mathbf{p}} \delta \bar{n}_{\mathbf{p}} \mathbf{v}_{\mathbf{p}}. \tag{1.89}$$

Our naive guess (1.84) is thus seen to be incorrect; in place of $\delta n_{\mathbf{p}}$ one finds the departure from *local* equilibrium $\delta \bar{n}_{\mathbf{p}}$.

We shall often use Eq. (1.89) as it stands. However, it is sometimes desirable to express $\mathbf{J}$ in terms of the usual distribution $\delta n_{\mathbf{p}}$. By inserting the expression, (1.30), for $\delta \bar{n}_{\mathbf{p}}$ into (1.89), we may write $\mathbf{J}$ in the form

$$\mathbf{J} = \sum_{\mathbf{p}} \delta n_{\mathbf{p}} \mathbf{j}_{\mathbf{p}}, \tag{1.90}$$

where $\mathbf{j}_{\mathbf{p}}$ is given by

$$\mathbf{j}_{\mathbf{p}} = \mathbf{v}_{\mathbf{p}} - \sum_{\mathbf{p}'} f_{\mathbf{p}\mathbf{p}'} \frac{\partial n^{\circ}}{\partial \epsilon_{\mathbf{p}'}} \mathbf{v}_{\mathbf{p}'}. \tag{1.91}$$

$\mathbf{j}_{\mathbf{p}}$ thus appears as the *current* carried by the quasiparticle $\mathbf{p}$.

According to Eq. (1.91), the *current* $\mathbf{j}_{\mathbf{p}}$ is different from the *velocity* $\mathbf{v}_{\mathbf{p}}$. This very important feature is characteristic of an interacting system. When a quasiparticle $\mathbf{p}$ is added to the system, it carries one extra particle at a group velocity $\mathbf{v}_{\mathbf{p}}$; the corresponding contribution to the current is obviously equal to $\mathbf{v}_{\mathbf{p}}$. But, that is not the whole story! Because of the particle interaction, the moving quasiparticle tends to drag part of the "medium" along with it, thus producing an extra current: $\mathbf{j}_p$ will thus be different from $\mathbf{v}_p$; the difference $(\mathbf{j}_p - \mathbf{v}_p)$ may be described as a *drag current*.

Such a picture is only correct for a homogeneous quasiparticle excitation, corresponding to a plane wave. Suppose we instead consider a localized wave packet, which moves at the group velocity $\mathbf{v}_p$. To the extent that the wave packet contains as a whole *one* extra particle, the total current is clearly equal to $\mathbf{v}_p$, not to $\mathbf{j}_p$. The current is thus considerably modified by the *localization* of the quasiparticle. The difference between $\mathbf{v}_p$ and $\mathbf{j}_p$ may then be interpreted from a new vantage point: we may regard the localized quasiparticle as carrying a current $\mathbf{j}_p$, which is modified by the return flow of the other particles around the moving excitation. The difference $(\mathbf{v}_p - \mathbf{j}_p)$ thus appears as a *backflow current*, arising from the interaction of the moving wave packet with the surrounding fluid. This interpretation is complementary to that given earlier for a plane wave excitation. Note that the backflow current is opposite to the "drag current."

A similar analysis may be used to obtain the momentum and energy currents in the liquid. On multiplying the transport equation respectively by $\mathbf{p}$ and ϵ_p, and noting that the collisions conserve the total momentum and energy, one finds the following expressions for the momentum current tensor $\Pi_{\alpha\beta}$ and the energy current $\mathbf{Q}$:

$$\begin{aligned} \Pi_{\alpha\beta} &= \sum_{\mathbf{p}} \delta\bar{n}_{\mathbf{p}} p_\alpha v_{\mathbf{p}\beta}, \\ \mathbf{Q} &= \sum_{\mathbf{p}} \delta\bar{n}_{\mathbf{p}} \epsilon_p \mathbf{v}_{\mathbf{p}}. \end{aligned} \tag{1.92}$$

Equations (1.92) constitute an obvious extension of (1.89). Again, one may calculate the momentum and energy currents of a given quasiparticle $\mathbf{p}$. They are different from the single values $p_\alpha v_{\mathbf{p}\beta}$ and $\epsilon_p \mathbf{v}_{\mathbf{p}}$; this difference represents the flux of momentum and energy carried by the drag current. The detailed calculation is left as a problem to the reader.

QUASIPARTICLE CURRENT IN A TRANSLATIONALLY INVARIANT SYSTEM

The total current operator $\mathbf{J}$ is given by

$$\mathbf{J} = \sum_i \frac{\mathbf{p}_i}{m}, \tag{1.93}$$

where $\mathbf{p}_i$ is the momentum of the ith particle, and m its bare mass. If the system remains invariant in *any* translation, the total momentum $\mathbf{P}$ of the system is a good quantum number. We then have

$$\mathbf{J} = \mathbf{P}/m. \tag{1.94}$$

(Equation (1.94) does not hold for a real metal, since in that nontranslationally invariant case, the lattice may supply momentum in arbitrary multiples of the reciprocal lattice vectors, $\hbar/\mathbf{K}$.) Let us consider the state containing a single excited quasiparticle of momentum $\mathbf{p}$, which carries a current $\mathbf{j}_\mathbf{p}$. By applying Eq. (1.94) to that special case, we find that

$$\mathbf{j}_\mathbf{p} = \mathbf{p}/m. \tag{1.95}$$

For a translationally invariant system, the current $\mathbf{j}_\mathbf{p}$ is thus the same as in the absence of interaction; again, (1.95) does *not* hold for electrons in metals.

For a translationally invariant system, we may obtain the expression (1.91) for the current $\mathbf{j}_\mathbf{p}$ by a method which shows clearly the physical origin of the drag current. Let us first write the total current $\mathbf{J}$ as the expectation value of the current operator

$$\mathbf{J} = \left\langle \varphi \left| \sum_i \frac{\mathbf{p}_i}{m} \right| \varphi \right\rangle.$$

We now consider the effect of a translation of the entire system, Fermi surface as well as excited quasiparticles, by a constant amount $\mathbf{q}$; such a translation is equivalent to looking at the system from a moving frame of reference which has a constant velocity $(-\mathbf{q}/m)$. In this moving frame, the interaction energy remains unchanged, while the kinetic energy operator increases by an amount

$$\sum_i \left\{ \frac{\mathbf{q}\cdot\mathbf{p}_i}{m} + \frac{q^2}{2m} \right\}.$$

We next assume that $\mathbf{q}$ is small, and use elementary perturbation theory to obtain the first-order correction to the energy E:

$$\begin{aligned} \delta E &= \left\langle \varphi \left| \sum_i \frac{\mathbf{q}\cdot\mathbf{p}_i}{m} \right| \varphi \right\rangle + O(q^2) \\ &= \mathbf{q}\cdot\mathbf{J} + O(q^2). \end{aligned}$$

$\mathbf{J}$ thus appears as the first derivative of the energy with respect to $\mathbf{q}$, an arbitrary component J_α being given by

$$J_\alpha = dE/dq_\alpha. \tag{1.96a}$$

(We use the total derivative symbol d to emphasize that the derivative involves a translation of the *whole* system.) Equation (1.96a) is a direct consequence of the Galilean invariance of the system under translations.

In the ground state $|\varphi_o\rangle$, the current is zero by symmetry: dE_o/dq_α therefore vanishes. Let us consider the excited state $|\varphi_\mathbf{p}\rangle$ containing a single excited quasiparticle $\mathbf{p}$. According to Eq. (1.96a), the corresponding current $\mathbf{j}_\mathbf{p}$ may be written as

$$j_{p\alpha} = d\epsilon_p/dq_\alpha. \tag{1.96b}$$

In other words, $\mathbf{q} \cdot \mathbf{j}_p$ is the change in ϵ_p when *both* $\mathbf{p}$ and the Fermi surface are translated by an amount $\mathbf{q}$.

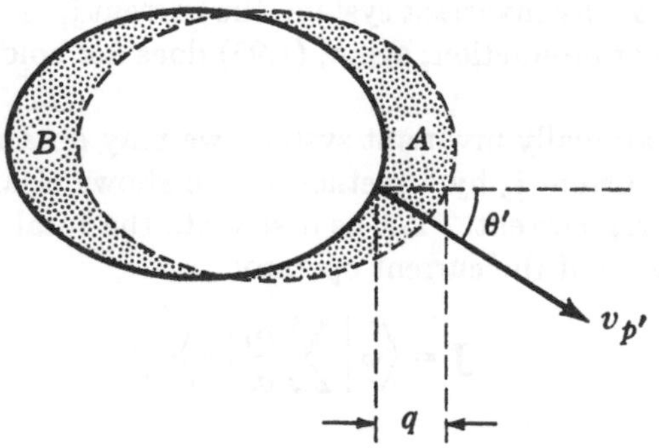

FIGURE 1.3. *When the Fermi surface is translated by an amount* $\mathbf{q}$, *quasiparticles are added in the shaded area A, and removed from the shaded area B.*

Let us first translate $\mathbf{p}$ alone, without touching the Fermi surface. The energy ϵ_p varies by an amount $\mathbf{q} \cdot \nabla\epsilon_p = \mathbf{q} \cdot \mathbf{v}_\mathbf{p}$. The corresponding contribution to the current is just $\mathbf{v}_\mathbf{p}$, as would be expected in the absence of interactions. We now translate the Fermi surface: this produces a change $\delta n_{\mathbf{p}'}$ in the distribution function. There is a further shift in the energy because of the interaction of the quasiparticle $\mathbf{p}$ with the distribution $\delta n_{\mathbf{p}'}$. We may write

$$\delta E = \mathbf{q} \cdot \mathbf{j}_\mathbf{p} = \mathbf{q} \cdot \mathbf{v}_\mathbf{p} + \sum_{\mathbf{p}'} f_{\mathbf{p}\mathbf{p}'} \delta n_{\mathbf{p}'}. \tag{1.97}$$

This shows clearly that the drag current, $(\mathbf{j}_\mathbf{p} - \mathbf{v}_\mathbf{p})$, is due to the interaction of the quasiparticle $\mathbf{p}$ with the medium.

The situation is illustrated on Fig. 1.3: the original Fermi surface (heavy line) has been shifted by an amount $\mathbf{q}$. The shift creates quasiparticles in the regions A, quasiholes in the region B. The corresponding $\delta n_{\mathbf{p}'}$ may be written

$$\delta n_{\mathbf{p}'} = -\mathbf{q} \cdot \nabla_{\mathbf{p}'} n^{\circ}_{\mathbf{p}'} = -\mathbf{q} \cdot \mathbf{v}_{\mathbf{p}'} (\partial n^{\circ}/\partial \epsilon_{p'}). \tag{1.98}$$

Let us insert Eq. (1.98) into (1.97): we recover the expression of ($\mathbf{j}_{\mathbf{p}} - \mathbf{v}_{\mathbf{p}}$) derived from Eq. (1.91).

On comparing Eqs. (1.91) and (1.95), we see that we have, for a translationally invariant system,

$$\mathbf{j}_{\mathbf{p}} = \mathbf{p}/m = \mathbf{v}_{\mathbf{p}} - \sum_{\mathbf{p}'} f_{\mathbf{pp}'} (\partial n^{\circ}/\partial \epsilon_{p'}) \mathbf{v}_{\mathbf{p}'}. \tag{1.99}$$

Equation (1.99) is a condition *imposed* on $f_{\mathbf{pp}'}$ by Galilean invariance. In the case of an isotropic system, such as liquid ^{3}He, $\mathbf{j}_{\mathbf{p}}$ and $\mathbf{v}_{\mathbf{p}}$ are both parallel to $\mathbf{p}$; Eq. (1.99) involves only the spin symmetric $\ell = 1$ component of $f_{\mathbf{pp}'}$. By using the addition theorem for spherical harmonics, and the definitions (1.11) and (1.23), we transform (1.99) into the relation

$$\frac{m^*}{m} = 1 + \frac{F_1{}^s}{3}. \tag{1.100}$$

Since m^* may be derived from specific heat measurements, (1.100) provides additional information on the interaction $f_{\mathbf{pp}'}$. We note that $F_1{}^s$ must be larger than -3; otherwise the system becomes unstable, since for a negative effective mass it is energetically favorable to excite quasiparticles across the Fermi surface.

1.6. LOCALIZED QUASIPARTICLE EXCITATIONS

We consider now the Landau transport equation in the absence of an external field and collision terms. We seek to determine the self-consistent, spatially varying solutions of the resulting homogeneous equation. These solutions correspond to the natural (resonant) frequencies for excitations in the Fermi liquid. The excitations are of two kinds: localized quasiparticles and collective modes. The corresponding solutions are very different in their structure. In the present section we take up localized quasiparticles, postponing consideration of collective modes until the following section.

QUALITATIVE STRUCTURE OF AN INHOMOGENEOUS EXCITATION

Assume that a single quasiparticle with momentum $\mathbf{p}_0$ is added to the ground state. The response of the system differs according to whether the added quasiparticle is spread out uniformly throughout the system, or is localized. In the former case, the state of the system remains homogeneous; the interaction energy of the added quasiparticle with the ground state distribution is uniform. There is no force tending to excite other quasiparticles. On the contrary, if the added quasiparticle is localized, the local energy $\tilde{\epsilon}_{\mathbf{p}}$ of other quasiparticles varies in space. This in turn gives rise to a "polarizing force," which tends to excite quasiparticles other than $\mathbf{p}_0$. In order to build a *localized wave packet* for the quasiparticle $\mathbf{p}_0$, we must find the equilibrium-distribution of the other quasiparticles around it.

We first consider excitations which are periodic in space and time, with wave vector $\mathbf{q}$ and frequency ω. They are characterized by the distribution function,

$$\delta n_{\mathbf{p}}(\mathbf{r}, t) = \delta n_{\mathbf{p}}(\mathbf{q}, \omega)e^{i(\mathbf{q}\cdot\mathbf{r}-\omega t)} + \text{c.c.} \tag{1.101}$$

(As usual, we denote all quantities by their amplitude.) We shall see later how to relate $\mathbf{q}$ and ω; for the moment, we consider them as independent variables. We assume that ω is much larger than the collision frequency, which means that we study the motion of quasiparticles over times short compared to the collision time τ. We can then ignore collisions.

The free flow of excited quasiparticles is governed by the transport equation (1.80), which in our simple plane wave geometry may be written as

$$-\omega\delta n_{\mathbf{p}}(\mathbf{q}, \omega) + \mathbf{q}\cdot\mathbf{v}_{\mathbf{p}}\delta\bar{n}_{\mathbf{p}}(\mathbf{q}, \omega) = 0. \tag{1.102}$$

Let us replace $\delta\bar{n}_{\mathbf{p}}$ by its expression (1.30). Equation (1.102) becomes an integral equation with respect to $\delta n_{\mathbf{p}}$:

$$(\mathbf{q}\cdot\mathbf{v}_{\mathbf{p}} - \omega)\delta n_{\mathbf{p}} - \mathbf{q}\cdot\mathbf{v}_{\mathbf{p}}(\partial n^0/\partial\epsilon_p)\sum_{\mathbf{p}'} f_{\mathbf{p}\mathbf{p}'}\delta n_{\mathbf{p}'} = 0. \tag{1.103}$$

This homogeneous equation will only have solutions for certain "eigenvalues" of the frequency ω; the solutions will provide the elementary excitations of the system which vary in space with a wave vector $\mathbf{q}$. (Equation (1.103) applies both to the "dressed" quasiparticles discussed in the present section and to the collective modes we shall discuss in Section 1.7.)

Consider an excitation of the form (1.101) which would involve a single added quasiparticle with momentum $\mathbf{p}_0$. The corresponding distribution function would be

$$\delta n_{\mathbf{p}} = \delta_{\mathbf{p},\mathbf{p}_0}.$$

In fact, we have seen that such a state does not represent a real excitation of the system. The corresponding disturbance acts to polarize the medium; as a result, the "bare" quasiparticle is "dressed" with an induced polarization cloud of other excited quasiparticles. The equilibrium solution for this problem corresponds to some distribution function, $\delta n_{\mathbf{p}}$, which is determined by Eq. (1.103), and takes the form

$$\delta n_{\mathbf{p}} = \delta_{\mathbf{p},\mathbf{p}_o} + \xi_{\mathbf{p}}. \tag{1.104}$$

$\xi_{\mathbf{p}}$, which describes the polarization cloud, is of order $1/N$. Nonetheless the total effect of the polarization cloud may be appreciable, since the number of values of $\mathbf{p}$ is of order N. We shall call a solution such as (1.104), in which one component $\mathbf{p}_o$ plays the leading role, an *individual excitation* of the system.

Let us apply the transport equation (1.103) to the special case $\mathbf{p} = \mathbf{p}_o$. Since $f_{\mathbf{pp'}}$ is of order $1/N$, we find within corrections of that order

$$\mathbf{q} \cdot \mathbf{v}_{\mathbf{p}_o} - \omega = 0. \tag{1.105}$$

Equation (1.105) fixes the frequency ω in terms of $\mathbf{q}$. We see that the plane wave moves along $\mathbf{q}$ with a velocity $v_{\mathbf{p}_o} \cos\theta$, where θ is the angle between $\mathbf{q}$ and $\mathbf{p}_o$. $\mathbf{v}_{\mathbf{p}_o}$ is therefore the *group velocity* of the quasiparticle, in agreement with our earlier remarks.

We consider next the components $\mathbf{p} \neq \mathbf{p}_o$ of Eq. (1.103). On collecting terms of order $1/N$, we find the following equation for $\xi_{\mathbf{p}}$:

$$(\mathbf{q} \cdot \mathbf{v}_{\mathbf{p}} - \omega)\xi_{\mathbf{p}} - \mathbf{q} \cdot \mathbf{v}_{\mathbf{p}}(\partial n^o/\partial \epsilon_p) \sum_{\mathbf{p'}} f_{\mathbf{pp'}}\xi_{\mathbf{p'}}$$
$$= +\mathbf{q} \cdot \mathbf{v}_{\mathbf{p}}(\partial n^o/\partial \epsilon_p) f_{\mathbf{pp}_o}. \tag{1.106}$$

If $q \equiv 0$, Eq. (1.106) has the trivial solution $\xi_{\mathbf{p}} = 0$: the excitation corresponds to a single "bare" quasiparticle $\mathbf{p}_o$, such as we defined in Section 1.1. The corresponding state is homogeneous. If, however, $\mathbf{q}$ is finite, the interaction with the "bare" quasiparticle $\mathbf{p}_o$ acts as a *driving force* on the right-hand side of the transport equation (1.106), and the resulting value of $\xi_{\mathbf{p}}$ will be nonvanishing.

We remark that the distribution $\xi_{\mathbf{p}}$ has the same wave vector $\mathbf{q}$ and frequency $\omega = \mathbf{q} \cdot \mathbf{v}_{\mathbf{p}_o}$ as the bare quasiparticle $\mathbf{p}_o$: it propagates through the system at a velocity $\mathbf{v}_{\mathbf{p}_o}$, not $\mathbf{v}_{\mathbf{p}}$. In other words, while the component $\mathbf{p}_o$ undergoes a *free* oscillation, the motion of the other components is *forced* by the bare quasiparticle $\mathbf{p}_o$. This distinction is important; it is characteristic of an individual excitation.

It would be nice to be able to solve Eq. (1.106) for $\xi_{\mathbf{p}}$, since we could thereby obtain a detailed description of the quasiparticle dressed with its polarization cloud. Unfortunately, a straightforward solution of Eq. (1.106) does not exist, apart from certain trivial cases in which $f_{\mathbf{pp'}}$ assumes a simple "separable" form. Indeed, although one may obtain a formal solution for $\xi_{\mathbf{p}}$ in terms of a suitably defined scattering amplitude (and we shall do this in the following section), such a solution is not especially helpful in obtaining an explicit expression for $\xi_{\mathbf{p}}$.

Let us turn, therefore, to the "gross" properties of the dressed quasiparticle $\mathbf{p}_o$, namely the total particle density $\rho_{\mathbf{p}_o}(\mathbf{q}, \omega)$ and current $\mathbf{J}_{\mathbf{p}_o}(\mathbf{q}, \omega)$ that it carries. Using Eq. (1.104), we may write

$$\begin{aligned} \rho_{\mathbf{p}_o}(\mathbf{q}, \omega) &= \sum_{\mathbf{p}} \delta n_{\mathbf{p}}(\mathbf{q}, \omega) = 1 + \sum_{\mathbf{p}} \xi_{\mathbf{p}}, \\ \mathbf{J}_{\mathbf{p}_o}(\mathbf{q}, \omega) &= \sum_{\mathbf{p}} \mathbf{j}_{\mathbf{p}} \delta n_{\mathbf{p}}(\mathbf{q}, \omega) = \mathbf{j}_{\mathbf{p}_o} + \sum_{\mathbf{p}} \mathbf{j}_{\mathbf{p}} \xi_{\mathbf{p}}. \end{aligned} \tag{1.107}$$

The last terms on the right-hand side of these two equations correspond to the charge and current of the polarization cloud carried by the dressed quasiparticle $\mathbf{p}_o$. They represent a sizable correction to the first terms, 1 and $\mathbf{j}_{\mathbf{p}_o}$, which correspond to the *bare* quasiparticle charge and current.

The current $\mathbf{J}_{\mathbf{p}_o}$ may also be written as

$$\mathbf{J}_{\mathbf{p}_o} = \sum_{\mathbf{p}} \mathbf{v}_{\mathbf{p}} \delta \bar{n}_{\mathbf{p}}(\mathbf{q}, \omega). \tag{1.108}$$

Let us sum the original transport equation (1.102) over $\mathbf{p}$, and compare the result with Eqs. (1.107) and (1.108). We see at once that

$$\omega \rho_{\mathbf{p}_o} = \mathbf{q} \cdot \mathbf{J}_{\mathbf{p}_o}. \tag{1.109}$$

Equation (1.109) expresses the law of particle conservation which governs the motion of the dressed quasiparticle. If we recall that $\omega = \mathbf{q} \cdot \mathbf{v}_{\mathbf{p}_o}$, we see that the component of $\mathbf{J}_{\mathbf{p}_o}$ parallel to $\mathbf{q}$ is equal to the density $\rho_{\mathbf{p}_o}$ multiplied by the parallel component of $\mathbf{v}_{\mathbf{p}_o}$. Along $\mathbf{q}$, the total current is thus the particle density times the group velocity, a result which is perhaps obvious *a priori*. We note that there are no inherent contradictions in a theory in which the current carried by a quasiparticle, $\mathbf{j}_{\mathbf{p}}$, is different from the group velocity, $\mathbf{v}_{\mathbf{p}}$. The current $\mathbf{j}_{\mathbf{p}}$ is that carried by a *bare* quasiparticle distributed uniformly in space. For this homogeneous situation, conservation laws are of no importance. If we deal with an inhomogeneous situation, corresponding to a localized quasiparticle, the latter must be *dressed*; the longitudinal part of the current (given by the conservation laws) then corresponds to the motion of the total particle density at the group velocity (such arguments do not apply to the transverse part of $\mathbf{J}_{\mathbf{p}_o}$, which is perpendicular to $\mathbf{q}$; there is no simple expression for the latter).

The preceding discussion is somewhat academic, since the physically interesting situation corresponds to a localized wave packet rather than a plane wave. We need to make a Fourier transform in order to pass from one to the other. In practice, this proves difficult. The backflow current forms a complicated pattern inside the wave packet. The only simple result is that the total current is equal to the total particle density times the drift velocity $\mathbf{v}_{\mathbf{p}_0}$ of the wave packet—a fairly obvious result indeed! One may picture the bare quasiparticle as surrounded by a "*backflow*" current of excited quasiparticles, which changes the current from its original value, $\mathbf{j}_\mathbf{p}$, to the value $\mathbf{v}_\mathbf{p}$ characteristic of a localized excitation.

Perhaps the most important lesson to be learned from the preceding discussion is that a bare localized quasiparticle $\mathbf{p}_0$ does not represent an eigenstate of the system. It necessarily surrounds itself with a cloud of other quasiparticles, the distribution of which depends on the geometry of the wave packet. The cloud carries particle density and current: the dressed quasiparticle thus has properties quite different from those of the bare quasiparticles studied in Section 1.1. Our discussion sheds a new light on the difference between the *velocity* of a quasiparticle and its *current*, and shows in detail how the conservation of particle number is brought about.

This discussion may be extended to configurations containing several excited "dressed" quasiparticles. If their number is kept small, the excitations "dress" independently, and move at their own group velocity. They still interact between themselves, via an energy which has the same order of magnitude as the interaction $f_{\mathbf{pp}'}$ between bare quasiparticles, but which is more complicated, because of the interaction with and between the dressing clouds. In problems involving a cooperative motion of a large number of quasiparticles at a given frequency ω and wave vector $\mathbf{q}$, the concept of a "dressed" quasiparticle loses its meaning, and it is much simpler to work directly with the bare quasiparticles: one cannot obtain a complete description of the system in terms of dressed quasiparticles.

DEFINITION OF THE SCATTERING AMPLITUDES

We now set up a formal solution of (1.106) in terms of the so-called "scattering amplitudes." In and of itself, Eq. (1.106) scarcely merits such an elegant solution. However, we shall make use of the concept of scattering amplitudes in Section 1.8: we therefore choose to introduce them now.

We begin by changing variables in the transport equation from $\xi_\mathbf{p}$ to $x_\mathbf{p}$, according to

$$\xi_\mathbf{p} = -\frac{\mathbf{q}\cdot\mathbf{v}_\mathbf{p}}{\mathbf{q}\cdot\mathbf{v}_\mathbf{p} - \omega}\frac{\partial n^0}{\partial \epsilon_p}x_\mathbf{p}. \tag{1.110}$$

Equation (1.106) then takes the form of a Fredholm linear integral equation:

$$x_\mathbf{p} - \sum_{\mathbf{p}'} f_{\mathbf{pp}'}\frac{\mathbf{q}\cdot\mathbf{v}_{\mathbf{p}'}}{\mathbf{q}\cdot\mathbf{v}_{\mathbf{p}'} - \omega}\frac{\partial n^0}{\partial \epsilon_{p'}}x_{\mathbf{p}'} = -f_{\mathbf{pp}_0}. \tag{1.111}$$

In order to solve it, we introduce the so-called "resolvent" operator, $A_{\mathbf{pp}'}(\mathbf{q}, \omega)$, defined by the equation

$$A_{\mathbf{pp}'}(\mathbf{q}, \omega) - \sum_{\mathbf{p}''} f_{\mathbf{pp}''} \frac{\mathbf{q} \cdot \mathbf{v}_{\mathbf{p}''}}{\mathbf{q} \cdot \mathbf{v}_{\mathbf{p}''} - \omega} \frac{\partial n^0}{\partial \epsilon_{p''}} A_{\mathbf{p}''\mathbf{p}'}(\mathbf{q}, \omega) = f_{\mathbf{pp}'}. \qquad (1.112)$$

It is clear from the definition (1.112) that

$$A_{\mathbf{pp}'}(0, \omega) = f_{\mathbf{pp}'}.$$

For $q = 0$, $A_{\mathbf{pp}'}(\mathbf{q}, \omega)$ therefore does not depend on the direction of $\mathbf{q}$. On comparing Eq. (1.111) with (1.112), we see that $x_{\mathbf{p}}$ and $\xi_{\mathbf{p}}$ are simply expressed in terms of $A_{\mathbf{pp}_0}$:

$$x_{\mathbf{p}} = -A_{\mathbf{pp}_0}(\mathbf{q}, \omega), \qquad (1.113)$$

$$\xi_{\mathbf{p}} = +\frac{\mathbf{q} \cdot \mathbf{v}_{\mathbf{p}}}{\mathbf{q} \cdot \mathbf{v}_{\mathbf{p}} - \omega} \frac{\partial n^0}{\partial \epsilon_p} A_{\mathbf{pp}_0}(\mathbf{q}, \omega). \qquad (1.114)$$

We have thus obtained a formal solution for the polarization induced by the localized bare quasiparticle. We note that since the frequency ω of the dressed quasiparticle is equal to $\mathbf{q} \cdot \mathbf{v}_{\mathbf{p}_0}$, $\xi_{\mathbf{p}}$ depends only on $A_{\mathbf{pp}_0}(\mathbf{q}, \mathbf{q} \cdot \mathbf{v}_{\mathbf{p}_0})$.

One may show by a microscopic calculation that $A_{\mathbf{pp}'}(\mathbf{q}, \omega)$ represents the *scattering amplitude* for a process in which two quasiparticles with momenta $\mathbf{p}$ and $\mathbf{p}'$ exchange momentum $\hbar\mathbf{q}$ and $\hbar\omega$. Since $\mathbf{q}$ and ω are very small, the corresponding scattering is *nearly forward*. $A_{\mathbf{pp}'}(\mathbf{q}, \omega)$ may equivalently be regarded as the scattering amplitude for the process in which a quasiparticle-quasihole pair, with respective momenta $(\mathbf{p} + \hbar\mathbf{q}/2)$ and $(\mathbf{p} - \hbar\mathbf{q}/2)$, is scattered into the states $(\mathbf{p}' + \hbar\mathbf{q}/2)$ and $(\mathbf{p}' - \hbar\mathbf{q}/2)$. The defining equation (1.112) is closely related to the Bethe-Salpeter equation which describes multiple scattering of the quasiparticle-quasihole pair [Nozières (1963), Chapter 6].

In any real scattering event, the initial and final states of the colliding quasiparticles must lie within a distance κT of the Fermi surface, because of the exclusion principle. Thus $\hbar\omega$ is bound to be $\lesssim \kappa T$. Let us consider a moderately large value of $\mathbf{q}$, such that

$$\kappa T \ll q v_F \ll \mu. \qquad (1.115)$$

We then can neglect ω as compared to $\mathbf{q} \cdot \mathbf{v}_{\mathbf{p}''}$ in the denominator of Eq. (1.112). The corresponding scattering amplitude is given by the equation

$$A_{\mathbf{pp}'}(\mathbf{q}, 0) = \mathsf{f}_{\mathbf{pp}'} = f_{\mathbf{pp}'} - \sum_{\mathbf{p}''} f_{\mathbf{pp}''}(\partial n^0/\partial \epsilon_{p''})\mathsf{f}_{\mathbf{p}''\mathbf{p}'}. \qquad (1.116)$$

(Note that $\mathsf{f}_{\mathbf{pp}'}$ does not depend on the direction of $\mathbf{q}$). We shall see in Section 1.9 that $\mathbf{q}$ is proportional to the deflection angle of the quasiparticles. According to Eq. (1.115), $\mathsf{f}_{\mathbf{pp}'}$ is the scattering amplitude of the quasiparticles $\mathbf{p}$ and $\mathbf{p}'$ if one first lets the temperature go to zero, and then the scattering angle. If T is kept finite, the limit of $A_{\mathbf{pp}'}(\mathbf{q}, 0)$ when $\mathbf{q} \to 0$ is not well defined, as q/ω may take any value.

The scattering amplitude $\mathsf{f}_{\mathbf{pp}'}$, which turns out to be very useful in obtaining the collision integral for quasiparticles, may be calculated explicitly if the system is isotropic. In that case, it depends only on the angle ξ between $\mathbf{p}$ and $\mathbf{p}'$. By analogy with Eqs. (1.20) and (1.23), we may write

$$\mathsf{f}_{\mathbf{pp}'} = \frac{\pi^2\hbar^3}{m^* p_F} \sum_\ell \{\mathsf{F}_\ell^s \pm \mathsf{F}_\ell^a\} P_\ell(\cos\xi), \tag{1.117}$$

where the coefficients F_ℓ^s and F_ℓ^a are dimensionless. By inserting the expansion (1.117) into (1.116), and making use once more of the addition theorem for spherical harmonics, we find

$$\mathsf{F}_\ell^s = \frac{F_\ell^s}{1 + F_\ell^s/(2\ell+1)}, \tag{1.118}$$

$$\mathsf{F}_\ell^a = \frac{F_\ell^a}{1 + F_\ell^a/(2\ell+1)}. \tag{1.119}$$

There is thus a simple relationship between $f_{\mathbf{pp}'}$ and $\mathsf{f}_{\mathbf{pp}'}$.

We may further demonstrate, by comparing (1.116) with the definition (1.30) of $\delta\bar{n}_\mathbf{p}$, that

$$\sum_{\mathbf{p}'} \mathsf{f}_{\mathbf{pp}'}\,\delta\bar{n}_{\mathbf{p}'} = \sum_{\mathbf{p}'} f_{\mathbf{pp}'}\delta n_{\mathbf{p}'}. \tag{1.120}$$

With the help of (1.120), we can invert the relation (1.30) giving $\delta\bar{n}_\mathbf{p}$ in terms of $\delta n_\mathbf{p}$: it becomes

$$\delta n_\mathbf{p} = \delta\bar{n}_\mathbf{p} + \frac{\partial n^0}{\partial \epsilon_p} \sum_{\mathbf{p}'} \mathsf{f}_{\mathbf{pp}'}\delta\bar{n}_{\mathbf{p}'}. \tag{1.121}$$

Equation (1.121) is useful if one wishes to perform an expansion of the transport equation in the vicinity of $\omega = 0$.

In conclusion, we return to the question of the structure of a dressed quasiparticle, characterized by Eq. (1.114). For wave vectors $\mathbf{q}$ that are perpendicular to $\mathbf{p}_0$, we may readily obtain an explicit solution, since in this case $\mathbf{q} \cdot \mathbf{v}_{\mathbf{p}_0} = 0$, and $A_{\mathbf{pp}_0}$ reduces to $\mathsf{f}_{\mathbf{pp}_0}$. It is not difficult to find the total particle density and current associated with that solution: the detailed calculation is left as an exercise to the reader.

1.7. COLLECTIVE MODES

Collective modes represent a second kind of possible elementary excitation for the Fermi liquid. Physically they involve a coherent motion of the system as a whole. As we have seen, any given quasiparticle is subject to the field of the surrounding particles in the medium. In equilibrium, under homogeneous conditions, the corresponding force averages to zero. If, however, a distortion of the quasiparticle distribution takes place as a result of some internal fluctuation of the system, the average interaction force no longer vanishes. Instead, it acts to return the distribution toward equilibrium, and thereby serves as a restoring force for a collective oscillation about the equilibrium state. In other words, a collective mode involves a *cooperative* motion of the system, governed by the global interaction between the particles. We may picture the system as moving in its own self-consistent field; in such a mode, the individual particles have lost their meaning.

In the present section we assume the Fermi liquid to be in its ground state. This case is somewhat special, in that one is automatically in a *collisionless* regime, in which collisions between thermally excited quasiparticles play no role. At finite temperatures such collisions act to disrupt the self-consistent fields responsible for collective modes, and hence act to damp the waves. We shall discuss the effect of such collisions in Secton 1.9. Here we merely remark that if ν characterizes the collision frequency for the quasiparticles, then as long as

$$\omega \gg \nu$$

one is effectively in a collisionless regime for which the considerations of the present section are valid.

REDUCTION OF THE TRANSPORT EQUATION

Collective modes, like the individual excitations studied in the preceding section, represent free oscillations of the liquid, and are characterized by a nonvanishing fluctuation, $\delta n_{\mathbf{p}}(\mathbf{q}, \omega)$, of the distribution function. At low values of $\mathbf{q}$ and ω, for which the Landau theory is applicable, their structure is determined by the homogeneous transport equation, (1.103), which, for convenience, we reproduce here:

$$(\mathbf{q} \cdot \mathbf{v}_{\mathbf{p}} - \omega)\delta n_{\mathbf{p}} - \mathbf{q} \cdot \mathbf{v}_{\mathbf{p}}(\partial n^{\circ}/\partial \epsilon_{p}) \sum_{\mathbf{p}'} f_{\mathbf{p}\mathbf{p}'} \delta n_{\mathbf{p}'} = 0. \quad (1.103)$$

The collective solutions of Eq. (1.103) differ in nature from the individual excitations described in the preceding section. In the latter, a single component, say $\mathbf{p}_{o}$, acts to force the motion of all the other particles. The core of the excitation is a single quasiparticle; it is surrounded by an appropriate polarization cloud. By contrast, in a collective mode all particles play an equal role; the distribution function extends smoothly over the entire Fermi surface. Such smooth solutions of the transport equation, (1.103), only exist for certain discrete eigenvalues: the frequency ω_{q} of a given collective mode is therefore a well-defined function of its wave vector $\mathbf{q}$. The corresponding eigenvectors provide the detailed structure of the collective mode. We note that Eq. (1.103) depends only on the ratio q/ω. The eigenvalues will thus be proportional to q, so that the collective modes will have a well-defined phase velocity, which of course varies from one mode to another. Such a property is characteristic of a neutral Fermi liquid; we shall see in Chapter 3 that matters are otherwise for a charged Fermi liquid.

At zero temperature, a collective mode may be pictured as an *oscillation of the Fermi surface.* Instead of characterizing the mode by the distribution function $\delta n_{\mathbf{p}}$, we shall use a more physical quantity, namely the normal displacement $u_{\mathbf{p}}$ of the Fermi surface at point $\mathbf{p}$. Let us assume for simplicity that the system is isotropic; we may then write

$$\delta n_{\mathbf{p}} = \delta(\epsilon_p - \mu) v_F u_{\mathbf{p}}. \tag{1.122}$$

In terms of $u_{\mathbf{p}}$, the transport equation (1.103) becomes

$$(\mathbf{q} \cdot \mathbf{v_p} - \omega) u_{\mathbf{p}} + \mathbf{q} \cdot \mathbf{v_p} \sum_{\mathbf{p}'} f_{\mathbf{pp}'} \delta(\epsilon_{p'} - \mu) u_{\mathbf{p}'} = 0. \tag{1.123}$$

Let us introduce polar coordinates (θ, φ) around the axis $\mathbf{q}$; $u(\theta, \varphi, \sigma)$ will be the displacement of the Fermi surface of spin σ in the direction (θ, φ). Using the "reduced" interaction F defined by Eq. (1.23), we may write (1.123) as

$$(\cos\theta - \lambda) u(\theta, \varphi, \sigma) + (\cos\theta/8\pi) \sum_{\sigma} \int d\Upsilon'\, F(\xi, \sigma, \sigma') u(\theta', \varphi', \sigma') = 0, \tag{1.124}$$

where ξ is the angle between the directions (θ, φ) and (θ', φ'). The dimensionless quantity

$$\lambda = \frac{\omega}{q v_F} \tag{1.125}$$

is the ratio of the phase velocity of the wave to the Fermi velocity. The eigenvalues of Eq. (1.124) will clearly correspond to discrete values of λ: the collective modes of a normal fluid thus have a *constant velocity*, in agreement with our above statement.

Collective modes in which the spins move in phase are decoupled from those in which spins move out of phase. Thus we may write the displacement $u(\theta, \varphi, \sigma)$ as

$$u(\theta, \varphi, \pm) = u^s(\theta, \varphi) \pm u^a(\theta, \varphi), \tag{1.126}$$

where u^s and u^a are respectively the spin symmetric and spin antisymmetric amplitudes of the oscillation. These amplitudes satisfy the independent equations

$$\begin{aligned} (\cos\theta - \lambda) u^s(\theta, \varphi) + (\cos\theta/4\pi \int d\Upsilon'\, F^s(\xi) u^s(\theta', \varphi') &= 0, \\ (\cos\theta - \lambda) u^a(\theta, \varphi) + (\cos\theta/4\pi \int d\Upsilon'\, F^a(\xi) u^a(\theta', \varphi') &= 0. \end{aligned} \tag{1.127}$$

Formal solutions of Eqs. (1.127) may be obtained by expanding u in a series of spherical harmonics $Y_{\ell m}(\theta, \varphi)$. It follows at once that m is a good "quantum number": different values of m are completely decoupled. We can therefore classify the solutions as "longitudinal" ($m = 0$), "transverse" ($m = 1$), "quadrupolar" ($m = 2$), etc. Unfortunately, the different ℓ values are coupled by the equation (1.127). The collective modes thus correspond to a single m, but to a mixture of ℓ's. This renders the actual solution of Eq. (1.127) rather difficult.

Among the collective modes perhaps the most important is the *longitudinal* symmetric mode. It is the only mode which involves fluctuations of the particle density, and thus represents the high frequency ($\omega \gg \nu$) counterpart of ordinary sound. For that reason, it has been called by Landau zero sound. Another important mode is the longitudinal antisymmetric mode, which involves spin density fluctuations. It may be viewed as a spin wave, according to the usual magnetic picture.

A SIMPLE MODEL FOR ZERO SOUND

We illustrate this general discussion with a simple example, which permits an explicit solution of Eq. (1.127). We assume that the interaction $f_{\mathbf{pp}'}$ is constant, independent of both $\mathbf{p}$ and $\mathbf{p}'$. In this case the only component of F which does not vanish is $F_0{}^s$, which we shall abbreviate as F_0: the only collective mode solution of (1.127) is therefore zero sound. The form of the eigenvector for a zero sound mode is clear by inspection. One finds that

$$u(\theta, \varphi) = C \frac{\cos\theta}{\lambda - \cos\theta}, \tag{1.128a}$$

where C is a constant; Eq. (1.128a) describes the distortion of the Fermi surface in the course of such an oscillation. The dispersion relation for zero sound is obtained by substituting Eq. (1.128a) into (1.127), and carrying out the integration over solid angles. It is left as a problem for the reader to show that one thereby obtains

$$\frac{\lambda}{2} \log \frac{\lambda + 1}{\lambda - 1} - 1 = \frac{1}{F_0}. \tag{1.128b}$$

The solutions, $\lambda = \lambda_0$, of Eq. (1.128b), yield the velocity of zero sound. If $F_0 > 0$ (corresponding to a repulsive quasiparticle interaction) there is one real root, such that $\lambda > 1$. In this case one has an undamped zero sound mode, with a phase velocity which is greater than the velocity of a particle on the Fermi surface, v_F. If, on the other hand, one has a weak attraction between the quasiparticles, such that

$$-1 < F_0 < 0,$$

the root of Eq. (1.128b) is complex, and the solution represents a damped zero sound oscillation. Finally, if there is a strong attraction between the quasiparticles, such that

$$F_0 < -1,$$

the zero sound mode is found to be unstable. The latter two possibilities are discussed later on in this section.

One may easily solve Eq. (1.128b) in the limiting cases of a strong and weak repulsive interaction between the quasiparticles. If the coupling is very strong ($F_0 \gg 1$), one obtains $\lambda \sim \sqrt{F_0/3}$: the motion of the Fermi surface reduces to a translation along **q**. In the limit of weak coupling ($F_0 \ll 1$), λ goes to 1, while the displacement u reduces to a small bump on the Fermi surface in the direction of **q**: the collective mode involves only a small number of quasiparticles, and propagates at essentially their velocity.

The present simple model also enables us to find an explicit solution for the scattering amplitude $A_{\mathbf{pp}'}(\mathbf{q}, \omega)$ defined by the integral equation (1.112), since that equation reduces to a simple algebraic equation. We express $A_{\mathbf{pp}'}$ in reduced units by setting

$$A_{\mathbf{pp}'}(\mathbf{q}, \omega) = \frac{\pi^2 \hbar^3}{m^* p_F} F(\mathbf{q}, \omega). \tag{1.129}$$

The solution of (1.112) may then be written as

$$F(\mathbf{q}, \omega) = \frac{F_0}{1 + F_0[1 - (\lambda/2) \log \{(\lambda + 1)/(\lambda - 1)\}]} \tag{1.130}$$

On comparing (1.130) with the dispersion equation (1.128), we see that $F(\mathbf{q}, \omega)$ is infinite when λ is equal to the velocity λ_0 of zero sound. In other words, $A_{\mathbf{pp}'}(\mathbf{q}, \omega)$ has a *pole* when ω is equal to the frequency of zero sound at wave vector **q**.

This conclusion is not specific to our simple model. Very generally, any collective mode of the system will show up as a *pole of the scattering amplitude* $A_{\mathbf{pp}'}(\mathbf{q}, \omega)$. This follows at once from the properties of linear integral equations. Let us write $\delta n_{\mathbf{p}}$ as

$$\delta n_{\mathbf{p}} = -\frac{\mathbf{q} \cdot \mathbf{v}_{\mathbf{p}}}{\mathbf{q} \cdot \mathbf{v}_{\mathbf{p}} - \omega} \frac{\partial n^0}{\partial \epsilon_p} x_{\mathbf{p}}. \tag{1.131}$$

The transport equation (1.103) becomes

$$x_{\mathbf{p}} - \sum_{\mathbf{p}'} f_{\mathbf{pp}'} \frac{\mathbf{q} \cdot \mathbf{v}_{\mathbf{p}'}}{\mathbf{q} \cdot \mathbf{v}_{\mathbf{p}'} - \omega} \frac{\partial n^{\circ}}{\partial \epsilon_{\mathbf{p}'}} x_{\mathbf{p}'} = 0 \tag{1.132}$$

[compare with Eq. (1.111)]. The kernel of (1.132) is the same as that which appears in the definition of $A_{\mathbf{pp}'}(\mathbf{q}, \omega)$, (1.112). A collective mode corresponds to an *eigenvalue* of the homogeneous equation (1.132). The solution of the *inhomogeneous* equation (1.112) will accordingly be infinite.

Since the scattering amplitude is singular at the collective mode frequencies, the latter clearly correspond to a *resonance* of the system. A pole in $A_{\mathbf{pp}'}(\mathbf{q}, \omega)$ may indeed be considered as arising from a "resonant exchange" of collective mode quanta between the colliding quasiparticles.

LANDAU DAMPING

Our explicit solution for the amplitude of zero sound, (1.128a), is obviously divergent if $\lambda = \cos\theta$, that is, when $\omega = \mathbf{q} \cdot \mathbf{v}_{\mathbf{p}}$. More generally, it is clear from the transport equation (1.124) that for any collective mode, $u_{\mathbf{p}}$ will be infinite if $(\mathbf{q} \cdot \mathbf{v}_{\mathbf{p}} - \omega)$ vanishes. Such a singularity gives rise to damping of the collective modes, as we shall now demonstrate.

Let us consider a quasiparticle $\mathbf{p}$, such that $\mathbf{q} \cdot \mathbf{v}_{\mathbf{p}} = \omega$: its velocity in the direction of $\mathbf{q}$ is just equal to the *phase velocity* ω/q of the collective mode. The phase "seen" by the moving quasiparticle is then stationary, which leads to a steady *energy transfer* between the quasiparticle and the running wave. If the quasiparticle goes slightly faster than the wave, it will be slowed down, and will thus give energy to the collective mode. On the contrary, if it is slightly slower, it will receive energy from the collective mode. Under equilibrium conditions, the distribution of quasiparticles is a decreasing function of their velocity. The net balance of energy corresponds to an energy transfer from the collective mode to the individual quasiparticles, i.e., to a *damping* of the collective mode.

Such a damping mechanism resembles that first proposed by Landau (1946) for the damping of a classical plasma oscillation and is known as *Landau damping*. It corresponds to a *coherent* interaction of the collective mode with those particles which "surf-ride" on the crests of the running wave. (Surf-riding is only possible if the phase velocity is smaller than v_F, i.e., if $\lambda < 1$). Such a resonant interaction may be pictured as a real transition, in which a collective mode decays by exciting a *single* quasiparticle-quasihole pair. In general, once Landau damping becomes possible, the collective mode has such a short lifetime that it no longer represents a well-defined excitation of the system.

Landau damping differs from the damping mechanism arising from collisions between quasiparticles. Collision damping, which we mentioned earlier, corresponds to an essentially incoherent scattering of the collective mode; it exists for any value of λ, and, at low temperatures, provides a small damping of the collective mode when $\lambda > 1$ (and $\omega \gg \nu$). If $\lambda < 1$, collision damping is negligible compared to Landau damping.

The calculation of the damped frequency, $\omega = \omega_1 - i\omega_2$, is not easy. First of all, we need a definite prescription for dealing with the singularities of the transport equation. Such a prescription may be derived from a study of the boundary conditions governing transport equations like (1.123). We shall see in Chapter 2 that the appropriate boundary condition corresponds to turning on adiabatically the fluctuation responsible for the collective mode. We are thus led to the prescription that, in Eq. (1.123), ω is to be replaced by $\omega + i\eta$, where η is positive and infinitesimal. We next solve Eq. (1.123), assuming ω to be real, and obtain a dispersion equation for ω. In order to find the frequency of damped collective modes, we then analytically continue the dispersion equation, obtained for real ω, into the lower half of the complex plane, where its roots will be found.

In order to show clearly how this procedure works, let us change ω into $(\omega + i\eta)$, and set

$$(\mathbf{q} \cdot \mathbf{v_p} - \omega - i\eta)u_\mathbf{p} = y_\mathbf{p}. \tag{1.133}$$

In terms of $y_\mathbf{p}$, the transport equation becomes

$$y_\mathbf{p} + \mathbf{q} \cdot \mathbf{v_p} \sum_{\mathbf{p}'} f_{\mathbf{pp}'}\delta(\epsilon_{p'} - \mu)y_{\mathbf{p}'}/(\mathbf{q} \cdot \mathbf{v}_{\mathbf{p}'} - \omega - i\eta) = 0. \tag{1.134}$$

Since η is infinitesimal, we may write, for *real values* of ω,

$$\frac{1}{\mathbf{q} \cdot \mathbf{v}_{\mathbf{p}'} - \omega - i\eta} = P\left(\frac{1}{\mathbf{q} \cdot \mathbf{v}_{\mathbf{p}'} - \omega}\right) + i\pi\delta(\mathbf{q} \cdot \mathbf{v}_{\mathbf{p}'} - \omega) \tag{1.135}$$

(where P stands for the Cauchy principal part). In order to find the damped collective mode, we insert Eq. (1.135) into (1.134), and perform the integration over $\mathbf{p}'$; we then continue the *final* result into the *lower* half of the complex ω-plane.

As an example, let us return to the simple model in which $f_{\mathbf{pp}'}$ is constant. In the dispersion equation (1.128b), we replace λ by $(\lambda + i\eta)$. For $\lambda > 1$, there is no change; on the other hand, for $\lambda < 1$, the logarithm becomes

$$\log\left[\frac{\lambda + i\eta + 1}{\lambda + i\eta - 1}\right]. \tag{1.136}$$

For *real* values of λ, Eq. (1.136) is equal to

$$\log\left|\frac{\lambda + 1}{\lambda - 1}\right| + i\pi.$$

We note that the imaginary part of (1.136) is now important, being comparable to the real part.

In replacing λ by $(\lambda + i\eta)$ in Eq. (1.128b), we fix the definition of the multivalued logarithm. For moderately negative values of F_0, the dispersion equation may then be shown to possess one root $(\omega_1 - i\omega_2)$, which corresponds to a damped collective mode (if F_0 is too negative, there appear unstable solutions, which we consider later in this section). Since on the real axis the imaginary part of (1.136) is large, we expect ω_2 to be roughly comparable to ω_1. Damping is thus important; the collective mode scarcely exists as such, being "dissolved" in the continuum of single quasiparticle excitations.

We can likewise calculate the scattering amplitude $A_{\mathbf{pp}'}(\mathbf{q}, \omega)$, for real values of the frequency ω. We find that Landau damping smoothes out the poles associated with the collective modes. This effect is best seen by considering the imaginary part of $A_{\mathbf{pp}'}(\mathbf{q}, \omega)$. In the absence of Landau damping $(\lambda > 1)$, the collective mode showed up as a discrete peak. With Landau damping $(\lambda < 1)$, the peak is considerably flattened, and appears more as a broad bump on the individual quasiparticle spectrum (this may be easily verified with the simple model for zero sound: the calculation is left as a problem to the reader). In the latter case, it is no longer appropriate to speak of a distinct collective mode; it might be better described as a weak resonance of the quasiparticles.

STABILITY AGAINST COLLECTIVE OSCILLATIONS

Under suitable conditions, the transport equation (1.123) may possess two pure imaginary roots, which are complex conjugates. One of these necessarily leads to an exponentially growing wave: the system is thus *unstable*. In such a case, the state which we assumed to be the ground state spontaneously evolves toward another state, containing *permanent fluctuations* of the quantity associated with the unstable collective mode.

In order to illustrate the phenomenon, we consider once more the simple model for zero sound, characterized by the dispersion equation

(1.128b). We look for a pure imaginary root, such that $\lambda = i\alpha$, where α is real. By paying due attention to the determination of the logarithm, we may write Eq. (1.128b) in the form

$$\alpha \tan^{-1}(1/\alpha) - 1 = 1/F_0. \tag{1.137}$$

It may easily be verified that Eq. (1.137) possesses two real solutions of opposite signs if $1/F_0$ lies within the range $(-1, 0)$. The condition for instability of zero sound is thus

$$F_0 < -1. \tag{1.138}$$

In other words, if the interaction is too attractive, damped zero sound is replaced by an unstable mode: the system builds up permanent density fluctuations, limited only by nonlinear effects.

Such instabilities also exist in the general case, corresponding to the transport equation (1.123). They may affect any collective mode whatsoever. There will thus exist an infinite number of stability criteria, analogous to Eq. (1.138), one for each type of collective mode.

In order to find the stability conditions, we note that the frequency ω vanishes at the *onset* of any instability, since at the threshold value of $f_{\mathbf{pp}'}$, the two complex conjugate roots merge together at the origin. The threshold condition may thus be obtained by putting $\omega = 0$ in the general equation (1.123). We find

$$u_{\mathbf{p}} + \sum_{\mathbf{p}'} f_{\mathbf{pp}'}\delta(\epsilon_{p'} - \mu)u_{\mathbf{p}'} = 0. \tag{1.139}$$

Equation (1.139) is easily solved if the system is assumed to be isotropic. We split u into spin symmetric and antisymmetric waves, according to (1.126), and further expand $u^s(\theta, \varphi)$ and $u^a(\theta, \varphi)$ into a series of spherical harmonics. Upon using the addition theorem for spherical harmonics, we find that Eq. (1.139) possesses a solution if one of the following conditions is met:

$$\begin{aligned} F_\ell^s &= -(2\ell + 1), \\ F_\ell^a &= -(2\ell + 1). \end{aligned} \tag{1.140}$$

In order for the system to be stable against the spontaneous growth of any collective mode, one must therefore satisfy the whole series of conditions:

$$\begin{aligned} F_\ell^s &> -(2\ell + 1), \\ F_\ell^a &> -(2\ell + 1). \end{aligned} \tag{1.141}$$

Too strong an attraction always leads to an instability of the normal state.

The first few of the conditions (1.141) may be interpreted in simple physical terms. On recalling Eqs. (1.58) and (1.67), we see that the stability conditions for $\ell = 0$ insure that the compressibility and spin susceptibility are positive. This is just what is required to prevent the growth of density or spin fluctuations in the system (if for instance χ_P were negative, the system would no longer be paramagnetic, but would instead turn to a ferromagnetic configuration). Similarly, the condition $F_1^s > -3$ means that the effective mass m^*, given by (1.100) is positive. In other words, the stability conditions (1.141), which we have just derived from a "dynamic" point of view, may also be obtained from "equilibrium" considerations.

In order to set up systematically such a "static" approach to the stability criteria, we note that the conditions (1.141) do not depend on the wave vector $\mathbf{q}$ (as long as the latter remains small). The corresponding instability will therefore show up when $\mathbf{q} = 0$, a case corresponding to an *homogeneous* deformation of the system. Such an instability will appear as a *spontaneous deformation* of the Fermi surface S_F, which is only possible if it corresponds to a decrease of the total free energy. In order to rule out homogeneous instabilities, we are thus led to require that the ground state free energy be a *minimum* with respect to deformations of the Fermi surface. By expressing this variational requirement, we should be able to reproduce the stability conditions (1.141) [Pomeranchuk (1958)].

Let us displace the Fermi surface of spin σ in the direction (θ, φ) by a small amount $u(\theta, \varphi, \sigma)$ (we consider an isotropic system). The corresponding change in free energy may be written, up to the second order in u, as

$$(F - F_0) = \frac{1}{(2\pi\hbar)^3} \sum_{\sigma} \int d\Upsilon \int_{p_F}^{p_F+u} (\epsilon_p - \mu) p^2 \, dp + \frac{1}{2(2\pi\hbar)^6} p_F^4 \sum_{\sigma,\sigma'} \int d\Upsilon \, d\Upsilon' \, u(\theta, \varphi, \sigma) u(\theta', \varphi', \sigma') f(\xi, \sigma, \sigma'), \quad (1.142)$$

where $d\Upsilon = \sin\theta \, d\theta \, d\varphi$ is the differential element of solid angle, and ξ the angle between the directions (θ, φ) and (θ', φ'). In the last term of Eq. (1.142), we have performed the integration over p and p', neglecting terms of order u^3 or higher.

The first-order term of Eq. (1.142) vanishes if $\epsilon_p = \mu$ everywhere on the Fermi surface. We know that this condition is always met—it prevents the spontaneous creation of quasiparticle-quasihole pairs. Let us now turn to the second-order terms. A simple integration yields

$$(F - F_0) = \frac{1}{(2\pi\hbar)^3} \frac{p_F^3}{2m^*} \sum_{\sigma} \int d\Upsilon \, u^2(\theta, \varphi, \sigma) + \frac{1}{(2\pi\hbar)^6} \frac{p_F^4}{2} \sum_{\sigma,\sigma'} \int d\Upsilon \, d\Upsilon \, u(\theta, \varphi, \sigma) u(\theta', \varphi', \sigma') f(\xi, \sigma, \sigma'). \quad (1.143)$$

This is a quadratic form in u. The system will be stable if this form remains positive, whatever the choice of $u(\theta, \varphi, \sigma)$.

In order to diagonalize the quadratic form (1.143), we write $u(\theta, \varphi, \sigma)$ in the form (1.126), and expand u^s and u^a in a series of normalized spherical harmonics:

$$\begin{aligned} u^s(\theta, \varphi) &= \sum_{\ell m} u^s_{\ell m} Y_{\ell m}(\theta, \varphi), \\ u^a(\theta, \varphi) &= \sum_{\ell m} u^a_{\ell m} Y_{\ell m}(\theta, \varphi). \end{aligned} \tag{1.144}$$

By using the addition theorem and the reduced form (1.23) of the interaction, we may cast Eq. (1.143) in the form

$$(F - F_0) = \frac{\Omega}{(2\pi\hbar)^3} \frac{p_F{}^3}{m^*} \sum_{\ell m} \left\{ |u^s_{\ell m}|^2 \left(1 + \frac{F_\ell{}^s}{2\ell + 1}\right) + |u^a_{\ell m}|^2 \left(1 + \frac{F_\ell{}^a}{2\ell + 1}\right) \right\}. \tag{1.145}$$

Since the $u^s_{\ell m}$ and $u^a_{\ell m}$ are all independent variables, $(F - F_0)$ will be positive definite if the whole set of inequalities (1.141) is satisfied. We therefore recover our previous stability conditions from another point of view.

SURVEY OF THE POSSIBLE SOLUTIONS FOR COLLECTIVE MODES

We have discussed essentially three possibilities for every collective mode:

(i) *No damping.* If the interaction is repulsive, and sufficiently strong, the collective mode has a phase velocity larger than the Fermi velocity of the particles. It is not Landau damped; it is subject only to a small collision damping, which is negligible except at very low frequencies.

(ii) *Strong Landau damping.* For weak repulsion or weak attraction, one generally finds a mode immersed in the continuum of single quasiparticle excitations. Such a collective mode is subject to very strong Landau damping, and no longer represents an independent, well-defined excitation of the system.

(iii) *Instability.* If the interaction is too attractive, the mode becomes unstable, and gives rise to the appropriate exponentially growing fluctuations in the system.

The threshold of the unstable region is given by one of the conditions (1.140). On the other hand, the transition from case (i) to case (ii) cannot be located exactly, as it requires an explicit solution of the integral equation (1.103). This solution can only be found for simple models in which the interaction $f_{\mathbf{pp}'}$ is separable.

It should be emphasized that the present discussion is restricted to *macroscopic* (i.e., long wavelength) collective modes, according to the general conditions of applicability of the Landau theory. Clearly, there may also exist "microscopic" collective modes, with a wavelength of atomic size. The latter lie outside the framework of the Landau theory, and must be studied by means of a suitable approximate theory, such as the random phase approximation discussed in Chapter 5; we consider their properties briefly.

Microscopic collective modes are markedly more complicated than the macroscopic modes studied in the present section. First of all, the frequency ω_q of a collective mode is no longer a linear function of $\mathbf{q}$ (the phase velocity thus varies with $\mathbf{q}$). Second, Landau damping becomes more efficient as one goes to microscopic values of $\mathbf{q}$ and ω. Finally, new damping mechanisms may come into play.

Landau damping may be viewed as a decay of the collective mode into individual quasiparticle excitations. At low $\mathbf{q}$, the only decay process of interest involves the excitation of a single quasiparticle-quasihole pair, with respective momenta $(\mathbf{p} \pm \hbar\mathbf{q}/2)$. Up to first order in $\mathbf{q}$, the corresponding excitation energy is simply $\mathbf{q} \cdot \mathbf{v}_\mathbf{p}$, which explains why Landau damping did not occur when $\omega > qv_F$. When $\mathbf{q}$ is finite, this linear approximation is no longer valid. The excitation energy of the single pair increases faster than $\mathbf{q} \cdot \mathbf{v}_\mathbf{p}$; for some sufficiently large value of $\mathbf{q}$, it will become equal to the frequency ω of the collective mode; beyond that point, the latter will be highly damped.

Furthermore for finite values of $\mathbf{q}$ the decay into *several* excited quasiparticle-quasihole pairs becomes important (this point will be discussed in Chapter 2). Such a "multipair" configuration may have an arbitrarily large energy, so that it will give rise to a weak damping of even a long wavelength collective mode. "Multipair" damping increases with increasing values of $\mathbf{q}$; whether the "effective" cut-off at which the collective mode ceases to be well defined is determined by Landau damping or multipair damping depends on the particular case under study.

Microscopic collective modes may also give rise to an instability at some finite value of $\mathbf{q}$: an example of such an instability is the giant spin density fluctuation considered by Overhauser (1962). At which wavelength a given instability first occurs obviously depends on the detailed properties of the system.

A complete analysis of the collective modes in an interacting system is thus seen to be difficult. It is fortunate that the Landau theory offers a simple treatment of the macroscopic limit of these excitations.

1.8. QUASIPARTICLE COLLISIONS

Thus far we have ignored the collisions between quasiparticles, since we have dropped the collision integral in the transport equation. Such a simplification is only valid at frequencies higher than the collision frequency ν—or at wavelengths short as compared to the mean free path. We now proceed to study the collision integral $I(n)$; we shall thus be able to extend the transport equation into the domain of frequencies lower than ν.

Detailed calculations of quasiparticle collisions turn out to be rather complicated. Since our purpose is only to set forth the main physical concepts, we shall frequently sketch the mathematical developments—and often simply state the result. It should, however, be realized that such calculations, although lengthy, do not pose any basic questions of principle.

STRUCTURE OF THE COLLISION INTEGRAL

At finite temperatures, the quasiparticles may undergo real collisions, similar to those described in the usual kinetic theory of gases. The simplest process is a *binary* collision, specified by

$$(\mathbf{p}_1) + (\mathbf{p}_2) \rightarrow (\mathbf{p}_1') + (\mathbf{p}_2'). \tag{1.146}$$

Suppose $\mathbf{p}_1$ is a thermally excited quasiparticle, with a free energy of the order of κT. In order to satisfy both the exclusion principle and energy conservation, $\mathbf{p}_2$, $\mathbf{p}_1'$, and $\mathbf{p}_2'$ must also lie within (κT) of the Fermi surface. When $T \rightarrow 0$, the transition probability therefore vanishes, since there are no final states available; we shall see later that the corresponding collision frequency is of order T^2. In addition to the single binary collision (1.146), there may occur more complicated processes, involving the simultaneous collision of 3, 4, etc. quasiparticles. Since all the particles involved in such a process must lie within κT of the Fermi surface, the corresponding transition probability will be of higher order in T, and is thus negligible: we need consider only the simple process (1.146).

The corresponding collision integral may be written as

$$I(n_1) = - \sum_{\text{Final states}} W(\mathbf{p}_1\sigma_1 \cdots \mathbf{p}_2'\sigma_2')\delta(\mathbf{p}_1 + \mathbf{p}_2 - \mathbf{p}_1' - \mathbf{p}_2')\delta_{\sigma_1+\sigma_2,\sigma_1'+\sigma_2'}$$
$$\delta(\tilde{\epsilon}_1 + \tilde{\epsilon}_2 - \tilde{\epsilon}_1' - \tilde{\epsilon}_2')\{n_1n_2(1 - n_1')(1 - n_2') - (1 - n_1)(1 - n_2)n_1'n_2'\}, \tag{1.147}$$

where $\tilde{\epsilon}_i$, n_i stand respectively for $\tilde{\epsilon}_{\mathbf{p}_i}$, $n_{\mathbf{p}_i\sigma_i}$. (We introduce the spin indices explicitly, for the sake of clarity.) $I(n_1)$ represents the rate of change of $n_{\mathbf{p}_1\sigma_1}$ as a consequence of collisions. $W(\mathbf{p}_1\sigma_1 \cdots \mathbf{p}_2'\sigma_2')$ is the transition probability for the process (1.146), from which we have split away the three δ-functions expressing the conservation of spin, momentum, and energy. We note that the quantity which should be conserved is the total energy of the system. Energy conservation must therefore involve the exact energy of quasiparticles in the excited medium, that is, the local energy $\tilde{\epsilon}_{\mathbf{p}}$ rather than the equilibrium energy $\epsilon_{\mathbf{p}}$. This remark, which was taken into account in writing Eq. (1.147), will prove to be very important for dealing with transport properties. Finally, within the brackets of Eq. (1.147), we find the usual occupation factors, corresponding to scattering in and out of the state $(\mathbf{p}_1\sigma_1)$, taking due account of the exclusion principle.

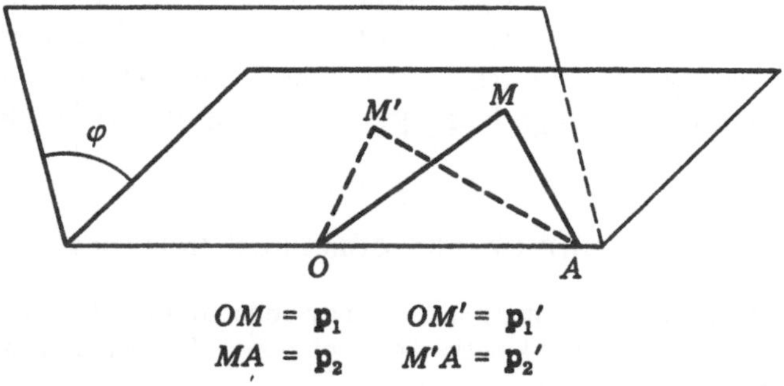

FIGURE 1.4. *Relative position of the momenta before and after a collision between two quasiparticles.*

When performing the sum over "final states" in (1.147), we must be careful to count each of these states only once. Let us first assume that $\sigma_1 = \sigma_2 = \sigma$; the conservation of spin implies $\sigma_1' = \sigma_2' = \sigma$: all the spins are fixed. In that case, the final state is unchanged by a permutation of $\mathbf{p}_1'$ and $\mathbf{p}_2'$. We shall take this into account by adding a factor $\frac{1}{2}$ to Eq. (1.147); we may then sum over the momenta $\mathbf{p}_1'$ and $\mathbf{p}_2'$ independently. We now turn to the other case $\sigma_2 = -\sigma_1$, which implies $\sigma_2' = -\sigma_1'$; we may then assume that $\sigma_1 = \sigma_1'$, $\sigma_2 = \sigma_2'$, and sum over $\mathbf{p}_1'$ and $\mathbf{p}_2'$ without restrictions: we thus avoid counting final states twice.

We next consider in some detail the "geometry" of the collision. We consider only an isotropic system, and follow closely the analysis of Abrikosov and Khalatnikov (1959), Momentum conservation requires that

$$\mathbf{p}_1 + \mathbf{p}_2 = \mathbf{p}_1' + \mathbf{p}_2' = \mathbf{OA}. \tag{1.148}$$

This situation is illustrated on Fig. 1.4, which shows the disposition of the various momenta, together with the angle φ between the planes $(\mathbf{p}_1\mathbf{p}_2)$ and $(\mathbf{p}_1'\mathbf{p}_2')$. We may simplify the figure by rotating the plane $OM'A$ onto the plane OMA. We thus obtain Fig. 1.5, where N is the new position of M'. In order to return to Fig. 1.4, one need merely rotate N by an angle $-\varphi$ around OA.

Because of the exclusion principle, all four momenta ($\mathbf{p}_1\mathbf{p}_2\mathbf{p}_1'\mathbf{p}_2'$) must lie within κT of the Fermi surface. Since we are dealing with a degenerate gas ($\kappa T \ll \mu$), the length of these momenta will be nearly equal to p_F. The triangles OMA and ONA are thus nearly isoceles, the displacement $MN = \lambda$ being very small. In other words, the angles between $\mathbf{p}_1$ and $\mathbf{p}_2$ on one hand, $\mathbf{p}_1'$ and $\mathbf{p}_2'$ on the other, are practically equal to the same value θ, and essentially independent of the energies associated with the four momenta ($\mathbf{p}_1 \cdots \mathbf{p}_2'$). This *decoupling* of the angular and energy variables, which makes the calculation much easier, is characteristic of a *degenerate* system; it is due to the exclusion principle which forces all the colliding quasiparticles to remain very close to the Fermi surface.

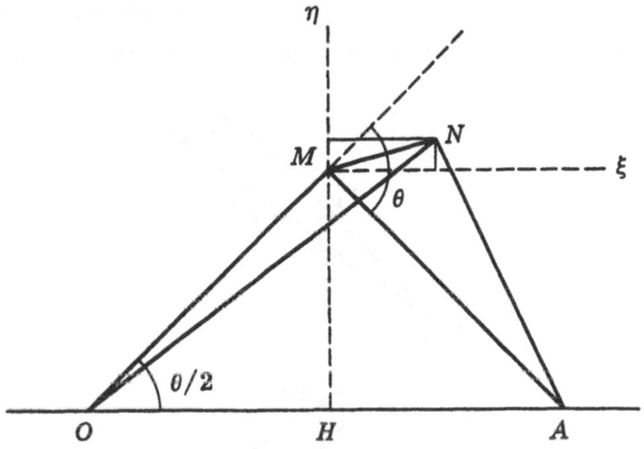

FIGURE 1.5. *Geometrical parameters of a particle-particle collision.*

Let $W_{\uparrow\uparrow}(\mathbf{p}_1 \cdots \mathbf{p}_2')$ and $W_{\uparrow\downarrow}(\mathbf{p}_1 \cdots \mathbf{p}_2')$ be the transition probabilities for incoming particles with parallel and antiparallel spins respectively. They will be practically independent of the *lengths* of the vectors ($\mathbf{p}_1 \cdots \mathbf{p}_2'$) (which remain very close to p_F); they depend only on their *relative orientation*, i.e., on the two angles (θ, φ). The collision integral is thus completely determined by two functions $W_{\uparrow\uparrow}(\theta, \varphi)$ and $W_{\uparrow\downarrow}(\theta, \varphi)$.

In order to calculate the collision integral explicitly, we need a suitable set of coordinates to describe the momenta $\mathbf{p}_2$ and $\mathbf{p}_1'$ ($\mathbf{p}_2'$ is then fixed by momentum conservation). We may for instance characterize $\mathbf{p}_1'$ by the variables $\lambda_\parallel$, $\lambda_\perp$, and φ (see Figs. 1.4 and 1.5). $\mathbf{p}_2$ is best defined by polar coordinates around the axis $\mathbf{p}_1$, i.e., the length p_2, the angle θ, and an azimuthal angle φ_2. It is then straightforward to write the volume element in phase space. It is actually more convenient to replace the variables $\lambda_\parallel$ and $\lambda_\perp$ by the lengths p_1' and p_2' (which is easily done if λ is assumed to be small). The lengths of the four vectors ($\mathbf{p}_1 \cdots \mathbf{p}_2'$) may in turn be replaced by the corresponding energies. After a straightforward calculation, one finds the following relation:

$$\int d^3\mathbf{p}_2 \int d^3\mathbf{p}_1' \int d^3\mathbf{p}_2' \, \delta(\mathbf{p}_1 \cdots - \mathbf{p}_2') = \frac{(m^*)^3}{2} \int \frac{\sin\theta}{\cos\theta/2} \, d\theta \, d\varphi \, d\varphi_2 \, d\epsilon_1' \, d\epsilon_2' \, d\epsilon_2. \quad (1.149)$$

The collision integral (1.147) may then be written explicitly. Let us assume for instance that $n_\mathbf{p}$ is spin independent; we thus obtain

$$I(n_1) = \frac{(m^*)^3}{2(2\pi\hbar)^6} \int \{W_{\uparrow\downarrow}(\theta, \varphi) + \tfrac{1}{2}W_{\uparrow\uparrow}(\theta, \varphi)\}\{n_1 n_2(1 - n_1')(1 - n_2') - n_1' n_2'(1 - n_1)(1 - n_2)\} \, \delta(\bar{\epsilon}_1 + \bar{\epsilon}_2 - \bar{\epsilon}_1' - \bar{\epsilon}_2') \frac{\sin\theta}{\cos\theta/2} \, d\theta \, d\varphi \, d\varphi_2 \, d\epsilon_1' \, d\epsilon_2' \, d\epsilon_2. \quad (1.150)$$

This rather unappealing expression may in fact be integrated easily.

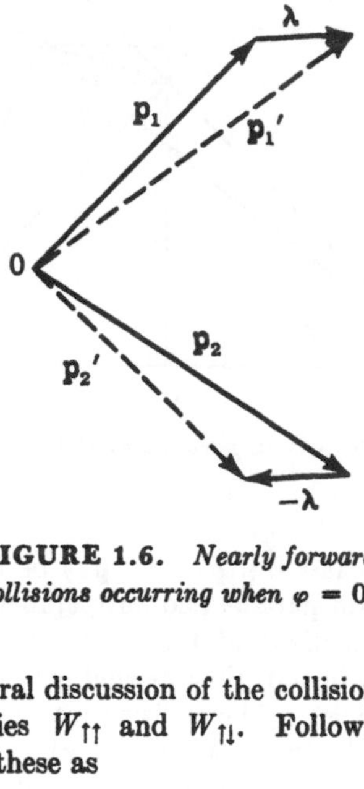

FIGURE 1.6. *Nearly forward collisions occurring when* $\varphi = 0$.

Before leaving the general discussion of the collision integral, let us consider the transition probabilities $W_{\uparrow\uparrow}$ and $W_{\uparrow\downarrow}$. Following the usual theory of scattering, we may write these as

$$W_{\uparrow\uparrow}(\theta, \varphi) = \frac{2\pi}{\hbar} |A_{\uparrow\uparrow}(\theta, \varphi)|^2,$$
$$W_{\uparrow\downarrow}(\theta, \varphi) = \frac{2\pi}{\hbar} |A_{\uparrow\downarrow}(\theta, \varphi)|^2, \quad (1.151)$$

where $A_{\uparrow\uparrow}$ and $A_{\uparrow\downarrow}$ are the corresponding *scattering amplitudes.*

Let us consider the case $\varphi = 0$. In that case, the vector $\mathbf{MN} = \boldsymbol{\lambda} = \hbar\mathbf{q}$ defined on Fig. 1.5 represents simply the momentum transfer in the collision. This process is shown more explicitly on Fig. 1.6. We note that $\sigma_1 = \sigma_1'$,

$\sigma_2 = \sigma_2'$; since λ is small, the scattering of the two quasiparticles $\mathbf{p}_1$ and $\mathbf{p}_2$ is almost completely in the *forward* direction. The transition also corresponds to a small energy transfer $\hbar\omega = \hbar\mathbf{q}\cdot\mathbf{v}_{\mathbf{p}_1} = -\hbar\mathbf{q}\cdot\mathbf{v}_{\mathbf{p}_2}$. The corresponding scattering amplitude $A_{\mathbf{p}_1\mathbf{p}_2}(\mathbf{q}, \omega)$ is then related to the interaction $f_{\mathbf{pp}'}$ by Eq. (1.112), according to the general discussion of Section 1.6.

When φ is identically zero, the energy transfer ω is of the order of qv_F. The scattering amplitudes then depend on the ratio $\mathbf{q}/\omega$, i.e., on the direction of $\mathbf{q}$: they cannot be uniquely defined. Let us instead assume that φ is very small, yet larger than $\kappa T/\mu$. The energy transfer $\hbar\omega$ remains of the order of κT, while the momentum transfer $\hbar\mathbf{q}$, although small, becomes much larger than $\kappa T/v_F$. In that case, $\mathbf{q}$ is practically perpendicular to $\mathbf{p}_1$ and $\mathbf{p}_2$, and the scattering amplitude is equal to the quantity, $\mathsf{f}_{\mathbf{p}_1\sigma_1,\mathbf{p}_2\sigma_2}$, defined by Eq. (1.116). Making use of the definition (1.117), we can thus write

$$
\left.
\begin{aligned}
A_{\uparrow\uparrow}(\theta, \varphi) &\to \frac{\pi^2\hbar^3}{m^* p_F}\,[\mathsf{F}^s(\theta) + \mathsf{F}^a(\theta)]\\
A_{\uparrow\downarrow}(\theta, \varphi) &\to \frac{\pi^2\hbar^3}{m^* p_F}\,[\mathsf{F}^s(\theta) - \mathsf{F}^a(\theta)]
\end{aligned}
\right\}
\qquad (\kappa T/\mu \ll \varphi \ll 1), \qquad (1.152)
$$

where θ is the angle between $\mathbf{p}_1$ and $\mathbf{p}_2$. Equations (1.152) are not valid when $\varphi \lesssim \kappa T/\mu$; this restriction is of no importance, as the corresponding region in phase space has a negligible volume.

We may use Eq. (1.118) to deduce the f's from the interaction energy $f_{\mathbf{pp}'}$. In principle, the scattering amplitude for $\varphi \to 0$ may thus be derived from the equilibrium properties of the liquid. Actually, such a procedure is not especially useful, since we know little about $f_{\mathbf{pp}'}$. Furthermore, it would tell nothing about the case $\varphi \neq 0$. Nevertheless, we can obtain an order of magnitude estimate of $A_{\uparrow\uparrow}$ and $A_{\uparrow\downarrow}$ by assuming these two quantities to be *constant*, independent of θ and φ. Using (1.152), we are thus led to write

$$
\begin{aligned}
A_{\uparrow\uparrow} &\sim \frac{\pi^2\hbar^3}{m^* p_F}\,\{\mathsf{F}_0^s + \mathsf{F}_0^a\},\\
A_{\uparrow\downarrow} &\sim \frac{\pi^2\hbar^3}{m^* p_F}\,\{\mathsf{F}_0^s - \mathsf{F}_0^a\}.
\end{aligned}
\qquad (1.153)
$$

The coefficients F_0^s and F_0^a, given by Eq. (1.118), may be obtained from sound velocity and spin susceptibility data. We can then use Eq. (1.153) to evaluate the *average* transition probabilities $W_{\uparrow\uparrow}$ and $W_{\uparrow\downarrow}$.

LIFETIME OF QUASIPARTICLES

Let us suppose that at time $t = 0$, we introduce an extra quasiparticle with momentum $\mathbf{p}$ into a degenerate Fermi liquid in an equilibrium state at temperature T. We may use the transport equation to study the rate of decay of the quasiparticle. We shall see that the distribution

$n_{\mathbf{p}}$ decays according to the usual law

$$\frac{\partial n_{\mathbf{p}}}{\partial t} = -\frac{n_{\mathbf{p}}}{\tau_p}, \tag{1.154}$$

where τ_p is the quasiparticle *lifetime*.

Again, the simplest decay process corresponds to the scheme (1.146), in which the quasiparticle $\mathbf{p}$ decays into two quasiparticles $\mathbf{p}_1'$ and $\mathbf{p}_2'$, and one quasihole $\mathbf{p}_2$. Obviously, we may imagine more complicated processes, which would produce $(m + 1)$ quasiparticles, and m quasiholes. Fortunately, the latter processes are negligible when $\mathbf{p}$ is close enough to the Fermi surface. Because of the exclusion principle, the free energy of an elementary excitation varies roughly between $-\kappa T$ and $+\infty$. In order to ensure energy conservation, each of the $(2m + 1)$ final elementary excitations must have a free energy lying in the range $(-\kappa T, +(\epsilon_p - \mu))$, where $(\epsilon_p - \mu)$ is the free energy of the initial quasiparticle. This leads to a transition probability of order $(\kappa T)^{2m}$ if $|\epsilon_p - \mu| \ll \kappa T$, of order $(\epsilon_p - \mu)^{2m}$ if $|\epsilon_p - \mu \gg \kappa T$. The higher-order decay processes are thus negligible if we work at low temperature $(\kappa T \ll \mu)$ *and* close enough to the Fermi surface $(|\epsilon_p - \mu| \ll \mu)$. We assume this to be the case, and therefore consider only the process (1.146).

At time $t = 0$, $1/\tau_p$ represents the rate of change of $n_{\mathbf{p}}$ due to collisions. The system is then characterized by the distribution

$$n_{\mathbf{p}'} = \begin{cases} 1 & \text{if} \quad \mathbf{p}' = \mathbf{p}, \\ n^0_{\mathbf{p}'}(T) & \text{if} \quad \mathbf{p}' \neq \mathbf{p}, \end{cases} \tag{1.155}$$

where $n_{\mathbf{p}}{}^0(T)$ is given by Eq. (1.37). Within corrections of order $1/N$, we can replace the local energy $\bar{\epsilon}_{\mathbf{p}}$ by its equilibrium value ϵ_p. By analogy with Eq. (1.147), we write

$$\frac{1}{\tau_p} = \sum_{\text{Final states}} W(\mathbf{p}\sigma \cdots \mathbf{p}_2'\sigma_2')\delta(\mathbf{p} \cdots - \mathbf{p}_2')\delta_{\sigma+\sigma_2,\sigma_1'+\sigma_2'}\delta(\epsilon_p \cdots - \epsilon_2')$$
$$\times\, n_2{}^0[1 - n^0_{1'}][1 - n^0_{2'}]. \tag{1.156}$$

Equation (1.156) has the same structure as the collision integral (1.147), and may be reduced to a form similar to (1.150).

We define

$$2W(\theta, \varphi) = W_{\uparrow\downarrow}(\theta, \varphi) + \tfrac{1}{2}W_{\uparrow\uparrow}(\theta, \varphi), \tag{1.157}$$

and introduce a dimensionless free energy, according to

$$x = \frac{\epsilon_p - \mu}{\kappa T}. \tag{1.158}$$

In terms of x, the equilibrium distribution may be written as

$$n_p^{\circ}(T) = n^{\circ}(x) = \frac{1}{1+e^x}. \tag{1.159}$$

By analogy with Eq. (1.150), we may write the expression (1.156) for $1/\tau_p$ in the form

$$\frac{1}{\tau_p} = \frac{(m^*)^3(\kappa T)^2)}{(2\pi\hbar)^6}\int \frac{\sin\theta\, W(\theta,\varphi)}{\cos\theta/2}\,d\theta\,d\varphi\,d\varphi_2 \int dx_2\,dx_1'\,dx_2'\,\delta(x+x_2-x_1'-x_2')$$
$$\times\, n_o(x_2)[1-n_o(x_1')][1-n_o(x_2')]. \tag{1.160}$$

The integration over energy variables may be done exactly [Morel and Nozières (1962), Appendix A]. One finally obtains

$$\frac{1}{\tau_p} = \frac{(m^*)^3(\kappa T)^2}{16\pi^4\hbar^6}\left\langle\frac{W}{\cos\theta/2}\right\rangle(\pi^2+x^2)[1-n^{\circ}(x)] \tag{1.161}$$

(where the symbol $\langle\ \ \rangle$ denotes the average over the whole sphere). In order to make this result more explicit, we replace x by its value (1.158). We thus obtain the important result,

$$\frac{1}{\tau_p} = \frac{(m^*)^3}{16\pi^4\hbar^6}\left\langle\frac{W}{\cos\theta/2}\right\rangle\frac{(\pi\kappa T)^2+(\epsilon_p-\mu)^2}{1+\exp[(\mu-\epsilon_p)/\kappa T]}, \tag{1.162}$$

which is rigorous in the limit $|\epsilon_p-\mu| \ll \mu$.

There is much to say about Eq. (1.162). We shall distinguish three cases of interest.

(*i*) $|\epsilon_p-\mu| \ll \kappa T$ This is the condition for *thermal* behavior of the system; all elementary excitations of interest lie within a layer of width κT around the Fermi surface. τ_p is then nearly independent of p and is proportional to T^{-2}; it gives a qualitative measure of the *collision time* of a thermal excitation.

(*ii*) $(\epsilon_p-\mu) \ll -\kappa T$ We are in the "ground state" regime, in which the quasiparticle distribution is effectively frozen by the lack of final states resulting from the exclusion principle. Under these circumstances, the exponential appearing in Eq. (1.162) is very large, so that $1/\tau_p$ is vanishingly small. Thus, the ground state particles are not subject to real collisions.

(*iii*) $(\epsilon_p-\mu) \gg \kappa T$ Thermal broadening of the distribution plays no role: we obtain the same result as for $T = 0$. We may write Eq. (1.162) as

$$\frac{\hbar}{\tau_p} = \frac{(\epsilon_p-\mu)^2}{\bar{\epsilon}}, \tag{1.163}$$

where $\bar{\epsilon}$, which has the dimension of an energy, is given by

$$\bar{\epsilon} = \frac{16\pi^4\hbar^5}{(m^*)^3\langle W/\cos\frac{1}{2}\theta\rangle} \tag{1.164}$$

We note that $\hbar/\tau_p$ depends quadratically on the free energy $(\epsilon_p - \mu)$. This result is likewise due to the Pauli principle, and the corresponding scarcity of final states near the Fermi surface. Equation (1.163) no longer holds when $(\epsilon_p - \mu)$ becomes comparable to μ: the higher-order scattering processes then give a sizable contribution to $1/\tau_p$, of the order of $(\epsilon_p - \mu)^{2m}$

According to Eq. (1.154), the probability of finding a particle **p** decreases exponentially, as e^{-t/τ_p}. The corresponding probability amplitude varies as $e^{-t/2\tau_p}$. This decay may be interpreted as arising from an *imaginary* part of the energy

$$i\Gamma_p = \frac{i\hbar}{2\tau_p} = i\,\frac{(\epsilon_p - \mu)^2}{2\bar{\epsilon}} \tag{1.165}$$

which is to be added to the usual real part ϵ_p. This is a devious way to say that a finite lifetime implies an uncertainty in energy. In the vicinity of the Fermi surface, Γ_p is negligible as compared to the free energy $(\epsilon_p - \mu)$: the concept of a quasiparticle then makes sense, and Landau's postulates are justified *a posteriori*. On the contrary, Γ_p is comparable to $(\epsilon_p - \mu)$ as soon as $(\epsilon_p - \mu) \gtrsim 2\bar{\epsilon}$: quasiparticles are so unstable that they lose any physical meaning. The "cut-off" value of the free energy, $2\bar{\epsilon}$, may be estimated from Eq. (1.164), using the average W deduced from equilibrium properties.

TRANSPORT COEFFICIENTS

Once we know the collision integral, we can study the usual transport properties of the system, such as viscosity, thermal conductivity, or spin diffusion. The calculation follows essentially the same path as that followed in the kinetic theory of gases. Here we shall only sketch the general method, and stress the new features brought about by the interaction between quasiparticles. A complete discussion can be found in the review article of Abrikosov and Khalatnikov (1959).

Let us apply to the system an inhomogeneous static perturbation, which may be a temperature gradient, a velocity gradient, or a spin magnetization gradient. The applied gradient sets up a flow of heat, momentum or spin, which is limited only by the collisions between quasiparticles. The induced flow is proportional to the applied gradient, the coefficient of proportionality being respectively the thermal conductivity $\mathbf{K}$, the viscosity η, or the spin diffusion coefficient $\mathbf{D}_\sigma$. We wish to calculate these transport coefficients.

At equilibrium, the system is characterized by a *time-independent* distribution function, $n_\mathbf{p}(\mathbf{r})$, which is determined by the transport equation obtained by adding the collision integral to (1.78):

$$\nabla_r n_\mathbf{p} \cdot \nabla_\mathbf{p} \bar{\epsilon}_\mathbf{p} - \nabla_\mathbf{p} n_\mathbf{p} \cdot \nabla_\mathbf{r} \bar{\epsilon}_\mathbf{p} = I(n_\mathbf{p}). \tag{1.166}$$

$\bar{\epsilon}_\mathbf{p}$ is the local energy, which depends on the position $\mathbf{r}$. We obtain the induced currents of interest by solving Eq. (1.166).

As a specific example of a calculation of a transport coefficient, we consider the process of thermal conduction. The other transport coefficients may be calculated in essentially the same way. Let us write $n_\mathbf{p}$ in the form

$$n_\mathbf{p}(\mathbf{r}) = n^\circ[\bar{\epsilon}_\mathbf{p}, T(\mathbf{r})] + \delta\bar{n}_\mathbf{p}(\mathbf{r}). \tag{1.167}$$

The first term of Eq. (1.167) depends on $\mathbf{r}$ both through $\bar{\epsilon}_\mathbf{p}$ and through T. If there were no temperature gradient, ∇T, the equilibrium solution would simply correspond to $n_\mathbf{p} = n^\circ$. The second term $\delta\bar{n}_\mathbf{p}$ is thus proportional to ∇T, and represents the departure from isotropy which gives rise to the heat current.

Let us substitute the expression (1.167) into the left-hand side of Eq. (1.166). The latter involves gradients in $\mathbf{r}$-space: in these terms we can neglect $\delta\bar{n}_\mathbf{p}$, within corrections of order $(\nabla T)^2$. We thus replace $n_\mathbf{p}$ by $n^\circ(\bar{\epsilon}_\mathbf{p}, T)$. The left-hand side of Eq. (1.166) becomes

$$\frac{\partial n_o}{\partial T} \nabla_\mathbf{r} T \cdot \nabla_\mathbf{p} \bar{\epsilon}_\mathbf{p} + \frac{\partial n_o}{\partial \bar{\epsilon}_\mathbf{p}} \nabla_\mathbf{r} \bar{\epsilon}_\mathbf{p} \cdot \nabla_\mathbf{p} \bar{\epsilon}_\mathbf{p} - \frac{\partial n_o}{\partial \bar{\epsilon}_\mathbf{p}} \nabla_\mathbf{p} \bar{\epsilon}_\mathbf{p} \cdot \nabla_\mathbf{r} \bar{\epsilon}_\mathbf{p}. \tag{1.168}$$

The last two terms of Eq. (1.168) cancel one another, as we had earlier noticed in Section 1.4. The left-hand side of the transport equation reduces to the single term

$$\frac{\partial n_o}{\partial T} \mathbf{v}_\mathbf{p} \cdot \nabla T \tag{1.169}$$

which acts as a driving force on $n_\mathbf{p}$.

We now consider the collision integral (1.150), in which we replace $n_\mathbf{p}$ by the form (1.167). The product of occupation factors may be expanded in powers of $\delta\bar{n}$. The zeroth-order term vanishes, because of the net *conservation of local energy*. The second- or higher-order terms are of order $(\nabla T)^2$, and thus negligible: we are only left with the first-order term. In order to write it, we find it convenient to introduce the quantity $\zeta_\mathbf{p}$, which is defined by the equation

$$\delta\bar{n}_\mathbf{p} = \frac{\partial n^\circ}{\partial \epsilon_\mathbf{p}} \zeta_\mathbf{p} = -\frac{n^\circ(1 - n^\circ)}{\kappa T} \zeta_\mathbf{p}. \tag{1.170}$$

It is then straightforward to cast Eq. (1.150) in the following form:

$$I(n_1) = -\frac{(m^*)^3}{2\kappa T(2\pi\hbar)^6}\int \frac{W(\theta, \varphi)\sin\theta}{\cos\theta/2} \\ n_1{}^{\circ}n_2{}^{\circ}[1 - n_{1'}^{\circ}][1 - n_{2'}^{\circ}]\{\zeta_1 + \zeta_2 - \zeta_1' - \zeta_2'\} \\ \times \delta(\epsilon_1 + \epsilon_2 - \epsilon_1' - \epsilon_2')\, d\theta\, d\varphi\, d\varphi_2\, d\epsilon_1'\, d\epsilon_2'\, d\epsilon_2. \quad (1.171)$$

(Once the equation is linearized, we are allowed to replace $\tilde{\epsilon}_p$ by ϵ_p.) Equation (1.171) holds for any spin-independent phenomenon (such as viscosity). For spin antisymmetric effects (such as spin diffusion), $I(n)$ involves a different combination of $W_{\uparrow\uparrow}$ and $W_{\uparrow\downarrow}$. Let us emphasize that the collision integral involves $\delta\tilde{n}$, instead of δn, essentially because $\tilde{\epsilon}_p$ is conserved rather than ϵ_p.

We next equate the two sides of the transport equation, (1.169) and (1.171). We thus get an integral equation which, in principle, allows us to calculate the unknown quantity ζ. From ζ, we get $\delta\tilde{n}_p$ from Eq. (1.170), and finally the free energy current, given by Eq. (1.92):

$$\mathbf{Q} = \sum_p \delta\tilde{n}_p(\epsilon_p - \mu)\mathbf{v}_p. \quad (1.172)$$

$\mathbf{Q}$ is proportional to ∇T, their ratio being the thermal conductivity $\mathbf{K}$.

The transport equation cannot be solved exactly. What one does is to postulate a simple form for ζ, and then find a corresponding approximate solution. In the case of thermal conduction, for instance, we may try the form

$$\zeta_p = \gamma\mathbf{v}_p \cdot \nabla T(\epsilon_p - \mu), \quad (1.173)$$

where γ is a constant to be determined. We substitute Eq. (1.173) into (1.171), and perform the necessary integrals. On combining the result with the transport equation, we obtain γ and finally, the thermal conductivity.

The calculation of $\boldsymbol{\eta}$ and $\mathbf{D}_\sigma$ proceeds in similar fashion. Since the calculations are lengthy, we quote here only the final results, which Abrikosov and Khalatnikov (1959) estimate to be correct to within ~10%:

$$\begin{aligned} \mathbf{K} &= \frac{4\pi^2}{3}\frac{(\hbar v_F)^3}{m^* T\langle W \sin^2 \frac{1}{2}\theta/\cos\frac{1}{2}\theta\rangle} \\ \boldsymbol{\eta} &= \frac{16}{45}\frac{m^*\hbar^3 v_F{}^5}{(\kappa T)^2\langle W \sin^4 \frac{1}{2}\theta \sin^2\varphi/\cos\frac{1}{2}\theta\rangle} \\ \mathbf{D}_\sigma &= \frac{16\pi^2}{3}\frac{\hbar^6 v_F{}^2(1 + F_0{}^a)}{(m^*)^3(\kappa T)^2\langle W_{\uparrow\downarrow} \sin^2 \frac{1}{2}\theta(1 - \cos\varphi)/\cos\frac{1}{2}\theta\rangle} \end{aligned} \quad (1.174)$$

We note that $\boldsymbol{\eta}$ and $\mathbf{D}_\sigma$ vary as T^{-2}, while $\mathbf{K}$ is of order T^{-1}. This difference may be understood by noticing that a quasiparticle carries finite momentum and spin, and a free energy of order T. We also remark that the angular averages are different, so that one does not obtain an *exact* relation between $\boldsymbol{\eta}$ and $\mathbf{K}$.

1.9. ZERO SOUND VS. FIRST SOUND

COLLISIONLESS VS. HYDRODYNAMIC REGIMES

We now consider the effect of quasiparticle collisions on the longitudinal collective modes of a neutral Fermi liquid. Suppose the frequency of the mode under consideration is ω, while the quasiparticle collision frequency is ν. We have already remarked that if

$$\omega \gg \nu, \tag{1.175}$$

then collisions are sufficiently infrequent that the contribution from the collision integral, (1.147), to the transport equation, (1.78), may safely be neglected. One is thereby in a *collisionless* regime. In this regime, provided $F_0 > 0$, zero sound is a well-defined collective mode. The corresponding restoring force is the averaged self-consistent field of a large number of particles. Collisions between quasiparticles act to disturb that field, and hence to damp zero sound; such damping is small as long as the condition (1.175) is met.

In the opposite limit, characterized by

$$\omega \ll \nu, \tag{1.176}$$

there are many quasiparticle collisions during the time, ω^{-1}, of interest. One is then in a collision-dominated, or *hydrodynamic regime*, in which the solution to the full transport equation (1.83) is effectively determined by the collision integral, (1.147). Under these circumstances too, organized density oscillations are possible, namely ordinary hydrodynamic or *first* sound waves. The frequent collisions between quasiparticles bring about local thermodynamic equilibrium; physically it is clear that such collisions provide the restoring force for a first sound wave, and that they must occur sufficiently often to restore equilibrium in a time short compared to the period of the motion. Hence the condition (1.176) is a necessary condition for the existence of a well-defined hydrodynamic first sound mode.

The transition between the above two regimes occurs when

$$\omega \sim \nu. \tag{1.177}$$

In that range of frequencies, collisions act often enough to disrupt thoroughly a zero sound mode, but not sufficiently often to make possible an undamped first-sound mode: matters are clearly least favorable for wave propagation. We may therefore anticipate that as one increases the frequency at a fixed temperature the transition from first sound to zero sound is marked by a maximum in sound wave attenuation.

We now proceed to discuss the solutions for the transport equation in the hydrodynamic regime, the effect of collisions on both zero sound and first sound, and the transition between the two modes of wave propagation.

FIRST SOUND

The velocity of first sound, corresponding to frequencies $\omega \ll \nu$, can be derived along the same lines as that of zero sound. Again the starting point is the transport equation (1.123), to which we must add a suitably defined "collision integral." We present here a qualitative account of the calculation, as it clearly displays the differences between first and zero sound. A more exact treatment can be set up along the lines of Sections 1.8 and 2.7: it is left as a problem to the reader.

The collisions between particles act to restore the equilibrium distribution n°; at frequencies $\omega \ll \nu$, they tend to maintain the displacement u of the Fermi surface at a negligibly small value. However there is one thing which collisions cannot do; that is, change the local density and current (since collisions between particles conserve the total number of particles and the total momentum). Put another way, if we expand u in a series of spherical harmonics, the $\ell = 0$ and $\ell = 1$ components are unaffected by the particle collisions. We can thus guess that in the very low-frequency limit, $\omega \ll \nu$, u will only contain $\ell = 0$ and $\ell = 1$ terms; for a longitudinal mode ($m = 0$), we shall have

$$u(\theta) = a + b \cos \theta, \tag{1.178}$$

where a and b are constants. The result (1.178) is substantiated by detailed calculation. Physically, it corresponds to a density oscillation superimposed on a uniform translation of the fluid: this is just what we expect to find for an ordinary acoustic wave.

In order to solve the transport equation (1.123), we expand it in a series of Legendre polynominals. For $\ell > 1$, we must add a collision term, which for low frequency dominates the system behavior, and which acts to damp completely such displacements. On the other hand, for $\ell = 0$ and $\ell = 1$, collisions are totally ineffective, since they must conserve the total charge and current. We may therefore write explicitly the first two terms of the expansion. On making use of (1.178), we find after some easy algebra

$$\begin{aligned} \ell = 0, \qquad & -\omega a + \frac{q v_F}{3}\left[1 + \frac{F_1{}^s}{3}\right] b = 0; \\ \ell = 1, \qquad & -\omega b + q v_F \left[1 + F_0{}^s\right] a = 0. \end{aligned} \tag{1.179}$$

In order that Eqs. (1.179) have a nonzero solution, ω and q must be such that

$$\omega^2 = \frac{q^2 v_F^2}{3}[1 + F_0^s]\left[1 + \frac{F_1^s}{3}\right]. \tag{1.180}$$

On referring to Eqs. (1.61) and (1.100), we see that Eq. (1.180) reduces to

$$\omega = sq, \tag{1.181}$$

where s is the sound velocity derived by macroscopic arguments. We thus obtain the usual *first sound* as the low-frequency propagating mode.

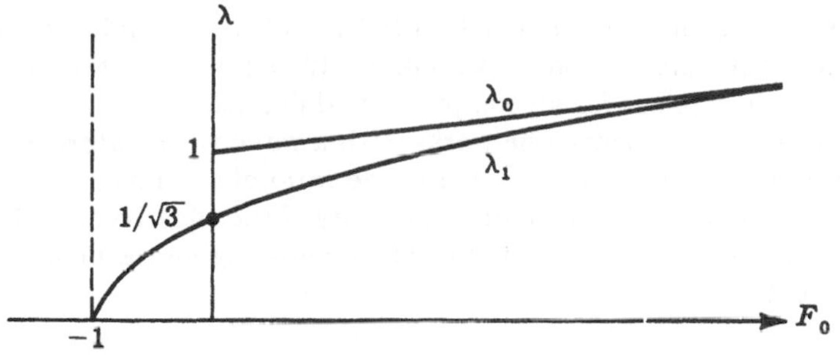

FIGURE 1.7. *The reduced velocities of first and zero sound as a function of interaction strength.*

Let us emphasize that in the very low-frequency limit, the collisions act to *truncate* the transport equation (1.123). While zero sound is obtained from the full equation, first sound follows only from the $\ell = 0$ and $\ell = 1$ components, all other components being "damped out" by the collisions. Clearly, these comments only apply to the extreme cases, $\omega \gg \nu$ or $\omega \ll \nu$. In the transition region, neither of the approaches is valid; a qualitative description of this region is presented below.

Let us consider the result (1.180) in the framework of our simple model, in which

$$F_0^s = F_0,$$
$$F_1^s = 0.$$

Using the same notation as for zero sound, we find that first sound possesses a "reduced" velocity

$$\lambda_1 = \frac{\omega}{qv_F} = \sqrt{\frac{1 + F_0}{3}}. \tag{1.182}$$

λ_1, characteristic of first sound, is plotted on Fig. 1.7 together with λ_0

(characteristic of zero sound), as a function of the interaction strength F_0. We see that the two solutions merge together in the limit of strong interaction. For $F_0 \cong 0$, λ_0 is larger than λ_1 by a factor of $\sqrt{3}$. In the region $-1 < F_0 < 0$, the zero sound solution is damped, while first sound still exists.

DAMPING OF COLLECTIVE MODES: TRANSITION FROM ZERO TO FIRST SOUND

Let us now consider the collision damping of longitudinal waves in a neutral Fermi liquid, as the frequency of the wave is varied from the collisionless regime, (1.175), to the hydrodynamic regime (1.176). This variation has been studied with the aid of a simplified collision integral by Abrikosov and Khalatnikov (1959); we refer the interested reader to their paper for all computational details.

In typical experiments, one works with a wave with *real* frequency ω, and complex wave vector $\mathbf{q}$. Arg q, the ratio of the imaginary part of q to its real part, furnishes a direct measure of the effectiveness of damping. Suppose first we are in the collisionless regime; a little thought shows that

$$\arg q \sim \frac{1}{\omega \tau_q} \tag{1.183}$$

where τ_q is the lifetime of the quasiparticles which effectively take part in the collective mode. Depending on the frequency ω, we can distinguish two regimes, corresponding to different collision mechanisms.

(i) When $\hbar\omega \ll \kappa T$ we are in a classical or *thermal* regime. The collective mode is sustained by the thermally excited quasiparticles, spread over an energy κT around the chemical potential μ. According to Eq. (1.162), the lifetime τ_q is controlled by collisions with other thermal quasiparticles, and is inversely proportional to T^2.

(ii) When $\hbar\omega \gg \kappa T$, we enter the quantum or *multipair* regime. The collective mode is then a coherent superposition of quasiparticle-quasihole pairs with total momentum $\mathbf{q}$. The quasiparticle state involved in such a mode extends over an energy ω on either side of the Fermi level. The corresponding collision time, τ_q, is given by Eq. (1.162). It is independent of temperature. In this regime, the collisions which limit τ_q are those with the ground state particles, in which the colliding quasiparticle excites a *further* quasiparticle-quasihole pair. Zero sound damping occurs when either component of the "coherent" quasiparticle-quasihole pair undergoes such a collision process. The corresponding damping mechanism may therefore be viewed as decay of the collective mode into *two* such pairs, hence the name "*multipair*" given to this regime.

As the frequency is reduced, and

$$\omega\tau_q \sim 1,$$

we pass from the collisionless to the hydrodynamic regime. In the *very low*-frequency regime, there is very little damping of the first-sound

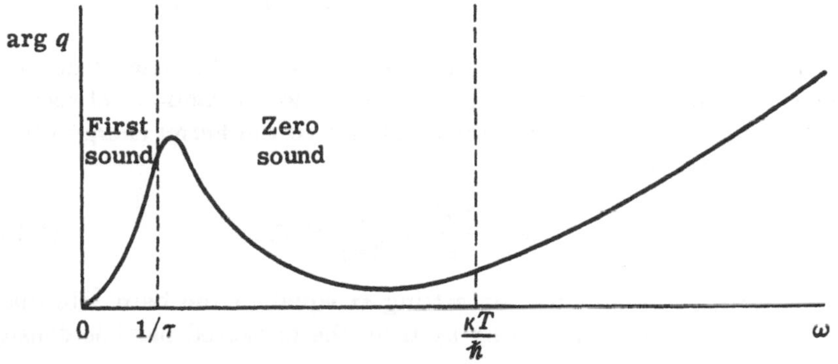

FIGURE 1.8. *Qualitative behavior of the damping of sound waves as a function of frequency.*

mode. What damping there is corresponds to the well-known viscous damping of first sound, for which one has

$$\text{Im}\, q = \frac{2\omega^2}{3Nms^3}\,\eta. \tag{1.184}$$

If one makes use of the expressions (1.174) for η and (1.162) for τ_q (in this thermal regime), one finds readily that

$$\arg q \cong \omega\tau_q. \tag{1.185}$$

In the hydrodynamic regime, the damping is proportional to τ_q, rather than to its inverse; it is thus inversely proportional to the temperature.

The results (1.183) and (1.185) may be combined in the simple formula

$$\arg q \cong \frac{\omega\tau_q}{1 + \omega^2\tau_q^{\,2}} \qquad (\omega \lesssim \mu), \tag{1.186}$$

which provides a simple interpolation formula linking the two physically different regimes. The main physical results in the general evolution of arg q are shown in a qualitative way in Fig. 1.8. There we see that the damping is large in two distinct regimes: on the high-frequency side, because of multipair decay, on the low-frequency side because of thermal quasiparticle collisions. Between these regimes one finds weakly damped zero sound propagation ($\omega\tau \gg 1$), while in the lowest frequency regime ($\omega\tau \ll 1$) one has first sound propagation.

1.10. PROPERTIES OF DEGENERATE ^{3}HE

APPLICABILITY OF LANDAU THEORY

Liquid ^{3}He, the only Fermi liquid found in nature, becomes degenerate at a very low temperature. If there were no interaction between the ^{3}He atoms, the temperature corresponding to the Fermi energy (in the ground state) would be

$$T_F = \frac{E_F^\circ}{\kappa} = \frac{p_F^2}{2m\kappa} \cong 5^\circ. \tag{1.187}$$

This estimate is obtained by setting m equal to the bare ^{3}He mass, $\sim 5 \times 10^{-24} g$, and calculating p_F from the measured particle density,

$$p_F = \hbar(3\pi^2\rho^{1/3}). \tag{1.188}$$

An order of magnitude estimate of the temperature at which typical quantum effects ($C_v \sim T$, χ_P independent of T) would be decisive is $(T_F/10)$, or roughly 0.5°K. In fact, such quantum behavior is found experimentally only at much lower temperatures, of the order of 0.1°K. This reduction in the degeneracy temperature arises as a consequence of the interaction between He atoms, which must indeed be strong in order to achieve such a large effect.

The Landau theory described above is only applicable to a degenerate liquid, i.e., at temperatures below 0.1°K. Until very recently, few experimental results were available in this region. Nowadays, experiments have been pushed down to a few thousandths of a degree: they mostly agree with the predictions of the Landau theory.

It would, of course, be highly desirable to extend the Landau theory to a higher range of temperatures—say between 0.1°K and 1°K. This would involve a systematic expansion of the free energy up to higher orders in the departure from equilibrium, $\delta n_{\mathbf{p}}$. We have seen that such an expansion, besides its mathematical complexity, raises difficulties of principle: it is not clear to what extent one can define quasiparticles in a momentum region in which they are appreciably damped. A formal definition of quasiparticles in this temperature domain has been given by Balian and de Dominicis (1964), using field-theoretical techniques;

however, the relationship of their "quasiparticles" to the actual, physical excitation is still not clear. Richards (1963) has assumed that such an expansion exists and has explored the next-order terms beyond the Landau approximation. At present, his calculation remains approximate, although the comparison with experiment is not unpromising. In what follows, we shall restrict our attention to the temperature range where the Landau theory is clearly valid, i.e., for $T \lesssim 0.1°K$.

Thus far we have implicitly assumed that liquid ^{3}He remains a *normal* Fermi liquid when the temperature is lowered. Actually the liquid may undergo a transition to a *superfluid* state, which resembles the superconducting state of a metal (we shall briefly describe this state in Chapter 7). The existence of the superfluid state is at present uncertain. Theoretical conjectures locate the transition around 0.02°K. In 1961, Anderson *et al.* (1961) found no evidence of the transition down to temperatures of the order of 0.008°K. More recently, Peshkov (1965) reported evidence, based on specific heat measurements, for a transition to the superfluid state at 0.0055°K, in contradiction with the experimental results of Wheatley and colleagues [Abel, Anderson, Black, and Wheatley (1965)]. Wheatley *et al.* measured the spin susceptibility, the spin diffusion coefficient being used as a thermometer (a very sensitive one, indeed, since $\mathbf{D}_\sigma$ varies as T^{-2}): no evidence for a superfluid transition was found down to 0.0036°K. Peshkov's result thus appears doubtful, although it may well be that his temperature scale is different from that of Wheatley. In any event, it is clear that between, say 0.005°K and 0.1°K, we may assume safely ^{3}He to be a *normal* Fermi liquid.

MACROSCOPIC PROPERTIES

By measuring various macroscopic properties, one obtains information on the effective mass m^* and the interaction $f_{\mathbf{pp}'}$. One can, moreover, study the dependence of these coefficients on the pressure applied to the system.

We first collect the relevant results found in the course of the present chapter. The specific heat C_v, sound velocity s, and spin susceptibility χ_P are given by

$$C_v = \frac{m^* p_F}{3\hbar^3} \kappa^2 T, \tag{1.43}$$

$$s^2 = \frac{p_F^2}{3mm^*} (1 + F_0^s), \tag{1.61}$$

$$\chi_P = \frac{m^* p_F}{\pi^2 \hbar^3} \frac{\beta^2}{1 + F_0^a}. \tag{1.67}$$

All these quantities are experimentally measured; we furthermore know p_F from Eq. (1.188). We may thus obtain m^* from Eq. (1.43), and subsequently F_0^s and F_0^a respectively from Eqs. (1.61) and (1.67). Since the system is translationally invariant, we can also make use of the relation

$$\frac{m^*}{m} = \frac{1 + F_1^s}{3} \tag{1.100}$$

to find the coefficient F_1^s.

In practice, such an approach raises difficulties. The "experimental" value of m^* is found to increase as the temperature is decreased [Wheatley (1966)]. Anderson (1965) has suggested that the present specific heat data would fit a $T \log T$ law better than the simple T dependence predicted by the Landau theory. A tentative explanation of this surprising result has been proposed by Balian and Fredkin (1965): the singularity in C_v would arise from the contribution of *long wavelength zero sound* to the self-energy operator (in analogy with the infrared divergence of quantum electrodynamics). At the moment, this aspect of ^{3}He remains very controversial. In what follows, we shall avoid the difficulty by noting that m^* is a *slowly* varying function of T. In a given temperature range (say ~0.01°K), one may assume m^* to be constant: the Landau theory is then approximately valid. In Table 1.1 we present a compilation of the latest experimental results for m^*/m (and F_1^s), F_0^s, and F_0^a, obtained from Wheatley (1966).

TABLE 1.1. *Parameters of the Fermi Liquid Theory for* ^{3}He

Pressure (atm)	m^*/m	F_1^s	F_0^s	F_0^a
0.28	3.1	6.3	10.8	−0.67
27.0	5.8	14.4	75.6	−0.72

We make the following comments:

(i) All the F's are large; we are in a region of strong coupling.
(ii) F_0^s increases rapidly with increasing pressure (much faster than does the density); this variation may be attributed to the hard-core part of the interaction between two ^{3}He atoms, the atoms being not far from a close-packed array.
(iii) The large effective mass lowers T_F, in agreement with experiment.
(iv) $F_0^a < 0$: the average interaction is thus larger between anti-

parallel spin particles than between those with parallel spin. This results from the Pauli principle which tends to keep parallel spins apart. As a consequence, particles of parallel spin sample less of a short-range interaction than do those of antiparallel spin. The effect on the energy is, however, small since $|F_0^a| \ll |F_0^s|$.

Knowing F_0^s and F_0^a, we can calculate the coefficients F_0^s and F_0^a defined in Eqs. (1.118) and (1.119). For liquid ^{3}He at zero pressure, we find

$$\begin{aligned} \mathsf{F}_0^s &= 0.91, \\ \mathsf{F}_0^a &= -2.0. \end{aligned} \tag{1.189}$$

According to Eq. (1.153), the combinations $\mathsf{F}_0^s \pm \mathsf{F}_0^a$ are proportional to the average *scattering amplitudes* of two quasiparticles, for parallel and antiparallel spins, respectively. On making use of Eqs. (1.189), we see that the scattering amplitude for parallel spins is about 40% of that for antiparallel spins. The difference is again a consequence of the exchange repulsion, which is here far more efficient than it is for the system energy.

From the numerical values (1.189), we can obtain a rough estimate of the transport coefficients. For that purpose, we first replace the scattering amplitudes by their value (1.153). We then calculate W and $W_{\uparrow\downarrow}$, given by Eqs. (1.151) and (1.157). We finally perform the integrations in Eq. (1.174). The results obtained at low pressure are compared to the experimental values in Table 1.2. In view of the crudeness of the theoretical estimates, the agreement is excellent (the nearly perfect agreement obtained for thermal conductivity is most certainly accidental!). It may, in fact, be possible to use the observed discrepancies to obtain information on the exact behavior of the two functions $W_{\uparrow\downarrow}(\theta, \varphi)$ and $W_{\uparrow\uparrow}(\theta, \varphi)$.

TABLE 1.2. *Comparison of Calculated and Measured Transport Coefficients for* ^{3}He

Quantity	Calculated	Measured
$\mathrm{K}T$ (erg cm^{-1} sec^{-1})	59	50
ηT^2 (poise °K^2)	1.4×10^{-6}	2.8×10^{-6}
$\mathrm{D}_\sigma T^2$ (cm^2 sec^{-1} °K^2)	6.3×10^{-6}	1.52×10^{-6}

The average scattering amplitudes can also be used to estimate the lifetime of quasiparticles, given by Eq. (1.162). In the thermal range

$\epsilon_p \sim \mu \sim \kappa T$ one finds that

$$\tau T^2 \sim 10^{-12}, \tag{1.190}$$

in which τ is expressed in seconds and T in °K. The thermal lifetime is thus very short, a fact which renders the observation of collective modes rather difficult. Away from the Fermi surface ($\epsilon_p - \mu \gg \kappa T$), the collision frequency is proportional to $(\epsilon_p - \mu)^2$. The energy $\bar{\epsilon}$ at which $\hbar/\tau_p$ is equal to $(\epsilon_p - \mu)$ is given by (1.164). For ^{3}He at low pressure, one finds

$$\bar{\epsilon} \sim 1\text{°K}. \tag{1.191}$$

It is thus impossible to extend the concept of well-defined quasiparticles to the entire Fermi sea: one must be careful to use quasiparticles only in the immediate vicinity of the Fermi surface.

The above results demonstrate the utility of the Landau theory; it provides a compact, coherent picture by means of which we can *link together* many experimental results. We can first use the theory on a semiphenomenological basis, the coefficients F_0^s and F_0^a being measured experimentally. We then obtain theoretical information on other phenomena, such as transport properties, etc. We thus arrive at a unified description of normal Fermi liquids.

EXPERIMENTAL INFORMATION ON ZERO SOUND

We discussed briefly the existence of collective modes in liquid ^{3}He. We cannot hope to obtain exact results, first because the integral equation (1.124) is not soluble in closed form; second, because we have little evidence concerning the quasiparticle interaction $f_{pp'}$. On the other hand, we can achieve a qualitative understanding of the problem by using approximate forms for the reduced interaction $F_{pp'}$. For instance, we have studied in Section 1.7 the simple model in which $F_{pp'}$ is replaced by a constant F_0^s: *zero sound* is found to exist as long as

$$F_0^s > 0 \tag{1.192}$$

(for $F_0^s < 0$, the collective mode is destroyed by Landau damping). The condition (1.192) is *not* exact, but probably represents a reasonable guess for the existence of zero sound. Similarly, longitudinal spin waves may be expected to exist if the following condition is met:

$$F_0^a > 0. \tag{1.193}$$

Finally by taking an interaction of the form

$$F_0^s + F_1^s \cos \xi, \tag{1.193a}$$

one may show that transverse ($m = 1$) spin symmetric modes exist as long as

$$F_1^s > 6 \tag{1.194}$$

(see Problem 1.5). Again, the conditions (1.193) and (1.194) are only approximate.

We now apply these results to liquid ^{3}He, making use of the results shown in Table 1.1. Since F_0^s is positive and rather large, there is no doubt that *zero sound* exists as a well-defined collective mode. On the other hand, F_0^a is negative: spin waves are likely to be destroyed by Landau damping. Finally, F_1^s, given by Eq. (1.100), is close to satisfying Eq. (1.194): transverse spin symmetric modes probably exist at comparatively low pressures. Of course, all these modes only exist when their frequency ω is larger than the thermal collision frequency $1/\tau$, given by Eq. (1.190).

Experimentally, it should be easiest to observe zero sound, since it involves density fluctuations which readily couple to an external probe. Experiments which would detect a spin wave—even less a transverse collective mode—are difficult to conceive, still more so to carry out. We shall thus focus our attention on zero sound. For an interaction of the form (1.193a), its velocity at zero pressure is found to be about 3.6 times the Fermi velocity: zero sound should thus be a well-defined collective mode in ^{3}He. Indeed the coupling between quasiparticles is so strong that the velocity of zero sound lies within a few percent of that of first sound (see Fig. 1.7).

The most direct way to observe zero sound is by studying ultrasonic propagation in liquid ^{3}He. As the frequency ω is increased, one should observe a maximum in the wave attenuation, obtained when ω is equal to the collision frequency $1/\tau$ (see Fig. 1.8). The maximum corresponds to the transition from first to zero sound. To the extent that the frequency can cover both sides of that maximum, one should observe the small difference in velocity between first and zero sound.

In practice, such an experiment is extremely difficult, in view of the high frequencies required: at 0.1°K, the frequency of the attenuation maximum is of the order of 10^{10}; this is the extreme limit at which presently available ultrasonic equipment functions. As yet, only the beginning of the curve has been observed, i.e., the increase of attenuation as ω approaches $1/\tau$. In order to observe the full transition from first to zero sound, it will be necessary to go to still lower temperatures ($<10^{-2}$°K), while continuing to work with the highest ultrasonic frequencies available.

Present ultrasonic methods have, nevertheless, provided indirect evidence for the existence of zero sound. Keen, Matthews, and Wilks

(1963) have measured the acoustic impedance, Z, of ^{3}He as a function of temperature. In the hydrodynamic regime, an elementary calculation shows that it is

$$Z_1 = Nms, \tag{1.195}$$

where s is the first sound velocity. Bekarevitch and Khalatnikov (1961) have solved the transport equation for quasiparticles which controls Z in the collisionless regime. Gavoret (1965) has recently shown that since one is, in fact, in a strong coupling regime, the "collisionless" acoustic impedance is dominated by zero sound propagation, and is given by

$$Z_0 = Nms_0, \tag{1.196}$$

where s_0 is the zero sound velocity. One thereby expects to see an increase in the acoustic impedance on passing to temperatures appropriate to the collisionless regime.

Such an increase has been observed by Keen *et al.*, who find that in the vicinity of 0.09°K, Z/mN rises abruptly from a value of ~181 m/sec to a value of ~200 m/sec. Gavoret (1965) calculates, on the basis of Eq. (1.196), a value of Z_0/mN of ~191 m/sec, in qualitative agreement with the experimental results. The lack of detailed agreement may well be traced to our inadequate knowledge of $F_{pp'}$.

Further indirect evidence for the existence of zero sound has been found in detailed study of the thermal boundary resistance between ^{3}He and an adjoining metal (the so-called "Kapitza effect"). We refer the reader to the original literature [Anderson, Connolly, and Wheatley (1964), Gavoret (1965)] for a discussion of this problem, which lies somewhat beyond the purview of our book.

In view of the difficulties with direct ultrasonic observation of zero sound, various other methods of observation have been proposed which involve the inelastic scattering of light, γ-rays, or neutrons by liquid ^{3}He [Abrikosov and Khalatnikov (1961); Akhiezer, Akhiezer, and Pomeranchuk (1961)]. We shall see in Chapter 2 that such experiments provide a direct measurement of the density fluctuation excitation spectrum. At temperatures and frequencies such that zero sound is a well-defined excitation, one expects it will dominate this spectrum, and so appear as a discrete peak in the spectrum of inelastically scattered "particles" at a given angle. Such measurements possess the advantage that they permit the direct observation of high-energy zero sound quanta (corresponding to $\omega \gg 10^{12}$/sec) and hence can be carried out at "moderate" temperatures of 0.1°K to 1°K without quasiparticle collisions posing a serious problem.

Of these probes, that which is most appealing at first sight is neutron scattering, which we shall see has proved so useful in the study of excitations in ^{4}He. Unfortunately, such an experiment is hardly feasible, as the cross section for neutron capture by an ^{3}He nucleus is far bigger than that for neutron scattering: the neutrons are absorbed by the liquid in a very short distance. Akhiezer *et al.* (1961) have suggested that one might avoid this trouble by studying neutron reflection on the surface of the liquid, under grazing incidence. Such an experiment is extremely difficult, and has not been achieved as yet. Assuming that it may be performed with polarized neutrons, it would also provide information on the spin excitations of the Fermi liquid, through the spin flip collisions of the incoming neutrons with ^{3}He atoms.

The next possibility is to study the scattering of x-rays, perhaps by making use of very narrow Mossbauer lines. As yet, no experiment has been tried in this direction. The search for zero sound in liquid ^{3}He remains one of the exciting problems of low-temperature physics.

1.11. CONCLUSION

We conclude this long chapter with a brief summary of the range of validity and main achievements of Landau's theory. The theory applies to *normal* Fermi liquids (apart from the possible logarithmic singularity in C_v). It is valid only if the following two conditions are met: (i) The system must be highly degenerate, such that $\kappa T \ll \mu$. (ii) Only macroscopic phenomena are considered, of wave vector $\mathbf{q}$ and frequency ω such that $\hbar q \ll p_F$, and $\hbar\omega \ll \mu$. The two conditions ensure that the number of quasiparticles involved in the problem, whether thermally excited or produced by an external field, remains very small compared to the total density N. In other words, Landau's theory corresponds to an expansion in powers of the fraction of excited particles.

Within these two restrictions, the theory is rigorous. It lends itself to a large number of problems: equilibrium properties, transport properties, response to long wavelength external perturbations. On the other hand, it does not permit one to treat the response to a microscopic perturbation. For instance, the distortion of the system around a point impurity involves large momentum transfers, and lies beyond the framework of the Landau theory. Similarly, the scattering of a fast charged particle by an electron gas involves large energy transfers, which again prevents the application of Landau's formalism. Therefore, despite its very broad field of application, Landau's theory is not a "universal panacea": its well-defined limitations should always be kept in mind. Moreover, it uses (and provides) far more information than any experiment will ever sample. For example, one never detects experimentally *one* quasiparticle; put another way, one never *measures* the quasiparticle distribution $\delta n_{\mathbf{p}}$. What is observed experimentally is always an *average* property of the system, its charge, current, magnetization, etc. A theory as detailed as the Landau theory is therefore not always needed for the interpretation of physical experiments.

For these two reasons—lack of generality and excessive information—there is need for another formulation, perhaps less ambitious in its scope, which would provide for any system a formal description of the relevant experimental results. The following chapter is devoted to building this formalism.

PROBLEMS

1.1. Calculate the spin susceptibility (1.67) without using the quantity $\delta\bar{n}$; that is, take into account explicitly the change in quasiparticle energy due to the magnetization.

1.2. Calculate, beginning with (1.92), the momentum and energy carried by a single quasiparticle with momentum $\mathbf{p}$.

1.3. Calculate the first-order terms of an expansion of the scattering amplitude $A_{\mathbf{pp'}}(\mathbf{q}, \omega)$ in powers of qv_F/ω or ω/qv_F.

1.4. Derive the dispersion relation for zero sound, Eq. (1.128b).

1.5. Study the transverse collective mode ($m = 1$) taking for the reduced interaction $F_{\mathbf{pp'}}$ the simplified form

$$F = F_0 + F_1 \cos \xi.$$

Find the dispersion law. Show that the mode is subject to Landau damping unless $F_1 > 6$.

1.6. Calculate, with the aid of the simple model of a constant $f_{\mathbf{pp'}}$, the influence of Landau damping on $A_{\mathbf{pp'}}(\mathbf{q}, \omega)$.

1.7. Derive along the same lines as those used for the thermal conductivity the expression (1.179) for the viscosity.

1.8. Derive the expression (1.179) for the spin diffusion coefficients. [First obtain the expression of the collision integral for a spin antisymmetric phenomenon, which is different from (1.155); then proceed along the same lines as for thermal conductivity.]

1.9. Study the transition from first to zero sound, and the damping of the oscillation in the two limits. For that purpose, replace the quasiparticle interaction by a constant F_0, and assume that the collision integral may be written as

$$I(n) = -\frac{[\delta\bar{n} - \delta\bar{n}_0 - \delta\bar{n}_1]}{\tau},$$

where n_0 and n_1 are the $\ell = 0$ and $\ell = 1$ components of n. (Note that by subtracting these terms, the conservation of charge and current in quasiparticle collisions is automatically guaranteed. The dispersion law may then be obtained in closed form.)

REFERENCES

Abel, W. R., Anderson, A. C., Black, W. L., and Wheatley, J. C. (1965), *Physics* **1**, 337.

Abrikosov, A. A. and Khalatnikov, I. M. (1959), *Rept. Progr. Phys.* **22**, 329.

Abrikosov, A. A. and Khalatnikov, I. M. (1961), *Sov. Phys. JETP* **14**, 389.

Akhiezer, A. J., Akhiezer, I. A., and Pomeranchuk, Y. A. (1961), *Sov. Phys. JETP* **14**, 343.

Anderson, A. C., Salinger, G. L., Steyert, W. A., and Wheatley, J. C. (1961), *Phys. Rev. Letters* **6**, 331.

Anderson, A. C., Connolly, J. I., and Wheatley, J. C. (1964), *Phys. Rev.* **135**, A910.

Anderson, P. W. (1965), *Physics*, **2**, 1.

Balian, R. and de Dominicis, C. (1964), *Physica* **30**, 1927.

Balian, R. and Fredkin, D. R. (1965), *Phys. Rev. Letters* **15**, 480.

Bekarevich, I. L. and Khalatnikov, I. M. (1961), *Sov. Phys. JETP* **12**, 1187.

Gavoret, J. (1965), *Phys. Rev.* **137**, A721.

Hone, D. (1962), *Phys. Rev.* **125**, 1494.

Hugenholtz, N. M. and Van Hove, L. (1958), *Physica* **24**, 363.

Kadanoff, L. P. and Prange, R. E. (1964), *Phys. Rev.* **134**, A566.

Keen, B. E., Matthews, P. W., and Wilks, J. (1963), *Phys. Letters* **5**, 5.

Landau, L. D., (1946), *J. Phys. USSR* **10**, 25.

Landau, L. D. (1956), *Sov. Phys. JETP* **3**, 920.

Landau, L. D. (1957), *Sov. Phys. JETP* **5**, 101.

Luttinger, J. M. and Nozières, P. (1962), *Phys. Rev.* **127**, 1423, 1431.

Morel, P. and Nozières, P. (1962), *Phys. Rev.* **126**, 1909.

Nozières, P. (1963), *Theory of Interacting Fermi Systems*, W. A. Benjamin, New York.

Overhauser, A. (1962), *Phys. Rev.* **128**, 1437.

Peshkov, V. (1965), *Sov. Phys. JETP* **21**, 663.

Pitaevskii, L. P. (1959), *Sov. Phys.* **10**, 1267.

Pomeranchuk, Y. A. (1958), *Sov. Phys. JETP* **8**, 361.

Richards, P. M. (1963), *Phys. Rev.* **132**, 1867.

Simkin, D. (1963), Ph. D. Thesis, University of Illinois, unpublished.

Wheatley, J. (1966), *Proc. Sussex Symposium on Quantum Fluids*, North-Holland, Amsterdam.

CHAPTER 2

RESPONSE AND CORRELATION IN NEUTRAL SYSTEMS

Any *experiment* conducted on a physical system involves the excitation of the system by some external probe, and the subsequent measurement of the system's response to that probe. If the interaction between the probe and the system is sufficiently weak, the system's response is *linear;* it is determined entirely by the properties that the system possesses in the absence of the probe. Consequently, "weak interaction" experiments offer the ideal method for studying the behavior of many-particle systems.

The *response* of the system is closely linked to the *correlations* which exist between the positions and momenta of different particles, so that it is often possible to measure various correlation functions directly. Moreover, under certain circumstances there is a further connection between the correlation functions and the spectrum of *elementary excitations* in the many-particle system. One can then pass from a knowledge of the correlation functions to a determination of the elementary excitation spectrum, and vice versa.

In this chapter we set up a general theory of linear response, and then discuss its application to various situations of physical interest. Such a theory is necessarily somewhat formal; it is, however, extremely useful. It is exact, and provides a number of exact results of great practical importance. Even more important, the theory establishes the *language* one should use in discussing the properties of quantum liquids, both in the macroscopic and the microscopic regimes. In the remainder of this book we shall rely again and again on the general theory presented here, essentially because the formalism helps one appreciate the underlying unit of all quantum liquids, a unity which is frequently hidden by a diversity of mathematical methods.

We begin by considering a simple physical example of probe-liquid interaction, the inelastic scattering of an external particle beam by a neutral quantum liquid. We show, in Section, 2.1, that one measures thereby the "dynamic form factor" of the system; in Section 2.2 we establish an important sum rule satisfied by this quantity. We establish the general theory of linear response in Section 2.3; there we discuss in detail the formal features of the theory, which underlie its subsequent applications. All of the above considerations are presented for a system that is in its ground state; the theory is extended to finite temperatures in Section 2.6.

The general theory developed in Sections 2.1–2.3 is applied to a particular example, that of density correlations in a neutral Fermi liquid, in Section 2.4. A qualitative discussion of the dynamic form factor is given; the intimate relation between the response functions and the elementary excitation spectrum, which exists in the macroscopic limit at $T = 0$, is exploited to determine the long wavelength excitations of physical interest. The Landau theory is used to carry out an explicit calculation of the density response function in Section 2.5; a link is thereby established between Chapters 1 and 2.

The final section of the chapter, Section 2.7, is concerned with the effect of particle collisions on the system's response to an external probe. At finite temperatures, as the frequency of the probe is reduced, one passes from the "collisionless" regime, in which the collisions play a negligible role, to the "hydrodynamic" regime in which the collisions bring about a state of local equilibrium. The general theory of linear response is of course valid in the hydrodynamic regime; we show that in this limit the response functions may be obtained from purely macroscopic arguments.

2.1. SCATTERING OF A PARTICLE IN THE BORN APPROXIMATION

DESCRIPTION OF THE SCATTERING EVENT

As a typical *microscopic* probe of the system behavior, we first consider the measurement of the *energy* and *angular distribution* of a beam of inelastically scattered particles. An example of such a probe is the scattering of slow neutrons in liquid ^{4}He. We shall see in Chapter 4 that similar experiments may be carried out for charged electron systems, by studying the scattering of a fast electron, or the inelastic scattering of light.

Let us consider an incoming "external" particle, with mass M_e, momentum $\mathbf{P}_e$, and energy $E_e = P_e^2/2M_e$. The particle scatters on the system, which we assume to be initially in its ground state $|0\rangle$, with energy E_0. After the collision, the external particle has momentum $(\mathbf{P}_e - \mathbf{q})$, while the system finds itself in some excited state $|n\rangle$, with energy E_n. If we assume the system to be translationally invariant, momentum conservation implies that the state $|n\rangle$ has total momentum $\mathbf{q}$. Furthermore, the conservation of energy in the collision requires that

$$\omega \equiv E_n - E_0 = \frac{\mathbf{q}\cdot\mathbf{P}_e}{M_e} - \frac{q^2}{2M_e}. \tag{2.1}$$

The collision may thus be characterized by a momentum transfer $\mathbf{q}$, and an energy transfer ω, from the particle to the system.*

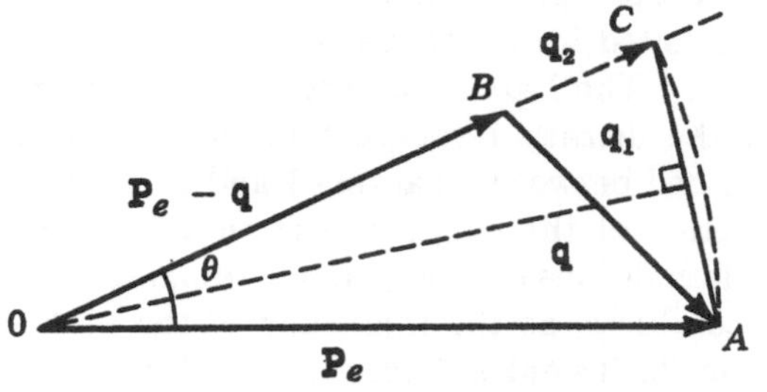

FIGURE 2.1. *Inelastic scattering of a fast particle with momentum $\mathbf{P}_e$. Relation between the scattering angle, energy transfer, and momentum transfer.*

For future reference, we first develop the geometry of the collision process. The momentum of the external particle before and after the collision is represented in Fig. 2.1. We see that the momentum transfer $\mathbf{q} = BA$ can be split into two parts:

(i) A part $\mathbf{q}_1 = CA$, which corresponds to an elastic scattering event, and which fixes the deflection angle θ, according to the relation

$$\frac{q_1}{P_e} = 2\sin\frac{\theta}{2}. \tag{2.2a}$$

* In this chapter, we shall use "atomic" units in which $\hbar$ is equal to 1. We shall also assume the system to be of unit volume.

(ii) Another term $\mathbf{q}_2 = BC$, which does not affect the scattering angle, but determines the energy transfer ω, via the relation

$$\omega = \mathbf{q}_2 \cdot \frac{\mathbf{P}_e}{M_e} - \frac{q^2}{2M_e}. \tag{2.2b}$$

These relations simplify if the momentum transfer $\mathbf{q}$ is much smaller than $\mathbf{P}_e$ (the corresponding scattering angle θ is then very small). In this case, $\mathbf{q}_1$ is nearly perpendicular to $\mathbf{P}_e$, while $\mathbf{q}_2$ is nearly parallel to it. Up to first order in $\mathbf{q}$, we have

$$\begin{aligned} q_1 &= q_\perp = P_e\theta, \\ q_2 &= q_\parallel = \frac{\omega M_e}{P_e}, \end{aligned} \tag{2.3}$$

so that q_1 and q_2 are completely decoupled.

Experimentally, one readily measures the deflection angle θ. It is much harder to obtain the energy transfer ω; the latter may be determined by a "time of flight" measurement of the velocity of the scattered neutron [Brockhouse (1960)]. To the extent that one measures *both* θ and ω, one can easily calculate the net momentum transfer, whose modulus is given by (see Fig. 2.1)

$$q^2 = q_1^2 + q_2^2 - 2q_1q_2 \sin\theta/2. \tag{2.4}$$

In the case of small angle scattering, to which Eqs. (2.3) apply, Eq. (2.4) reduces to

$$q = P_e\sqrt{\theta^2 + (\omega/2E_e)^2}. \tag{2.5}$$

SCATTERING CROSS SECTION: THE DYNAMIC FORM FACTOR

Let us assume that the probe and the system are coupled by a velocity-independent interaction potential

$$H_{\text{int}} = \sum_i \mathcal{V}(\mathbf{r}_i - \mathbf{R}_e), \tag{2.6}$$

where $\mathbf{r}_i$ and $\mathbf{R}_e$ are the system particle and probe positions respectively. It is convenient to Fourier analyze Eq. (2.6) as follows:

$$H_{\text{int}} = \sum_i \sum_{\mathbf{q}} \mathcal{V}_{\mathbf{q}} e^{i\mathbf{q}\cdot(\mathbf{r}_i - \mathbf{R}_e)} = \sum_{\mathbf{q}} \mathcal{V}_{\mathbf{q}} \rho_{\mathbf{q}}^{+} e^{-i\mathbf{q}\cdot\mathbf{R}_e}. \tag{2.7}$$

In Eq. (2.7), $\mathcal{V}_{\mathbf{q}}$ is the Fourier transform of $\mathcal{V}(\mathbf{r})$, while $\rho_{\mathbf{q}}$ is the Fourier transform of the particle density

$$\rho_{\mathbf{q}} = \int d^3r\, \rho(\mathbf{r})e^{-i\mathbf{q}\cdot\mathbf{r}} = \sum_i \int d^3r\, \delta(\mathbf{r} - \mathbf{r}_i)e^{-i\mathbf{q}\cdot\mathbf{r}}$$
$$= \sum_i e^{-i\mathbf{q}\cdot\mathbf{r}_i}. \tag{2.8}$$

The $\rho_{\mathbf{q}}$ describe the fluctuations of the particle density about its average value $\rho_o = N$.

We consider now the theoretical description of the scattering act considered above, in which the probe particle transfers momentum **q** and energy ω to the many-particle system. We shall make the important assumption that the probe particle is *weakly coupled* to the many-particle system, so that the scattering act may be described within the Born approximation. This approximation is valid for the cases we have mentioned above, those of the scattering of slow neutrons or fast electrons. Because the probe and the system are weakly coupled, the probe eigenfunctions before and after the scattering act are simply plane wave states, corresponding to the momenta $\mathbf{P}_e$ and $\mathbf{P}_e - \mathbf{q}$. The initial wave function for the many-particle system is its *exact* ground state wave function, $|0\rangle$; the matrix element for the system to make a transition into an *exact* excited state $|n\rangle$ of momentum **q**, is, from Eq. (2.7),

$$\langle n, \mathbf{P}_e - \mathbf{q}|H_{\text{int}}|0, \mathbf{P}_e\rangle = \mathcal{V}_{\mathbf{q}}\langle n|\rho_{\mathbf{q}}^{+}|0\rangle = \mathcal{V}_{\mathbf{q}}(\rho_{\mathbf{q}}^{+})_{no}. \tag{2.9}$$

Apart from $\mathcal{V}_{\mathbf{q}}$, which is presumed known, Eq. (2.9) depends only on the matrix element of the density fluctuations, $\rho_{\mathbf{q}}^{+}$, taken between the exact many-particle eigenstates in the absence of the external probe. The external particle thus acts as a direct probe of the *density fluctuations* in the system.

According to the "golden rule" of second-order perturbation theory, the probability per unit time $\mathcal{P}(\mathbf{q}, \omega)$ that the particle transfer momentum **q**, energy ω, to the system is given by

$$\mathcal{P}(\mathbf{q}, \omega) = 2\pi|\mathcal{V}_{\mathbf{q}}|^2 \sum_n |(\rho_{\mathbf{q}}^{+})_{no}|^2\delta(\omega - \omega_{no}). \tag{2.10}$$

In Eq. (2.10), we have set $\omega_{no} = E_n - E_o$; the delta function expresses the conservation of energy. The summation is over all states $|n\rangle$ coupled to the ground state $|0\rangle$ by the density fluctuation $\rho_{\mathbf{q}}^{+}$. In applying Eq. (2.10) it is important to realize that ω and **q** are not independent variables, but are related according to Eq. (2.1). One can formally allow for this relation by multiplying Eq. (2.10) by an additional factor

$$\delta\left(\omega - \mathbf{q}\cdot\frac{\mathbf{P}_e}{M_e} + \frac{q^2}{2M_e}\right),$$

which expresses the conservation of incoming particle energy.

The transition probability (2.10) may be written as

$$\mathcal{P}(\mathbf{q}, \omega) = 2\pi|\mathcal{V}_\mathbf{q}|^2 S(\mathbf{q}, \omega), \tag{2.11}$$

where the factor $S(\mathbf{q}, \omega)$ is defined by

$$S(\mathbf{q}, \omega) = \sum_n |(\rho_\mathbf{q}{}^+)_{no}|^2 \delta(\omega - \omega_{no}). \tag{2.12}$$

$S(\mathbf{q}, \omega)$ embodies all the properties of the many-particle system that are relevant to the scattering of the probe: it is known as the *dynamic form factor* of the system. $S(\mathbf{q}, \omega)$ represents the maximum information one can obtain about the system behavior in a particle scattering experiment. It furnishes a direct measure of the excitation spectrum of the density fluctuations, being proportional to the squared matrix element for each permissible excitation energy. Note that $S(\mathbf{q}, \omega)$ is real and positive, and that it vanishes for $\omega < 0$, since at $T = 0$ all the excitation frequencies, ω_{no}, are necessarily positive.

The dynamic form factor is closely related to the density correlations in the many-particle system. Let us consider the evolution in time of a density fluctuation, $\rho_\mathbf{q}{}^\dagger$. This may be described by the state vector

$$\rho_\mathbf{q}{}^+(t)|0\rangle = e^{iHt}\rho_\mathbf{q}{}^+ e^{-iHt}|0\rangle, \tag{2.13}$$

where H is the total system Hamiltonian. The *dynamic* correlation between a density fluctuation at time $t = 0$ and one at time τ is expressed by the correlation function,

$$S(\mathbf{q}, \tau) = \langle 0|\rho_\mathbf{q}(\tau)\rho_\mathbf{q}{}^+(0)|0\rangle. \tag{2.14}$$

On making use of Eq. (2.13), and, further, introducing the complete set, $|n\rangle$, of eigenstates of H, we may write Eq. (2.14) as

$$S(\mathbf{q}, \tau) = \sum_n \langle 0|e^{iH\tau}\rho_\mathbf{q} e^{-iH\tau}|n\rangle\langle n|\rho_\mathbf{q}{}^+|0\rangle = \sum_n |(\rho_\mathbf{q}{}^+)_{no}|^2 e^{-i\omega_{no}\tau}. \tag{2.15}$$

On comparing Eq. (2.15) with (2.12) we see that $S(\mathbf{q}, \tau)$ is the Fourier transform of $S(\mathbf{q}, \omega)$ with respect to time. The dynamic form factor thus determines the *correlations* between density fluctuations at different times, as well as the frequency distribution of the state $\rho_\mathbf{q}{}^+(t)|0\rangle$, i.e., its spectral density.

THE STATIC FORM FACTOR

In a "time of flight" neutron scattering experiment, one does not measure directly the momentum transfer $\mathbf{q}$, but rather the deflection angle θ and the energy transfer ω. $\mathbf{q}$ is then inferred from the values of θ and ω. Under these circumstances, the quantity of physical interest is the differential probability $\mathcal{P}(\theta, \omega)\, d\Upsilon\, d\omega$ that the particle be scattered into a solid angle $d\Upsilon$, while suffering an energy loss in the range $(\omega, \omega + d\omega)$. $\mathcal{P}(\theta, \omega)$ is given by

$$\mathcal{P}(\theta, \omega)\, d\Upsilon = \sum_{\mathbf{q}} \mathcal{P}(\mathbf{q}, \omega)\delta\left(\omega - \frac{\mathbf{q}\cdot\mathbf{P}_e}{M_e} + \frac{q^2}{2M_e}\right),$$

where the summation runs over those values of $\mathbf{q}$ for which the final particle momentum lies in the solid angle $d\Upsilon$. The explicit calculation of $\mathcal{P}(\theta, \omega)$ is rather complicated, unless q is much smaller than P_e. In such a case, we can use Eq. (2.4); we thus find

$$\mathcal{P}(\theta, \omega) = \frac{P_e^{\,2}}{(2\pi)^3}\int dq_{\parallel}\, \mathcal{P}(\mathbf{q}, \omega)\delta(\omega - q_{\parallel}V_e), \tag{2.16}$$

where V_e is the particle velocity. The momentum transfer $\mathbf{q}$ is a function of θ and ω, which in this limit is given by Eq. (2.5). Such an analysis has been applied to the scattering of fast electrons in metals; it will be considered in Section 4.4.

In the more general case in which $q \sim P_e$, the passage from $\mathcal{P}(\theta, \omega)$ to $\mathcal{P}(\mathbf{q}, \omega)$ is somewhat more complicated. Yet, it can be achieved using, for example, the ingenious automatic procedure devised by Brockhouse (1960). In this way, it has been possible to obtain the dynamic form factor $S(\mathbf{q}, \omega)$ of liquid ^{4}He from measurements of the scattering of slow neutrons. Such experiments will be discussed in Chapter 6.

In many cases, one is not able to analyze the energy of the outgoing particles: one only measures the differential cross section as a function of the scattering angle θ. The corresponding probability is

$$\mathcal{P}(\theta) = \int_0^\infty \mathcal{P}(\theta, \omega)\, d\omega.$$

When $\mathcal{P}(\theta, \omega)$ is replaced by its expression (2.16), $\mathcal{P}(\theta)$ appears as an integral of $\mathcal{P}(\mathbf{q}, \omega)$ over the frequency ω. In general, $\mathcal{P}(\mathbf{q}, \omega)$ depends implicitly upon ω through the wave vector $\mathbf{q}$, which renders such an integral rather complicated. In the most general case, it is no longer possible to split $\mathcal{P}(\theta)$ into two factors, characterizing the probe and the system respectively.

Matters are considerably simplified when the momentum transfer $\mathbf{q}$ is essentially independent of the energy transfer ω in the range where scattering is appreciable. For inelastic scattering, such a condition can be satisfied only if $q \ll P_e$, which ensures that Eq. (2.16) is valid. On making use of Eqs. (2.11) and (2.16), we may then write the probability $\mathcal{P}(\theta)$ as

$$\mathcal{P}(\theta) = \frac{M_e P_e}{(2\pi)^3} \int_0^\infty \mathcal{P}(\mathbf{q}, \omega)\, d\omega = \frac{M_e P_e}{(2\pi)^3} 2\pi |\mathcal{V}_\mathbf{q}|^2 N S_\mathbf{q}, \tag{2.17}$$

where $S_\mathbf{q}$ is defined by

$$S_\mathbf{q} = \frac{1}{N} \int_0^\infty d\omega\, S(\mathbf{q}, \omega). \tag{2.18}$$

$S_\mathbf{q}$ is called the *static form factor* of the system.

We consider in more detail the range of validity of Eq. (2.17). We have already remarked that a necessary condition for its validity is that $P_e \gg q$; in order that this condition be satisfied for all momentum transfers of interest, one must have $P_e \gg p_F$. According to Eq. (2.5), a second condition is that $\theta \gg \omega/E_e$. For a translationally invariant neutral system, we shall see that this second condition is easily satisfied for small q, since the corresponding energy transfer is of order qv_F. Under these circumstances, the last term of Eq. (2.5) is of order

$$\left[\frac{qv_F}{2E_e}\right]^2 = \left[\frac{q}{P_e}\frac{v_F}{V_e}\right]^2 \sim \theta^2 \left(\frac{v_F}{V_e}\right)^2. \tag{2.19}$$

Since this term is negligible with respect to θ^2, it follows that q is effectively independent of ω, and is equal to $P_e\theta$: it is then possible to measure $S_\mathbf{q}$ directly.

If the system is not translationally invariant, the energy transfer ω may be of the order of the chemical potential μ, even at very low $\mathbf{q}$. (In a real solid, for example, the incoming particle excites interband transitions.) In such a case, the last term of Eq. (2.5) is negligible only if θ is much larger than $\mu/E_e \sim (p_F/P_e)^2$. Since, as we have seen, one has necessarily $p_F \ll P_e$, this second restriction is of importance only for extremely small values of θ.

Let us again emphasize that for any system, the simple relation (2.17) breaks down when the particle velocity V_e is comparable to or smaller than v_F.

According to Eq. (2.18), $NS_\mathbf{q}$ is equal to the Fourier transform $S(\mathbf{q}, t)$ taken for $t = 0$. Indeed, one sees directly from Eq. (2.12) that

$$NS_\mathbf{q} = \langle 0 | \rho_\mathbf{q}^+ \rho_\mathbf{q} | 0 \rangle. \tag{2.20}$$

$NS_{\mathbf{q}}$ therefore gives a measure of the instantaneous density correlations in the system, that is, of the mean square density fluctuations. It is not surprising that the scattering of a *fast* particle should only involve $S_{\mathbf{q}}$. The time that a fast particle takes to go across the range of the interaction is very short; such a particle samples the correlations over a time scale that is much shorter than the characteristic periods of the system.

2.2. PARTICLE CONSERVATION: THE f-SUM RULE

Conservation laws represent one of the most valuable tools of the theoretical physicist, in that they provide a stringent test for any approximate description of a many-particle system. In this connection one of the simplest, and most useful, conservation laws is that of particle conservation. We shall see in Section 4.6 that particle conservation represents a necessary and sufficient condition for gauge invariance. Moreover, it yields an important relation, the f-sum rule, which the dynamic form factor $S(\mathbf{q}, \omega)$ must satisfy. In the present section, we shall first state the law of particle conservation. and then discuss its consequences.

CURRENT DENSITY FLUCTUATIONS

The operator which describes the current density at point $\mathbf{r}$ may be written in the form

$$\mathbf{J}(\mathbf{r}) = \frac{1}{2} \sum_i \left\{ \frac{\mathbf{p}_i}{m}\, \delta(\mathbf{r} - \mathbf{r}_i) + \delta(\mathbf{r} - \mathbf{r}_i)\, \frac{\mathbf{p}_i}{m} \right\}, \tag{2.21}$$

where $\mathbf{r}_i$ and $\mathbf{p}_i$ are the position and momentum of the ith particle in the system. The symmetrized expression is needed since $\mathbf{r}_i$ and $\mathbf{p}_i$ do not commute. The Fourier transform of $\mathbf{J}(\mathbf{r})$ with respect to $\mathbf{r}$ is equal to

$$\mathbf{J}_{\mathbf{q}} = \frac{1}{2} \sum_i \left\{ \frac{\mathbf{p}_i}{m}\, e^{-i\mathbf{q}\cdot\mathbf{r}_i} + e^{-i\mathbf{q}\cdot\mathbf{r}_i}\, \frac{\mathbf{p}_i}{m} \right\}. \tag{2.22}$$

$\mathbf{J}_{\mathbf{q}}$ represents a current density fluctuation with wave vector $\mathbf{q}$, and is the obvious analog of the density fluctuation, $\rho_{\mathbf{q}}$.

It is clear that $\rho_{\mathbf{q}}$ and $\mathbf{J}_{\mathbf{q}}$ must be related by a "continuity equation," which expresses the conservation of particle number in the system. In order to derive the exact form of the relation, we calculate the commutator $[\rho_{\mathbf{q}}, H]$, where H is the Hamiltonian of the system, which we write in the form

$$H = \sum_i \left[\frac{\mathbf{p}_i^2}{2m} + U(\mathbf{r}_i)\right] + \frac{1}{2}\sum_{i\neq j} V(\mathbf{r}_i - \mathbf{r}_j). \tag{2.23}$$

$U(\mathbf{r}_i)$ is an external potential applied to the particles (for instance, the periodic potential of the lattice in a metal); $V(\mathbf{r}_i - \mathbf{r}_j)$ is the interaction potential between a pair of particles. We make the important assumption that U and V are *velocity independent*, i.e., that both depend only on the positions $\mathbf{r}_i$ and $\mathbf{r}_j$ (and, perhaps, on spin). As a consequence, $\rho_{\mathbf{q}}$, which depends only on position, commutes with the potential and interaction terms. The only contribution to the commutator comes from the kinetic energy; one finds

$$[\rho_{\mathbf{q}}, H] = \frac{1}{2}\sum_i \left\{\frac{\mathbf{q}\cdot\mathbf{p}_i}{m}e^{-i\mathbf{q}\cdot\mathbf{r}_i} + e^{-i\mathbf{q}\cdot\mathbf{r}_i}\frac{\mathbf{q}\cdot\mathbf{p}_i}{m}\right\} = \mathbf{q}\cdot\mathbf{J}_{\mathbf{q}}. \tag{2.24}$$

Equation (2.24) is the desired *conservation law*. Let us emphasize that it is valid only when U and V are velocity independent. If this condition is not satisfied, the physical current is no longer given by Eq. (2.21): the conservation law takes a form different from Eq. (2.24).

In order to make explicit the relation between Eq. (2.24) and the conservation of particle number, we write $\rho_{\mathbf{q}}$ and $\mathbf{J}_{\mathbf{q}}$ in the Heisenberg representation. We then have

$$\frac{\partial\rho_{\mathbf{q}}}{\partial t} = -i[\rho_{\mathbf{q}}, H] = -i\mathbf{q}\cdot\mathbf{J}_{\mathbf{q}}, \tag{2.25}$$

which is the Fourier transform in space of the well-known equation

$$\frac{\partial\rho}{\partial t} + \operatorname{div}\mathbf{J} = 0. \tag{2.26}$$

Equation (2.26) serves to guarantee particle conservation. The operator relation (2.24) takes a simple form if we write its matrix elements between eigenstates of the real system,

$$\omega_{no}(\rho_{\mathbf{q}})_{no} = -[\mathbf{q}\cdot\mathbf{J}_{\mathbf{q}}]_{no}. \tag{2.27}$$

We shall make use of Eq. (2.27) to derive the *f*-sum rule in the following section.

THE f-SUM RULE

The dynamic form factor $S(\mathbf{q}, \omega)$ obeys the following sum rule,

$$\int_0^\infty \omega S(\mathbf{q}, \omega)\, d\omega = \frac{Nq^2}{2m}, \tag{2.28}$$

where N is the number of particles in the system, and m their bare mass. By using the definition (2.12) of $S(\mathbf{q}, \omega)$, we may write Eq. (2.28) as

$$\sum_n \omega_{no} |(\rho_{\mathbf{q}}^{+})_{no}|^2 = \frac{Nq^2}{2m}. \tag{2.29}$$

It is customary to define an "oscillator strength" f_{on} for each transition from the ground state $|0\rangle$ to the excited state $|n\rangle$:

$$f_{on} = \frac{2m}{q^2} \omega_{no} |(\rho_{\mathbf{q}}^{+})_{no}|^2. \tag{2.30}$$

In terms of the oscillator strengths, the sum rule (2.29) becomes

$$\sum_n f_{on} = N, \tag{2.31}$$

in which form it is known as the *f-sum rule.*

The f-sum rule (2.31) is equivalent to a longitudinal version of the well-known Thomas–Reich–Kuhn sum rule for atomic spectra: in its form (2.31), it was first determined by Placzek (1952). The derivation we present is due to the authors [Nozières and Pines (1958)].

Let us calculate the double commutator,

$$[[\rho_{\mathbf{q}}, H], \rho_{\mathbf{q}}^{+}]. \tag{2.32}$$

On making use of Eq. (2.24), we may write (2.32) as

$$[\mathbf{q} \cdot \mathbf{J}_{\mathbf{q}}, \rho_{\mathbf{q}}^{+}]. \tag{2.33}$$

The commutator, (2.33), may easily be calculated if we make use of the explicit expressions for $\mathbf{J}_{\mathbf{q}}$ and $\rho_{\mathbf{q}}^{+}$, (2.8) and (2.22); one finds

$$[[\rho_{\mathbf{q}}, H], \rho_{\mathbf{q}}^{+}] = \frac{Nq^2}{m}. \tag{2.34}$$

On the other hand, the expectation value of (2.33) in the ground state $|0\rangle$ may be found by inserting a complete set of eigenstates of H, $|n\rangle$, between the two factors of the commutator. On making use of Eq. (2.27), we find

$$\langle 0|[[\rho_{\mathbf{q}}, H], \rho_{\mathbf{q}}^{+}]|0\rangle = \sum_n \{\omega_{no}|(\rho_{\mathbf{q}}^{+})_{no}|^2 + \omega_{no}|(\rho_{\mathbf{q}})_{no}|^2\}. \tag{2.35}$$

We next make use of time reversal invariance, which tells us that we may associate to each state $|n\rangle$, with momentum $\mathbf{q}$, another state $|m\rangle$, degenerate with $|n\rangle$, which has a momentum $-\mathbf{q}$ and a complex conjugate wave function. Hence we may write:

$$\sum_n \omega_{no}|(\rho_{\mathbf{q}}^+)_{no}|^2 = \sum_m \omega_{mo}|(\rho_{\mathbf{q}})_{mo}|^2. \tag{2.36}$$

Equation (2.35) may therefore be written in the simpler form

$$\langle 0|[[\rho_{\mathbf{q}}, H], \rho_{\mathbf{q}}^+]|0\rangle = \sum_n 2\omega_{no}|(\rho_{\mathbf{q}}^+)_{no}|^2. \tag{2.37}$$

The f-sum rule follows at once from a comparison of Eqs. (2.34) and (2.37).

According to Eq. (2.28), the f-sum rule provides the first moment of the dynamic form factor $S(\mathbf{q}, \omega)$. In that respect, it complements the definition of the static form factor

$$NS_{\mathbf{q}} = \int_0^\infty S(\mathbf{q}, \omega)\, d\omega, \tag{2.18}$$

which provides the average value of $S(\mathbf{q}, \omega)$. We note that $S_{\mathbf{q}}$ can be written in terms of the oscillator strengths:

$$NS_{\mathbf{q}} = \sum_n |(\rho_{\mathbf{q}}^+)_{no}|^2 = \frac{q^2}{2m}\sum_n \frac{f_{on}}{\omega_{no}}. \tag{2.38}$$

Let us emphasize that the f-sum rule (2.28) is a direct consequence of particle conservation in the system, and represents a statement of that conservation law. Satisfaction of the f-sum rule is thus necessary to guarantee a gauge invariant theory, a point on which we shall elaborate in Section 4.6. The f-sum rule provides a very useful check on the consistency of any approximate theory. Furthermore, it may happen that a given class of excited states exhausts the sum rule: one thereby proves that other excited states have negligible oscillator strengths. We shall see in Chapters 4 and 6 that the use of particle conservation and sum rules sometimes permits a direct calculation of collective mode frequencies in the long wavelength limit.

STOPPING POWER FOR A FAST PARTICLE

Let us consider again the scattering of an external probe by the system, within the Born approximation. Such a scattering is characterized by the probability $\mathcal{P}(\theta, \omega)$ that the probe be deflected by an angle θ, while suffering an energy loss ω to the system. If the momentum transfer $\mathbf{q}$ is small as compared to the probe momentum $\mathbf{P}_e$, $\mathcal{P}(\theta, \omega)$ is given by Eq. (2.16). The total rate at which energy is transferred to the system by probe particles scattered in the direction θ is equal to

$$\begin{aligned} \frac{dE(\theta)}{dt} &= \int_0^\infty \mathcal{P}(\theta, \omega)\omega\, d\omega \\ &= \frac{P_e M_e}{(2\pi)^3} \int_0^\infty \mathcal{P}(\mathbf{q}, \omega)\omega\, d\omega. \end{aligned} \tag{2.39}$$

As we mentioned in Section 2.1, the integral appearing in Eq. (2.39) is not simple, as $\mathbf{q}$ is usually a function of both θ and ω. However, as we have seen, if the probe is fast enough $\mathbf{q}$ is essentially independent of ω, being equal to $P_e\theta$. On making use of Eq. (2.11) and the f-sum rule (2.38), we may then write

$$\frac{dE(\theta)}{dt} = \frac{P_e M_e}{(2\pi)^3} 2\pi |\mathcal{V}_\mathbf{q}|^2 \frac{Nq^2}{2m}. \tag{2.40}$$

The result (2.40) is independent of the detailed properties of the many-particle system under study, a perhaps unexpected result which illustrates the usefulness of the f-sum rule.

By summing (2.40) over all directions, we obtain the total energy loss of the probe particle to the system per unit time. As long as the scattering angle θ is small, the element of solid angle may be written as

$$d\Upsilon = 2\pi\theta\, d\theta.$$

Let us eliminate θ in favor of $q = P_e\theta$. On using Eq. (2.40), we obtain

$$\frac{dE}{dt} = \frac{1}{2\pi V_e} \int_0^\infty \frac{Nq^3}{2m} |\mathcal{V}_\mathbf{q}|^2\, dq. \tag{2.41a}$$

It is customary to define the *stopping power* of the system as the energy loss per unit *path* of the probe,

$$\frac{dE}{dx} = \frac{1}{V_e}\frac{dE}{dt} = \frac{1}{2\pi V_e^2} \int_0^\infty \frac{Nq^3}{2m} |\mathcal{V}_\mathbf{q}|^2\, dq. \tag{2.41b}$$

The expression (2.41a,b) of the stopping power for fast particles was obtained a long time ago by Bethe. We note that it is likewise not affected by the interaction between particles; a measurement of the stopping power provides little information on system properties.

2.3. LINEAR RESPONSE FUNCTION

In Section 2.1 we studied the real transitions in the many-particle system produced by an external probe of system behavior. Such transitions correspond to an *irreversible dissipative* process. In addition to the real processes, the probe will produce virtual transitions; such transitions represent a reversible deformation of the system, corresponding to its "polarization" by the external probe. The situation is analogous to that encountered for electrical circuits, in which the system's response to an applied potential contains both a dissipative and a reactive part. The complete description of the system's response to an external probe is straightforward as long as the probe-system coupling is weak. Under these circumstances the response of the system is linear, and may be calculated by a straightforward application of time-dependent perturbation theory.

In the most general case, the probe is coupled to some physical quantity A (density, current, spin density, etc.); one wishes to measure the linear response of some other quantity B to the probe. The response is conveniently described in terms of the "*B-A response function*": the first letter denotes the measured quantity; the second determines the form of the probe-system coupling. In the present section, we shall limit ourselves to a particular example, namely the *density-density response function,* which describes the density fluctuations induced by a density probe. Most of the results one obtains are easily transposed to other types of response functions.

RESPONSE TO A SCALAR POTENTIAL

We consider a scalar potential, $\varphi(\mathbf{r}, t)$, which acts on the system density through an interaction Hamiltonian,

$$H_e = \sum_{\mathbf{q}} \int \frac{d\omega}{2\pi} \rho_{\mathbf{q}}{}^{+} \varphi(\mathbf{q}, \omega) e^{-i\omega t}. \tag{2.42}$$

Here $\varphi(\mathbf{q}, \omega)$ is the Fourier transform in space and time of $\varphi(\mathbf{r}, t)$. Since the perturbing field is real, one has the relation

$$\varphi(\mathbf{q}, \omega) = \varphi^*(-\mathbf{q}, -\omega). \tag{2.43}$$

An example of a scalar probe is the particle with velocity $\mathbf{V}_e$, position $\mathbf{R}_e$ whose scattering we studied in Section 2.1: the corresponding value of $\varphi(\mathbf{q}, \omega)$ is

$$\varphi(\mathbf{q}, \omega) = 2\pi \mathcal{V}_{\mathbf{q}} e^{-i\mathbf{q}\cdot\mathbf{R}_e} \delta(\omega - \mathbf{q}\cdot\mathbf{V}_e) \tag{2.44}$$

in the approximation in which the recoil of the scattered particles is neglected.

We shall assume the external potential, φ, to be very weak, so that it is a good approximation to consider only the linear response of the system to the interaction (2.42). Under such conditions, each Fourier component of (2.42) acts independently. A given $\varphi(\mathbf{q}, \omega)$ induces its own density fluctuation in the system, with wave vector $\mathbf{q}$ and frequency ω. Without any loss of generality, we can restrict our discussion to a single Fourier component of Eq. (2.42). Since, however, we must preserve the reality condition (2.43), we are led to consider the perturbation

$$H_e = \rho_{\mathbf{q}}^{+}\varphi(\mathbf{q}, \omega)e^{-i\omega t} + \text{c.c.}, \tag{2.45}$$

which corresponds to a running wave with wave vector $\mathbf{q}$ and frequency ω.

There are two difficulties with an interaction of the form (2.45). First, it extends from $t = -\infty$ to $t = +\infty$, a situation which is, of course, unphysical. Second, a probe at a finite frequency ω will act to produce real system transitions and thereby transfer energy to the system. Over a sufficiently long time interval, no matter how weak the probe-system coupling may be, the probe will act to "heat up" the system to such an extent that the system deformation can no longer be regarded as small. Under such circumstances, nonlinear effects will play an essential role.

The above difficulties may be avoided by the proper choice of boundary conditions in specifying the probe-system coupling. We shall here adopt "adiabatic" boundary conditions, which correspond to turning on the interaction, H_e, *very slowly*. Mathematically, this may be accomplished if we multiply H_e by a factor $e^{\eta t}$, where η is positive and very small. It is clear that such a procedure is only meaningful if the system is *stable:* otherwise, one could not let it evolve over an infinite lapse of time. The present discussion thus cannot be applied to a study of unstable growing waves (such as laser oscillations).

We shall see later in this section that the adiabatic boundary condition insures that the response to the probe be *causal:* the system response follows (in time) the probe. Qualitatively, the energy transferred by the probe to the system is proportional to the potential squared, $\varphi_{\mathbf{q}}^2$, and to the time during which the coupling is set up, $1/\eta$. It will thus contain a factor $\varphi_{\mathbf{q}}^2/\eta$, where both $\varphi_{\mathbf{q}}$ and η are small quantities. Heating of the system by the probe is avoided if we first let the potential, φ, go to zero and only later the adiabatic switching rate η. In other words, we should first calculate the linear response, keeping η finite; the limit $\eta \to 0$ is then taken at the very end of the calculation.

The average density fluctuation at time t induced in the system by the perturbation (2.45) is given by

$$\langle \rho(\mathbf{q}, t) \rangle = \langle \psi(\mathbf{r}, t) | \rho_{\mathbf{q}} | \psi(\mathbf{r}, t) \rangle, \tag{2.46}$$

where $|\psi(\mathbf{r}, t)\rangle$ is the exact system wave function in the presence of the test particle. Since the perturbation is periodic in time, we may write

$$\langle \rho(\mathbf{q}, t) \rangle = \langle \rho(\mathbf{q}, \omega) \rangle e^{-i\omega t} e^{\eta t}, \tag{2.47}$$

where the Fourier transform $\langle \rho(\mathbf{q}, \omega) \rangle$ is given by

$$\langle \rho(\mathbf{q}, \omega) \rangle = \langle \psi(\mathbf{r}, t) | \rho_{\mathbf{q}} e^{i\omega t} | \psi(\mathbf{r}, t) \rangle e^{-\eta t}. \tag{2.48}$$

$\langle \rho(\mathbf{q}, \omega) \rangle$, which characterizes the response of the system, may be expressed in terms of the quantity

$$\chi(\mathbf{q}, \omega) = \frac{\langle \rho(\mathbf{q}, \omega) \rangle}{\varphi(\mathbf{q}, \omega)}; \tag{2.49}$$

$\chi(\mathbf{q}, \omega)$ is the *density-density response function.*

We shall see that χ depends only on the system properties in the absence of the probe. We remark that where χ depends on $\mathbf{q}$ and ω the response to the potential is nonlocal in space and time; thus the potential acting at a point $\mathbf{r}$, time t, gives rise to a change in the density at some other point $\mathbf{r}'$, and later time, t'. For translationally invariant systems, to which we restrict ourselves, that response depends only on $|\mathbf{r}' - \mathbf{r}|$ and $t' - t$.

We now calculate $\chi(\mathbf{q}, \omega)$. Since we are only interested in the *linear* response of the system, it suffices to treat the coupling Hamiltonian (2.45) by first-order perturbation theory. The system wave function $|\psi(t)\rangle$ satisfies the Schrödinger equation

$$i \frac{\partial}{\partial t} |\psi(t)\rangle = (H + H_e) |\psi(t)\rangle, \tag{2.50}$$

where H is the system Hamiltonian (2.23), and H_e the coupling to the probe (2.45). Equation (2.50) must be solved with the causal boundary condition that for $t \to -\infty$, the system is in its ground state $|0\rangle$.

According to the usual procedure of perturbation theory, we project $|\psi(t)\rangle$ onto the various eigenstates of H, $|n\rangle$, having an energy E_n; we thus write

$$|\psi(t)\rangle = \sum_n a_n(t) e^{-iE_n t} |n\rangle. \tag{2.51}$$

Our boundary condition corresponds to

$$a_n(-\infty) = \begin{cases} 1 & \text{if} \quad n = 0, \\ 0 & \text{if} \quad n \neq 0. \end{cases} \tag{2.52}$$

On substituting Eq. (2.51) into (2.50) and keeping only the terms of first order in φ, we find, after a straightforward integration,

$$a_n(t) = \left\{\frac{(\rho_{\mathbf{q}}{}^+)_{no}\varphi(\mathbf{q}, \omega)}{\omega - \omega_{no} + i\eta} e^{(-i\omega + i\omega_{no} + \eta)t} - \frac{(\rho_{\mathbf{q}})_{no}\varphi^+(\mathbf{q}, \omega)}{\omega + \omega_{no} - i\eta} e^{(i\omega + i\omega_{no} + \eta)t}\right\},$$
$$(n \neq 0) \quad (2.53)$$

where $(\rho_{\mathbf{q}}{}^+)_{no}$ and ω_{no} have the same meaning as in the preceding sections [cf. Eq. (2.27)]. $\langle\rho(\mathbf{q}, t)\rangle$ is then obtained from Eq. (2.46). Up to first order in the perturbation,

$$\langle\rho(\mathbf{q}, t)\rangle = \sum_n \{(\rho_{\mathbf{q}})_{on}a_n(t)e^{-i\omega_{no}t} + a_n{}^+(t)(\rho_{\mathbf{q}})_{no}e^{i\omega_{no}t}\} \quad (2.54)$$

(since $\langle 0|\rho_{\mathbf{q}}|0\rangle$ vanishes for a translationally invariant system). We substitute Eq. (2.53) into (2.54); the resulting calculation is simplified if we note that a given state $|n\rangle$ cannot be coupled to the ground state $|0\rangle$ *both* by $\rho_{\mathbf{q}}$ and by $\rho_{\mathbf{q}}{}^+$, since it has either momentum $\mathbf{q}$ *or* $-\mathbf{q}$. We find

$$\langle\rho(\mathbf{q}, t)\rangle = \varphi(\mathbf{q}, \omega)e^{(-i\omega+\eta)t} \sum_n \left\{\frac{|(\rho_{\mathbf{q}}{}^+)_{no}|^2}{\omega - \omega_{no} + i\eta} - \frac{|(\rho_{\mathbf{q}})_{no}|^2}{\omega + \omega_{no} + i\eta}\right\}. \quad (2.55)$$

On making use of Eq. (2.47), and of time reversal invariance [see Eq. (2.36)], we may write

$$\langle\rho(\mathbf{q}, \omega)\rangle = \varphi(\mathbf{q}, \omega) \sum_n |(\rho_{\mathbf{q}}{}^+)_{no}|^2 \left\{\frac{1}{\omega - \omega_{no} + i\eta} - \frac{1}{\omega + \omega_{no} + i\eta}\right\}, \quad (2.56)$$

from which we obtain the response function $\chi(\mathbf{q}, \omega)$, defined by Eq. (2.49):

$$\chi(\mathbf{q}, \omega) = \sum_n |(\rho_{\mathbf{q}}{}^+)_{no}|^2 \frac{2\omega_{no}}{(\omega + i\eta)^2 - \omega_{no}^2}. \quad (2.57)$$

Equation (2.57) is our desired result.

Results that express the linear response function in terms of the true eigenstates of the system [such as (2.57)] were first obtained by Kubo (1956, 1957). The method is quite general, and is not restricted to the density-density response function. For instance, we may consider a vector perturbation, such as an electromagnetic field applied to a charged system. The physical quantity of interest is then the *current* induced by this perturbation. We are thus led to define a "current-

current response tensor" in a way similar to Eq. (2.49). We shall consider this tensor in Chapter 4.

RELATION OF $\chi(\mathbf{q}, \omega)$ TO THE DYNAMIC FORM FACTOR

Let us compare Eq. (2.57) with the definition (2.12) of the dynamic form factor $S(\mathbf{q}, \omega)$. We see that we can write χ in the form

$$\chi(\mathbf{q}, \omega) = \int_0^\infty d\omega' \, S(\mathbf{q}, \omega') \left\{ \frac{1}{\omega - \omega' + i\eta} - \frac{1}{\omega + \omega' + i\eta} \right\}. \tag{2.58}$$

The response function may therefore be calculated from a knowledge of $S(\mathbf{q}, \omega)$; the dynamic form factor serves as a spectral density for the density-density response function.

If $\omega \to \infty$, in Eq. (2.58) we may expand the quantity in brackets. We find

$$\chi(\mathbf{q}, \omega) \to \frac{2}{\omega^2} \int_0^\infty S(\mathbf{q}, \omega')\omega' \, d\omega'. \tag{2.59}$$

On comparing Eq. (2.59) with the f-sum rule (2.28), we see that

$$\chi(\mathbf{q}, \omega) \to \frac{Nq^2}{m\omega^2} \qquad (\omega \to \infty). \tag{2.60}$$

The asymptotic form of χ is thus unaffected by the interaction between the system particles.

We next split χ into its real and imaginary parts,

$$\chi = \chi' + i\chi''.$$

On making use of the Dirac relation,

$$\lim_{\eta \to 0} \frac{1}{x - a + i\eta} = P \frac{1}{x - a} - i\pi\delta(x - a) \tag{2.61}$$

(where P stands for principal part), we find

$$\chi'(\mathbf{q}, \omega) = \int_0^\infty d\omega' \, S(\mathbf{q}, \omega') P \left(\frac{2\omega'}{\omega^2 - \omega'^2} \right), \tag{2.62}$$

$$\chi''(\mathbf{q}, \omega) = -\pi\{S(\mathbf{q}, \omega) - S(\mathbf{q}, -\omega)\}. \tag{2.63}$$

We note that χ' is an even function of ω, while χ'' is odd.

This relation is not surprising. We know that $S(\mathbf{q}, \omega)$ provides a measure of the *real* transitions induced by the probe, i.e., of the energy transfer from the probe to the system. Such a dissipative process involves the imaginary part of the response function, in the same way as dielectric losses correspond to the imaginary part of the dielectric constant (the "in phase," real part of χ corresponds instead to a reversible "reactive" response). On the other hand, that the relation between S and χ'' should be as simple as Eq. (2.63) is not obvious. Actually, Eq. (2.63) is valid only at $T = 0$; we shall see in Section 2.6 that at finite temperatures the relation between S and χ'' is more complicated.

In order to emphasize the relationship between χ'' and energy dissipation, we calculate the energy transfer per unit time from the probe to the system. The scalar potential (2.45) gives rise to a force field

$$\mathfrak{F}(t) = -i\mathbf{q}\varphi(\mathbf{q}, \omega)e^{i(\mathbf{q}\cdot\mathbf{r}-\omega t)} + \text{c.c.} \tag{2.64}$$

This force field produces an average particle current, with Fourier components $\langle \mathbf{J}(\mathbf{q}, \omega)\rangle$. The energy gained by the system per unit time is equal to

$$\frac{dE}{dt} = \mathfrak{F}\cdot\langle\mathbf{J}\rangle = \{-i\mathbf{q}\varphi(\mathbf{q}, \omega)e^{i(\mathbf{q}\cdot\mathbf{r}-\omega t)} + \text{c.c.}\}\cdot\{\langle\mathbf{J}(\mathbf{q}, \omega)\rangle e^{i(\mathbf{q}\cdot\mathbf{r}-\omega t)} + \text{c.c.}\}. \tag{2.65}$$

The terms of Eq. (2.65) which oscillate at frequency 2ω correspond to the reversible deformation of the system. The other terms,

$$-i\mathbf{q}\cdot\langle\mathbf{J}^{+}(\mathbf{q}, \omega)\rangle\varphi(\mathbf{q}, \omega) + \text{c.c.}, \tag{2.66}$$

correspond to a *steady* energy transfer. We need not calculate $\langle\mathbf{J}(\mathbf{q}, \omega)$ explicitly, but can instead obtain it from the particle conservation equation (2.26), which possesses the Fourier transform.

$$\langle\mathbf{q}\cdot\mathbf{J}(\mathbf{q}, \omega)\rangle = \omega\langle\rho(\mathbf{q}, \omega)\rangle. \tag{2.67}$$

Making use of Eq. (2.67) and of the definition (2.49) of χ, we may put Eq. (2.66) in the form

$$-i\omega|\varphi(\mathbf{q}, \omega)|^2\chi^{+}(\mathbf{q}, \omega) + \text{c.c.} = -2\omega|\varphi(\mathbf{q}, \omega)|^2\chi''(\mathbf{q}, \omega). \tag{2.68}$$

As predicted, the steady energy transfer from the probe to the system is seen to involve only χ''. Replacing χ'' by its expression (2.63) (and assuming $\omega > 0$), we find

$$\frac{dE}{dt} = 2\pi\omega|\varphi(\mathbf{q}, \omega)|^2S(\mathbf{q}, \omega). \tag{2.69}$$

Equation (2.69) is equivalent to our former result (2.11), to which it reduces if we replace φ by the value (2.44) appropriate to a single incoming particle.

In view of the relation between $\chi(\mathbf{q}, \omega)$ and $S(\mathbf{q}, \omega)$, it is clear that χ does not provide any more physical information than S. As soon as we know one, we can calculate the other. Is one more fundamental than the other? At $T = 0$, the quantity which presents the physical information in the most compact form is $S(\mathbf{q}, \omega)$; it gives the density fluctuation excitation spectrum as a positive definite function defined on the positive real axis. We can pass from S to χ'', by Eq. (2.63). The choice between S and χ'' is thus essentially a matter of taste. On the other hand, we shall see in Section 2.6 that at finite temperatures, χ'' offers a more convenient description than S (essentially because it is less temperature dependent). We shall therefore use $\chi''(\mathbf{q}, \omega)$ more extensively than $S(\mathbf{q}, \omega)$ whenever we discuss the behavior of a finite-temperature system.

CAUSALITY AND DISPERSION RELATIONS

The adiabatic boundary condition used to define the response function is equivalent to replacing the real frequency ω by a complex frequency $(\omega + i\eta)$. The frequency describing the exact time dependence of the probe thus lies in the upper half of the complex ω-plane, infinitesimally close to the real axis. To put it another way, the real axis is to be considered as part of the upper half of the plane.

In practice, one is often interested in the spontaneous decay of an excited state, characterized by a frequency $(\omega_q - i\Gamma_q)$ which lies in the lower half of the complex ω-plane. At first sight, the response function, which is defined in the upper half of the ω-plane, does not appear to be very helpful in this respect. In order to achieve a satisfactory understanding of damping phenomena, we must therefore extend the definition of $\chi(\mathbf{q}, \omega)$ into the lower half of the complex ω-plane. We are thus led to study in detail the analytic structure of $\chi(\mathbf{q}, \omega)$ as a function of the complex variable ω.

We see in Eq. (2.57) that $\chi(\mathbf{q}, \omega)$ possesses an infinite number of poles, located in the complex ω-plane at

$$\omega = \pm\omega_{no} - i\eta. \tag{2.70}$$

Usually, the spectrum of excited states $|n\rangle$ coupled by the probe to the ground state $|0\rangle$ will consist of two parts:

(i) a continuum obtained by exciting individual particles;
(ii) a discrete state obtained by exciting a collective mode with wave vector $\mathbf{q}$.

The continuous part of the excitation spectrum will give rise to a *branch cut*, extending on either side of the origin slightly below the real axis (as shown on Fig. 2.2). A long-lived collective mode will give rise to a discrete pole, also located below the real axis.

The function $\chi(\mathbf{q}, \omega)$ is separately analytic in the two regions of the complex ω-plane on either side of the cut, labeled I and II on Fig. 2.2. In particular, χ is analytic in the upper half of the complex ω-plane, *including* the real axis, which is entirely contained in region I. We shall see in a short while that such a property is closely related to causality. χ_{I} and χ_{II} are not equal on the cut, where they satisfy the relation

$$\chi_{\text{I}}(\omega) = [\chi_{\text{II}}(\omega)]^* \tag{2.71}$$

[(which may be checked directly using Eq. (2.57)]. According to Eq. (2.71), the response function χ has a discontinuity on the cut which is given by:

$$\chi_{\text{I}} - \chi_{\text{II}} = 2i \operatorname{Im} \chi_{\text{I}}. \tag{2.72}$$

On comparing Eq. (2.72) with (2.63), we see that the discontinuity of χ involves the dynamic form factor S.

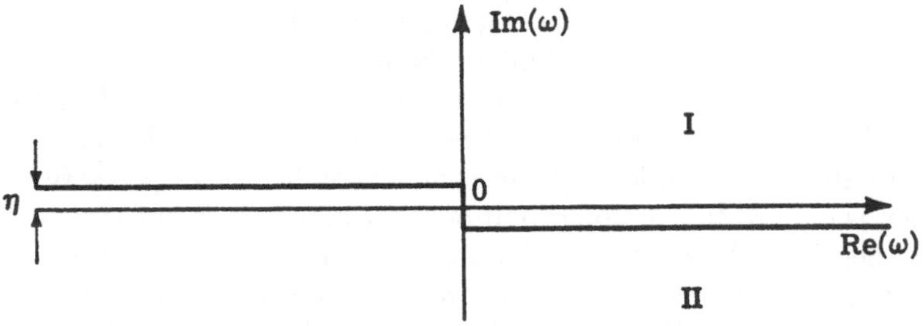

FIGURE 2.2. *The branch cut of $\chi(\mathbf{q}, \omega)$ in the complex ω-plane.*

As we pointed out above, we expect to obtain information on damped excitations by studying $\chi(\mathbf{q}, \omega)$ in the lower half of the complex ω-plane. Such insight cannot be derived from the function $\chi_{\text{II}}(\omega)$, since it is analytic in that region. On the other hand, the function obtained by analytically continuing χ_{I} into the lower half of the ω-plane is different from χ_{II}, and is no longer analytic. Since χ_{II} is analytic, the continuation of χ_{I} has the same poles as the difference $(\chi_{\text{I}} - \chi_{\text{II}})$, i.e., as $S(\mathbf{q}, \omega)$. A damped collective mode gives rise to a broad peak of S, centered at $\omega = \omega_q$, with a width Γ_q. The real function $S(\mathbf{q}, \omega)$ will accordingly possess two complex conjugate poles, located at $\omega = \omega_q \pm i\Gamma_q$. When continuing χ_{I} into the lower half of the complex ω-plane, we thus expect to find a pole at $\omega = \omega_q - i\Gamma_q$. We thereby justify the procedure used in Section 1.7 to study the Landau damping of collective modes.

Let us now use the methods of contour integration to calculate the

Fourier transform of $\chi(\mathbf{q}, \omega)$ with respect to time:

$$\chi(\mathbf{q}, t) = \frac{1}{2\pi} \int_{-\infty}^{+\infty} \chi(\mathbf{q}, \omega) e^{-i\omega t}\, d\omega. \tag{2.73}$$

When $t < 0$, we can close the integration contour at infinity by a half circle C lying *above* the real axis (Fig. 2.3). Since χ is analytic in the upper half of the complex ω-plane, and goes to zero at ∞, the integral (2.73) vanishes. We thus have

$$\chi(\mathbf{q}, t) = 0 \qquad \text{if} \quad t < 0. \tag{2.74}$$

If instead $t > 0$, we must close the contour by the half circle C' *below* the real axis (see Fig. 2-3). The integration contour then crosses the cut, so that Cauchy's theorem cannot be applied: χ is no longer zero.

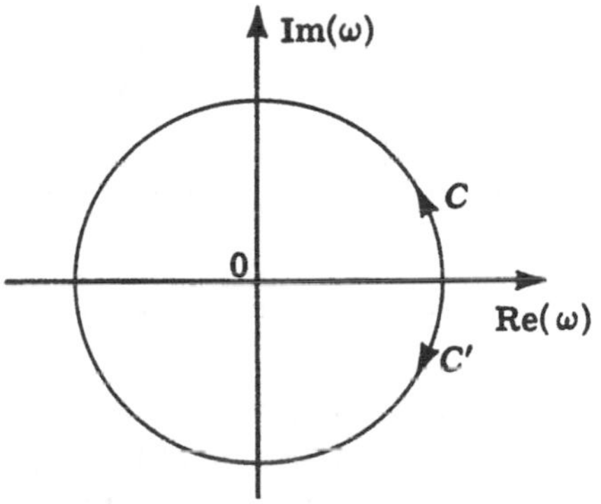

FIGURE 2.3. *Integration contours used to take the Fourier transform of* $\chi(\mathbf{q}, \omega)$.

According to its definition (2.49), $\chi(\mathbf{q}, \omega)$ represents the response to a scalar potential which does not depend on $\mathbf{q}$ or ω:

$$\varphi(\mathbf{q}, \omega) = \text{const.} \tag{2.75}$$

If we take the Fourier transform with respect to time, we see that the excitation (2.75) corresponds to a δ-function impulse applied at $t = 0$. $\chi(\mathbf{q}, t)$ represents the response of the system to that pulse. Our result (2.74) therefore means that the response of the system *follows* the excitation applied at $t = 0$. In other words, our system is *causal:* it is only distorted *after* the perturbation has been applied. Clearly this causality property follows directly from the analyticity of χ in the upper half of the complex ω-plane: they represent equivalent statements. From a physical point of view, it is obvious that the causality requirement must be satisfied. It is precisely in order to obtain a "causal theory" that we chose the adiabatic boundary conditions used throughout this section.

We now turn our attention to another consequence of the analytic properties of $\chi(\mathbf{q}, \omega)$. We consider the integral

$$\oint_{\Gamma} \frac{\chi(\mathbf{q}, \omega')}{\omega - \omega'}\, d\omega' = 0 \tag{2.76}$$

carried over the contour Γ indicated on Fig. 2.4. Since χ is analytic everywhere inside Γ, the integral (2.76) vanishes. We may separate

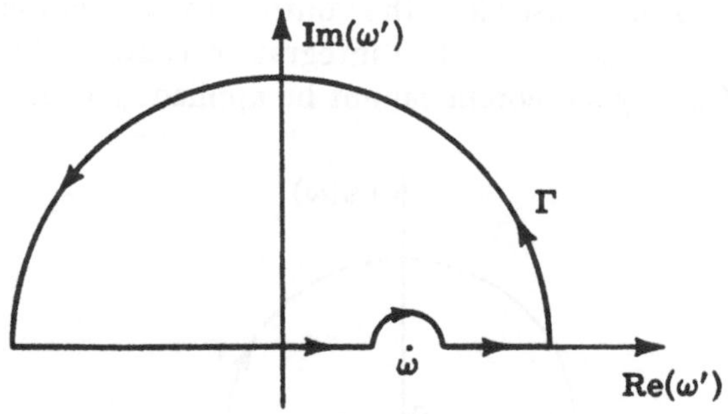

FIGURE 2.4. *Integration contour used to prove the Kramers–Kronig relations.*

the contributions along the contour Γ into three parts. The first, an integral over the half circle at infinity, vanishes, since in that limit χ is of order $1/\omega^2$. There remain two terms: first the integral over the two halves of the real axis (which, by definition, is the Cauchy principal part of the integral from $-\infty$ to $+\infty$), and second the integral over the small half circle surrounding ω, which is equal to $i\pi\chi(\mathbf{q}, \omega)$. Hence we find the relation

$$\int_{-\infty}^{+\infty} \chi(\mathbf{q}, \omega') P\left(\frac{1}{\omega - \omega'}\right) d\omega' + i\pi\chi(\mathbf{q}, \omega) = 0. \tag{2.77}$$

Let us again split χ into a real part χ' and an imaginary part χ''. By taking the real and imaginary parts of Eq. (2.77), we obtain the following two relations:

$$\chi'(\mathbf{q}, \omega) = \frac{-1}{\pi} \int_{-\infty}^{+\infty} \chi''(\mathbf{q}, \omega') P\left(\frac{1}{\omega - \omega'}\right) d\omega', \tag{2.78a}$$

$$\chi''(\mathbf{q}, \omega) = +\frac{1}{\pi} \int_{-\infty}^{+\infty} \chi'(\mathbf{q}, \omega') P\left(\frac{1}{\omega - \omega'}\right) d\omega'. \tag{2.78b}$$

These are the well-known "Kramers–Kronig relations" [Kramers (1927),

Kronig (1926)]. They are seen to be a direct consequence of the causal nature of the system response. They could have been established directly by using for χ its explicit expression (2.57). [Indeed, Eq. (2.78a) follows at once from a comparison of Eqs. (2.62) and (2.63).] More generally, all the properties of the linear response function are contained in its explicit expression (2.57), and can be found by inspection without any fancy mathematical analysis.

The equations (2.78) relate $\chi'(\mathbf{q}, \omega)$ to $\chi''(\mathbf{q}, \omega)$. If we can measure one of these quantities for all frequencies, we can thereby calculate the other. Such a relation remains valid for any response function. The Kramers–Kronig relations are therefore of great practical importance (especially in optics). In what follows, we shall often make use of them.

THE COMPRESSIBILITY SUM RULE

In the *long wavelength limit*, the static response function $\chi(\mathbf{q}, 0)$ represents the response of the system density to a static force field which varies slowly in space. It should therefore be related to the *compressibility* of the system. The connection is readily established by considering the *macroscopic* description of the response of the system to the scalar potential defined by Eq. (2.42).

The force field associated with the potential is given by Eq. (2.64). At zero frequency, it reduces to

$$\mathfrak{F}(\mathbf{r}) = -i\mathbf{q}\varphi(\mathbf{q}, 0)e^{i\mathbf{q}\cdot\mathbf{r}} + \text{c.c.} \tag{2.79}$$

$\mathfrak{F}(\mathbf{r})$ is the force felt by *one* particle. If it varies slowly over an atomic distance (i.e., if $qr_0 \ll 1$, where r_0 is the interparticle spacing), we can define the average force exerted on a unit volume of the system. The latter is equal to $\rho\mathfrak{F}(\mathbf{r})$; it can be replaced by $N\mathfrak{F}(\mathbf{r})$ to within corrections of higher order in φ. The force is longitudinal in nature, and acts to deform the system by creating a density fluctuation $\delta\rho(\mathbf{r})$.

The density fluctuation $\delta\rho(\mathbf{r})$ gives rise to an additional pressure distribution

$$\delta P(\mathbf{r}) = \frac{1}{\kappa N}\,\delta\rho(\mathbf{r}), \tag{2.80}$$

where κ is the macroscopic compressibility. $\delta P(\mathbf{r})$ is such that the corresponding pressure force, $-\,\mathbf{grad}\,P$, balances the external force. At equilibrium we therefore have

$$N\mathfrak{F} - \mathbf{grad}\,P = 0. \tag{2.81}$$

On making use of Eqs. (2.79) to (2.81) we easily obtain

$$\delta\rho = -N^2\kappa\varphi(\mathbf{q}, 0)e^{i\mathbf{q}\cdot\mathbf{r}} + \text{c.c.} \tag{2.82}$$

The result (2.82) provides the desired link between macroscopic definitions and the microscopic derivation of the system response. On comparing Eq. (2.82) with (2.49), we see that

$$\lim_{\mathbf{q}\to 0} \chi(\mathbf{q}, 0) = -N^2\kappa. \tag{2.83}$$

The relation (2.83) may be expressed in terms of the (isothermal) first sound velocity s, instead of κ. On making use of Eq. (1.49), we find

$$\lim_{\mathbf{q}\to 0} \chi(\mathbf{q}, 0) = -\frac{N}{ms^2}. \tag{2.84}$$

With the help of Eq. (2.62), we may transform Eq. (2.84) into a sum rule satisfied by the dynamic form factor $S(\mathbf{q}, \omega)$. We find

$$\lim_{\mathbf{q}\to 0} \int_0^\infty \frac{d\omega}{\omega} S(\mathbf{q}, \omega) = \frac{N}{2ms^2}. \tag{2.85}$$

Equation (2.85) is known as the "*compressibility sum rule*"; it provides an additional moment for $S(\mathbf{q}, \omega)$ in the long wavelength limit.

The compressibility sum rule, (2.85), will be valid as long as $\chi(\mathbf{q}, 0)$ exhibits no appreciable $\mathbf{q}$ dependence, that is, as long as the relation between the induced density and the external potential is a purely local one. In order to estimate the value of $\mathbf{q}$ at which it breaks down, we must know the $\mathbf{q}$ dependence of χ, which varies from one system to the other. For a neutral Fermi liquid, $\chi(\mathbf{q}, 0)$ remains constant until q is of the order of the Fermi momentum p_F. The compressibility sum rule will then be valid as long as $q \ll p_F$, i.e., for wavelengths much larger than the interparticle spacing. By contrast the f-sum rule is valid for all wave vectors.

In their present form, Eqs. (2.84) and (2.85) are valid only for a *neutral* system. For a charged system, complications arise, which are connected with the long-range character of the Coulomb interaction. There still exists a compressibility sum rule; however, it involves a "screened" dynamic form factor, in place of $S(\mathbf{q}, \omega)$. This problem is discussed in Section 4.2.

CORRELATIONS AND THE EXCITATION SPECTRUM

The response function formalism we have established is quite general; it applies to any quantum liquid, whether Bose or Fermi, charged or neutral, normal or superfluid. It provides a compact (and exact) way to describe experimental measurements involving weakly coupled external probes.

We see in Eq. (2.57) an example of the relationship which may exist between a response (or correlation) function of the quantum liquid and the appropriate elementary excitation spectrum. It is clear that a knowledge of the exact matrix elements, $|(\rho_{\mathbf{q}}{}^{+})_{no}|$, and excitation frequencies, ω_{no}, suffices to determine $\chi(\mathbf{q}, \omega)$. When all the excitations that contribute to an expression like (2.57) are long lived, we can pass directly from a knowledge of the relevant excitation spectrum (here the density fluctuation excitation spectrum) to the correlation function of interest. We shall see that at $T = 0$ this is the case for any quantum liquid in the long wavelength limit. Moreover, the compressibility sum rule, (2.85), and the f-sum rule, (2.28), prove of great assistance in sorting out the excitations of importance. At finite temperatures, similar connections exist, provided one is in a "collisionless" regime for which the mean free path of the excitations in question is long compared to the wavelengths of interest. In the opposite, "hydrodynamic" regime, the correlation functions are determined by macroscopic considerations, and are not simply related to the elementary excitations.

In the following two sections we apply the general theory to a specific example, that of the density-density response function (and density-fluctuation excitation spectrum) of a neutral Fermi liquid. We first discuss the qualitative behavior of the dynamic form factor, with particular emphasis on the contributions of various sorts of excitations to the sum rules as $\mathbf{q} \rightarrow 0$. We then apply the Landau theory to a direct calculation of $\chi(\mathbf{q}, \omega)$ in this limit, and establish thereby the detailed behavior of this correlation function. The corresponding hydrodynamic calculation is carried out in Section 2.7.

2.4. QUALITATIVE BEHAVIOR OF THE DYNAMIC FORM FACTOR FOR A NEUTRAL FERMI LIQUID

NONINTERACTING FERMI GAS

We first calculate $S(\mathbf{q}, \omega)$ for an especially simple example, that of a noninteracting Fermi gas. The ground state $|0\rangle$ of this system is obtained by filling all the plane wave states inside the Fermi sphere, of radius p_F. When acting on $|0\rangle$, the operator $\rho_{\mathbf{q}}{}^{+}$ produces single-particle transitions, in which a given particle is scattered from some state $\mathbf{p}$ to a state $(\mathbf{p} + \mathbf{q})$. This may be seen from Eq. (2.8), and still more clearly, if we express $\rho_{\mathbf{q}}{}^{+}$ in the second quantization representation:

$$\rho_{\mathbf{q}}{}^{+} = \sum_{\mathbf{p}\sigma} c^{+}_{\mathbf{p}+\mathbf{q},\sigma} c_{\mathbf{p},\sigma}, \tag{2.86}$$

where $c_{\mathbf{p}\sigma}^{+}$ and $c_{\mathbf{p}\sigma}$ are respectively the creation and destruction operators for a particle with wave vector **p** and spin σ. (Hereafter, we suppress the spin indices σ: it is understood that a summation over **p** also involves a summation over σ.) We may view $\rho_{\mathbf{q}}^{+}$ as representing a superposition of particle-hole pairs, each of net momentum **q** (since particle scattering is described equally well as creation of a particle-hole pair).

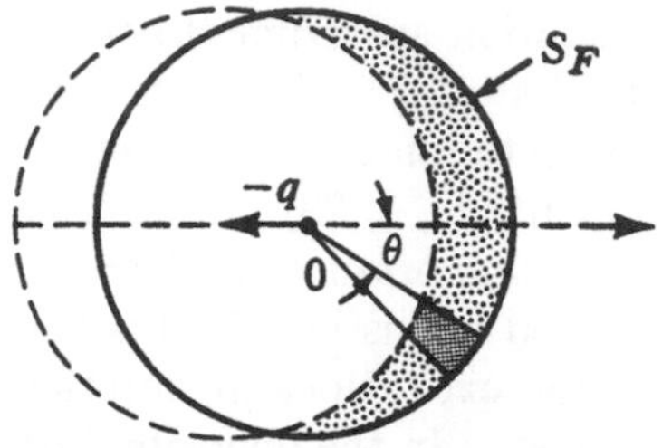

FIGURE 2.5. *The shaded crescent corresponds to the region of momentum space such that $p < p_F$, $|\mathbf{p} + \mathbf{q}| > p_F$, from which single particle hole pairs with total momentum $\mathbf{q}$ can be excited.*

In the ground state, because of the exclusion principle, a given pair can be excited only if the state **p** is filled and the state $(\mathbf{p} + \mathbf{q})$ empty. The allowed values of **p** correspond to the shaded crescent of Fig. 2.5, lying between the Fermi sphere, S_F, and the sphere obtained by translating S_F by an amount $-\mathbf{q}$. For $q < 2p_F$, we see that the Pauli exclusion principle reduces considerably the number of available final states.

The matrix element $(\rho_{\mathbf{q}}^{+})_{no}$, taken between two plane wave states, is equal to 1. The excitation energy ω_{no} corresponding to the transition $\mathbf{p} \rightarrow \mathbf{p} + \mathbf{q}$ is equal to

$$\omega_{\mathbf{pq}}^{0} = \frac{(\mathbf{p} + \mathbf{q})^2}{2m} - \frac{p^2}{2m} = \frac{\mathbf{q} \cdot \mathbf{p}}{m} + \frac{q^2}{2m}. \tag{2.87}$$

The dynamic form factor for the free Fermi gas, $S^{0}(\mathbf{q}, \omega)$, may therefore be written as

$$S^{0}(\mathbf{q}, \omega) = \sum_{\substack{p<p_F \\ |\mathbf{p}+\mathbf{q}|>p_F}} \delta(\omega - \omega_{\mathbf{pq}}^{0}), \tag{2.88}$$

The calculation of $S^{0}(\mathbf{q}, \omega)$ is thus reduced to a simple integration.

Without any calculation, it is clear that the excitation spectrum of particle-hole pairs with total momentum $\mathbf{q}$ will form a continuum, which lies between the following limits:

$$\begin{aligned} 0 \leqslant \omega^{\circ}_{\mathbf{pq}} &\leqslant \frac{qp_F}{m} + \frac{q^2}{2m} && \text{if} \quad q < 2p_F, \\ \frac{-qp_F}{m} + \frac{q^2}{2m} \leqslant \omega^{\circ}_{\mathbf{pq}} &\leqslant \frac{qp_F}{m} + \frac{q^2}{2m} && \text{if} \quad q > 2p_F. \end{aligned} \tag{2.89}$$

These allowed energies for pair excitation correspond to the shaded region of Fig. 2.6. Outside the finite range of frequency (2.89), $S^{\circ}(\mathbf{q}, \omega)$ vanishes. This feature is characteristic of a noninteracting system, and is no longer valid for real Fermi liquids.

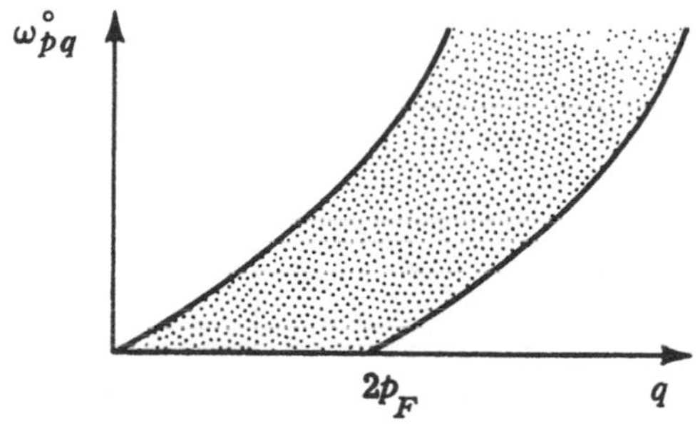

FIGURE 2.6. *The single-pair energy spectrum as a function of q.*

The explicit calculation of $S^{\circ}(\mathbf{q}, \omega)$ is straightforward, but somewhat lengthy. We leave the integration as an exercise to the reader, and quote here only the result

$$S^{\circ}(\mathbf{q}, \omega) = \begin{cases} \dfrac{\nu(0)}{2} \dfrac{\omega}{qv_F{}^{\circ}} & \text{if} \quad 0 \leqslant \omega \leqslant qv_F{}^{\circ} - \dfrac{q^2}{2m}, \\ \dfrac{\nu(0)}{2} \dfrac{p_F}{2q} \left\{ 1 - \left[\dfrac{\omega}{qv_F{}^{\circ}} - \dfrac{q}{2p_F} \right]^2 \right\} & \\ & \text{if} \quad qv_F{}^{\circ} - \dfrac{q^2}{2m} \leqslant \omega \leqslant qv_F{}^{\circ} + \dfrac{q^2}{2m}, \\ 0 & \text{if} \quad qv_F{}^{\circ} + \dfrac{q^2}{2m} \leqslant \omega. \end{cases} \tag{2.90}$$

In Eq. (2.90), $v_F{}^{\circ} = p_F/m$ is the free-particle Fermi velocity, while

$$\nu(0) = \frac{3N}{2\epsilon_F{}^{\circ}} = \frac{3mN}{p_F{}^2} \tag{2.91}$$

is the density of single-particle states per unit energy (and per unit volume) at the Fermi surface. We give in Fig. 2.7 plots of $S^o(\mathbf{q}, \omega)$ for $q < 2p_F$ and for $q > 2p_F$.

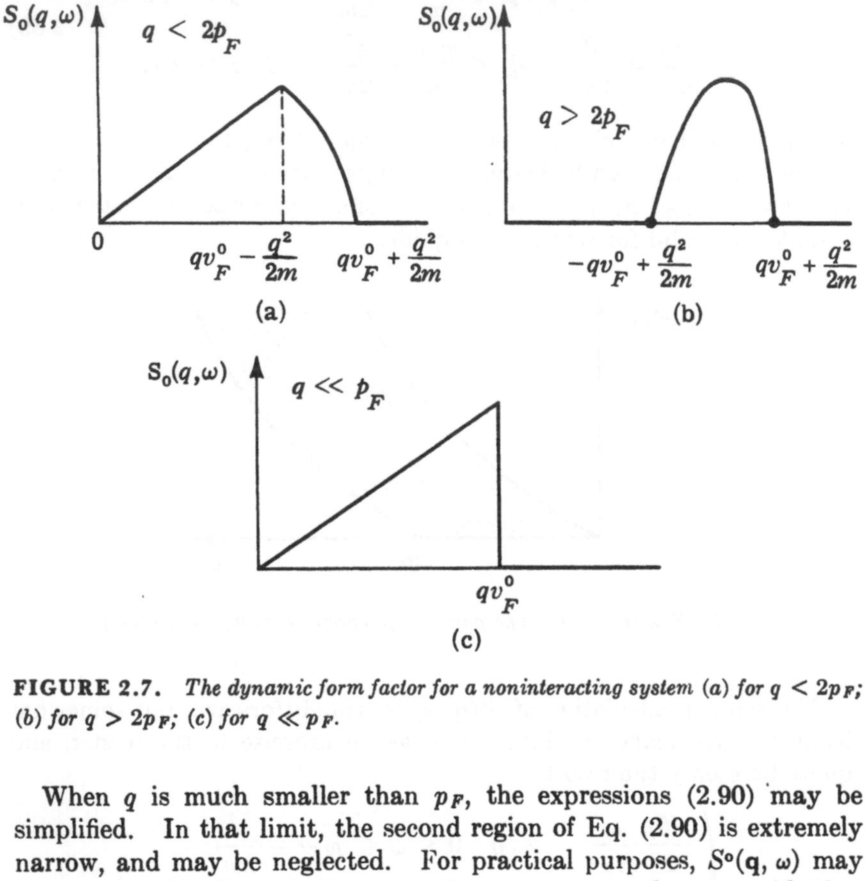

FIGURE 2.7. *The dynamic form factor for a noninteracting system (a) for $q < 2p_F$; (b) for $q > 2p_F$; (c) for $q \ll p_F$.*

When q is much smaller than p_F, the expressions (2.90) may be simplified. In that limit, the second region of Eq. (2.90) is extremely narrow, and may be neglected. For practical purposes, $S^o(\mathbf{q}, \omega)$ may be considered as linear in the range $0 < \omega < qv_F{}^o$, and zero outside that range (see Fig. 2.7c). This limit can in fact be studied directly, by noting that the crescent of Fig. 2.5 is then very narrow. Let us use polar coordinates with an axis parallel to $\mathbf{q}$. The number of states inside the crescent with polar angles comprised in the range $(\theta, \theta + d\theta)$ is

$$dn = \frac{2}{(2\pi)^3} 2\pi p_F{}^2 \sin\theta \cdot q \cos\theta \, d\theta \tag{2.92}$$

(within corrections which are of higher order in q/p_F). On the other hand, the excitation energy of the pair is nearly equal to

$$\omega = qv_F{}^o \cos\theta. \tag{2.93}$$

By replacing the summation in Eq. (2.88) by an integration over θ, one finds easily that

$$S^{\circ}(\mathbf{q}, \omega) = \begin{cases} \dfrac{m^2\omega}{2\pi^2 q}, & \omega < qv_F^{\circ}, \\ 0, & \omega > qv_F^{\circ}, \end{cases} \tag{2.94}$$

which, as can be seen, is equivalent to the rigorous result (2.90) in the limit $q \ll p_F$.

We note that for all values of q, $S^{\circ}(\mathbf{q}, \omega)$ vanishes when $\omega \to 0$, being proportional to ω (this region corresponds to the horns of the crescent in Fig. 2.5, where the width of the crescent vanishes). Such behavior is a direct consequence of the exclusion principle, and of the corresponding scarcity of low-energy excitations. This feature will therefore persist in the case of real, interacting, Fermi liquids, as we shall see later in this section.

Finally, let us calculate the static form factor $S_{\mathbf{q}}^{\circ}$ for a noninteracting gas. According to Eqs. (2.18) and (2.88), the static form factor $S_{\mathbf{q}}^{\circ}$ is given by

$$NS_{\mathbf{q}}^{\circ} = \sum_{\substack{p<p_F \\ |\mathbf{p}+\mathbf{q}|>p_F}} 1 \tag{2.95}$$

and is thus proportional to the area of the crescent in Fig. 2.5. A simple calculation yields

$$S_{\mathbf{q}}^{\circ} = \begin{cases} \dfrac{3q}{4p_F} - \dfrac{q^3}{16p_F^3} & \text{if } 0 < q \leqslant 2p_F, \\ 1 & \text{if } q \geqslant 2p_F. \end{cases} \tag{2.96}$$

The result (2.96) is not correct when $q \equiv 0$: since $\rho_0 = N$, it is clear from Eq. (2.20) that $S_{\mathbf{q}}^{\circ}$ is then equal to N. We observe that $S_{\mathbf{q}}^{\circ}$ vanishes when $q \to 0$, once again as a consequence of the exclusion principle.

INTERACTING FERMI LIQUID

Let us consider the matrix element $(\rho_{\mathbf{q}}^{+})_{n0}$ for a neutral Fermi liquid. What are the excited states $|n\rangle$ of the real liquid which are coupled to the ground state $|0\rangle$ by the density fluctuation $\rho_{\mathbf{q}}^{+}$? We first note that $|n\rangle$ must correspond to the same number of particles as $|0\rangle$, and have total momentum $\mathbf{q}$. Among the states $|n\rangle$ of interest, we shall therefore find configurations involving an equal number of individually excited quasiparticles and quasiholes, with total momentum $\mathbf{q}$. We may distinguish between *single* pair excitations and "*multipair*" excitations.

The single pair excitations involve a single quasiparticle-quasihole pair, while the multipair excitations involve two or more quasiparticle-quasihole pairs.

It should be noted that these "individual" excitations involve, in general, quasiparticles and quasiholes that are not close to the Fermi surface, and which thus correspond to highly damped excited states of the system. The quasiparticle concept is nonetheless suggestive, and may be used in a qualitative description of the behavior of $S(\mathbf{q}, \omega)$.

In addition to the excitation of individual quasiparticles and quasiholes, the density fluctuation $\rho_{\mathbf{q}}^{+}$ may also contain collective oscillations. Only *zero sound* can in fact be excited, since for an isotropic system none of the other collective modes gives rise to density fluctuations. Again, for a large enough wave vector $\mathbf{q}$, zero sound is appreciably damped. Where it does not represent a well-defined excitation, it may still prove useful in providing a qualitative guide to the behavior of $S(\mathbf{q}, \omega)$.

We shall now study in some detail each of these three types of excited states.

(*i*) *Single pair excitations* Such excitations are obtained by exciting a quasiparticle with momentum $\mathbf{p} < p_F$ to a state $(\mathbf{p} + \mathbf{q})$. Because of the exclusion principle (which holds for quasiparticles as well as for bare particles), $|\mathbf{p} + \mathbf{q}|$ must be larger than p_F. The available values of $\mathbf{p}$ are therefore restricted to the shaded crescent of Fig. 2.5, *exactly* as for a noninteracting Fermi gas. The single-pair excitation spectrum will therefore resemble closely that of the noninteracting system.

The excitation energy of the quasiparticle-quasihole pair is equal to

$$\epsilon_{\mathbf{p}+\mathbf{q}} - \epsilon_{\mathbf{p}}. \tag{2.97}$$

When $q < 2p_F$, we expect the corresponding spectrum to cover a finite range of energy, extending from 0 (when $\mathbf{p}$ lies on the horns of the crescent) to some maximum value (obtained when $\mathbf{p}$ is parallel to $\mathbf{q}$). When q is comparable to the Fermi momentum p_F, the energies $\epsilon_{\mathbf{p}}$ and $\epsilon_{\mathbf{p}+\mathbf{q}}$ are poorly defined for most $\mathbf{p}$ because of damping effects: the single-pair excitation spectrum is not likely to cut off sharply. If, however, $q \ll p_F$, the above excitations involve only well-defined quasiparticles, close to the Fermi surface. The single-pair excitation spectrum then extends over the finite range $0 < \omega < qv_F$, where v_F is the quasiparticle Fermi velocity: in that limit, we find the sharp cut-off characteristic of a noninteracting system.

One may obtain further information on the behavior of $S(\mathbf{q}, \omega)$ in the limit $\omega \ll qv_F$ (whatever the value of q). In this limit, for $q < 2p_F$, the single-pair contribution arises from the crescent horns in Fig. 2.5. We note that the corresponding quasiparticles are close to the Fermi surface, and thus well defined, even for finite q. It is clear from Fig. 2.5 that the width of the crescent vanishes when $\omega \rightarrow 0$: the density of single-pair excitations per unit energy is thus proportional to ω when $\omega \rightarrow 0$. (The detailed calculation is left as a problem to the reader.) Again, this feature is a consequence of the exclusion principle. We consequently expect $S(\mathbf{q}, \omega)$ to be proportional to ω in the vicinity of $\omega = 0$, as long as $q < 2p_F$. For $q > 2p_F$, the single-pair spectrum does not start from the origin $\omega = 0$, but from some finite value of ω.

We have emphasized the similarities between the single-pair excitation spectrum and that of a noninteracting system. Actually, there exist important differences in their contributions to $S(\mathbf{q}, \omega)$. For an interacting system, the matrix element $(\rho_{\mathbf{q}}^{+})_{n0}$ is no longer equal to 1. It will certainly depend on the state $|n\rangle$ under consideration, and therefore on the energy ω. We thus expect the single-pair contribution $S^{(1)}(\mathbf{q}, \omega)$ to the dynamic form factor to be a fairly complicated function of ω. For example, unlike the noninteracting fermion system, the linear behavior found for very small ω does not persist when ω is finite. Moreover, when $q \ll p_F$, we cannot carry out an explicit calculation of $S^{(1)}(\mathbf{q}, \omega)$, even though we know its range $(0 < \omega < qv_F)$. This difficulty, which is often overlooked, should be kept in mind.

(*ii*) *Multipair excitations* We now consider the contribution $S^{(n)}(\mathbf{q}, \omega)$ arising from the excitation of several quasiparticle-quasihole pairs. For the interacting fermion system, the density fluctuation $\rho_{\mathbf{q}}^{+}$ may couple such "higher configurations" to the ground state (such coupling did not occur for a noninteracting gas). Since momentum conservation involves only the total momentum, there is essentially no limitation on the momentum of any single quasiparticle or quasihole in such configurations. The multipair excitation energy will therefore vary over a broad range, extending from $\omega = 0$ to infinity, with a maximum density at a few times the chemical potential μ. The corresponding contribution, $S^{(n)}(\mathbf{q}, \omega)$, to the dynamic form factor possesses no particular structure and will appear as a broad background.

For very small frequencies, $\omega \ll \mu$, we can use the exclusion principle to estimate $S^{(n)}(\mathbf{q}, \omega)$. Because of energy conservation, each of the $2n$ quasiparticles or quasiholes excited in a multipair configuration must be within an energy ω of the Fermi surface. The density per unit energy of an n-pair configuration in the vicinity of ω is thus proportional to ω^{2n-1}. The corresponding contribution to $S^{(n)}$ will vary in the same way, being at least of order ω^3. In this limit it is thus negligible compared to the single-pair contribution, which is of order ω. The fact that the multipair excitations play a negligible role in low-frequency phenomena is reminiscent of the correspondingly small role played by the higher-order decay processes in determining the lifetime of a quasiparticle (cf. Section 1.8). Both contributions are small because of the Pauli principle.

We further note that the multipair excitations also play a negligible role in the long wavelength limit ($q \ll p_F$). We shall demonstrate later in this section that their contribution to $S(\mathbf{q}, \omega)$ is at least of order q^2.

(*iii*) *Collective excitations* In the zero sound excitation mode, the ground state $|0\rangle$ is coupled by $\rho_{\mathbf{q}}^+$ to a single state, containing one quantum of zero sound with wave vector $\mathbf{q}$ and energy ω_q. It is at first sight somewhat surprising that this single state may give a contribution to $S(\mathbf{q}, \omega)$ which is comparable in importance to that of the many individual pair excitations (whose number is of order N). How this happens is easily understood if we realize that a collective mode involves a *coherent* motion of all the quasiparticles. The matrix element $(\rho_{\mathbf{q}}^+)_{no}$ for excitation of a zero sound quantum appears as a sum of terms, one per quasiparticle involved in the collective mode; these contributions *interfere* constructively, with the result that $|(\rho_{\mathbf{q}}^+)_{no}|^2$ is N times bigger than the corresponding probability for any single individual pair excitation. The extra factor of N thus makes up for the fact that there is one final state, instead of a number of order N.

An undamped zero sound mode will contribute to the dynamic form factor a single *discrete peak:*

$$Z_q \delta(\omega - \omega_q), \tag{2.98}$$

where Z_q is a weight factor to be determined. As we have mentioned in Chapter 1, in fact the zero sound mode is always somewhat damped; the peak is correspondingly broadened, to an extent which depends on the system, and on the choice of $\mathbf{q}$. Where the zero sound excitation spectrum overlaps with that of single pairs, very strong Landau damping occurs; the corresponding peak of $S(\mathbf{q}, \omega)$ is very broad. For values of $\mathbf{q}$ such that Landau damping is impossible, there remains an overlap of the zero sound peak with the multipair excitation spectrum (since the latter extends from 0 to ∞). A resonant coupling between zero sound and multipair excitations is then energetically possible; it will act to damp the zero sound mode and broaden the "collective" peak of $S(\mathbf{q}, \omega)$. The broadening is small if $q \ll p_F$; it increases rapidly when q becomes comparable to p_F.

If the collective mode is strongly damped, it is no longer particularly useful to consider zero sound as a separate excitation of the system. The broad peak in $S(\mathbf{q}, \omega)$ is better interpreted as a weak resonance in the single-pair excitation spectrum. Under just what circumstances the collective mode is well enough "resolved" to be considered as a separate entity is obviously a matter of personal choice.

A rough sketch of the three types of contributions to $S(\mathbf{q}, \omega)$ for an interacting Fermi liquid is given in Fig. 2.8. Let us emphasize again the basic difference between the individual excitation *continuous* spectrum and the collective mode *discrete* spectrum. The former corresponds to an *incoherent* response to the density fluctuation $\rho_{\mathbf{q}}^{+}$, while the latter represents the *coherent*, resonant response to that same excitation, in agreement with the general picture of a collective mode which was presented in Section 1.7.

STRUCTURE OF $S(\mathbf{q}, \omega)$ WHEN $\mathbf{q} \to 0$

In the long wavelength limit, $\mathbf{q} \to 0$, it becomes possible to evaluate the order of magnitude of the various contributions to $S(\mathbf{q}, \omega)$ and thereby distinguish clearly between the different contributions. In the present

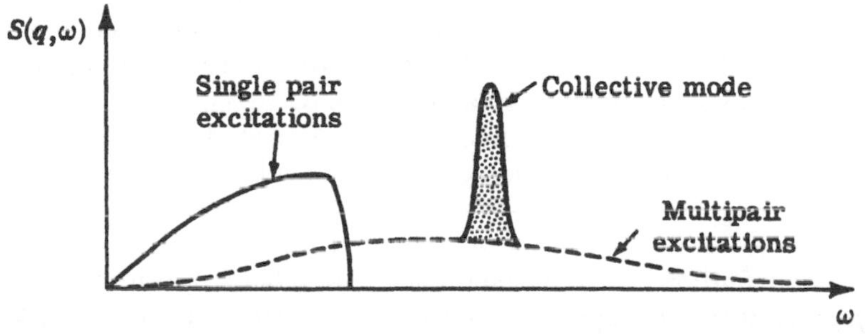

FIGURE 2.8. *A rough sketch of the various contributions to $S(\mathbf{q}, \omega)$ for an interacting Fermi liquid.*

section, we discuss the long wavelength response of a neutral system by relying on *qualitative* arguments. More detailed results may be obtained by using the Landau theory of Fermi liquids. We shall discuss the latter approach in Section 2.5: it will be seen to corroborate our qualitative results.

In what follows, we assume that we are in a *collisionless* regime. This implies that the collisions between thermally excited quasiparticles can be neglected, which is true when the frequency ω is much larger than the thermal collision frequency ν. It further implies that the lifetime of all excitations of physical interest is long compared to the period of the applied field. The latter requirement is satisfied for single-pair excitations in the long wavelength limit; it will not likely be so for multipair excitations: however, we shall show that these play a negligible role in most cases.

The collisionless regime is realized at $T = 0$, where there are no collisions. At finite temperatures, ν is finite: the collisionless regime gives way to a "hydrodynamic" regime when $\omega \ll \nu$. To avoid all complications, we focus our attention in what follows on the zero temperature case; the collisionless regime at finite temperatures, and the transition to the hydrodynamic regime are discussed in Section 2.7.

Let us consider the relative importance of the various contributions to $S(\mathbf{q}, \omega)$ described in the preceding section, in the limit $\mathbf{q} \to 0$, with $T = 0$. We first consider the *single-pair* excitations, obtained by exciting one quasiparticle with wave vector $\mathbf{p}$ to the state $(\mathbf{p} + \mathbf{q})$. When $\mathbf{q} \to 0$, we note that:

(i) The excitation energy $\omega_{no} \simeq \mathbf{q} \cdot \mathbf{v_p}$ is proportional to q. It varies between 0 and a maximum value qv_F.

(ii) Because of the exclusion principle, the wave vector $\mathbf{p}$ is restricted to the shaded crescent of Fig. 2.5. The number of states in that crescent is of order Nq/p_F. The summation over the intermediate states $|n\rangle$ will thus introduce a factor q/p_F.

(iii) Finally, the matrix element $(\rho_{\mathbf{q}}{}^{+})_{no}$ is likely to be of order 1, by analogy with the noninteracting gas (even though it may depend on the state $|n\rangle$ considered).

In the definition (2.12), the factor $\delta(\omega - \omega_{no})$ has a dimension $1/\omega_{no} \sim 1/qv_F$; we therefore expect the single-pair contribution to $S(\mathbf{q}, \omega)$ to be of order 1, over a narrow range of frequencies of width qv_F. Its exact dependence on ω is not simple, except when $\omega \ll qv_F$, in which case $S(\mathbf{q}, \omega)$ is proportional to ω.

Let us now study the *multipair* excitations. We first remark that when $\mathbf{q} \to 0$, the limit of a multipair configuration is a well-defined *excited state* of the system, which contains a number of excited quasiparticles and quasiholes, whose total momentum happens to be zero; its excitation energy is finite. It follows that:

(i) The average excitation energy $\omega_{no} = \bar{\omega}$ does not vanish when $\mathbf{q} \to 0$; it remains finite, equal to a few times the chemical potential μ. The corresponding spectrum is very broad.

(ii) There is no limitation (arising from the exclusion principle) on the number of available states $|n\rangle$. The order of magnitude of the corresponding contribution to $S(\mathbf{q}, \omega)$ thus depends entirely on the behavior of the matrix element $(\rho_{\mathbf{q}}{}^{+})_{no}$: we must find its limit when $\mathbf{q} \to 0$.

For that purpose, we use the fact that the limit of a multipair excitation when $\mathbf{q} \rightarrow 0$ is an excited state $|n\rangle$, which is *orthogonal* to the ground state. Consequently, if we expand $\rho_{\mathbf{q}}^{+}$ in powers of $\mathbf{q}$, the first term of the expansion, equal to a constant, does not contribute to the matrix element. The corresponding matrix element $(\rho_{\mathbf{q}}^{+})_{no}$ is thus *at least of order* q when $\mathbf{q} \rightarrow 0$. Such a conclusion is valid whenever the limit of $|n\rangle$ is orthogonal to $|0\rangle$ (it applies, for instance, to interband transitions in real solids). Clearly, such an argument cannot be applied to a single-pair "intraband" excitation, since in that case, the limit of $|n\rangle$ when $\mathbf{q} \rightarrow 0$ is the ground state itself.

For a translationally invariant system, the matrix element $(\rho_{\mathbf{q}}^{+})_{no}$ is even smaller, and is such that

$$\lim_{\mathbf{q}\rightarrow 0} (\rho_{\mathbf{q}}^{+})_{no} \sim q^{1+\alpha} \qquad (\alpha > 0). \tag{2.99}$$

In order to demonstrate (2.99), we make explicit use of both particle conservation and translational invariance. The continuity equation, (2.27), relates $(\rho_{\mathbf{q}}^{+})_{no}$ to the matrix element of the current, $(\mathbf{J}_{\mathbf{q}}^{+})_{no}$. When $\mathbf{q} \rightarrow 0$, the current density fluctuation $\mathbf{J}_{\mathbf{q}}$ becomes equal to the *total* current $\mathbf{J}$. For a translationally invariant system, $\mathbf{J}$ is a good quantum number, which commutes with the Hamiltonian. In such a case, we therefore have

$$\lim_{\mathbf{q}\rightarrow 0} (\mathbf{J}_{\mathbf{q}}^{+})_{no} = (\mathbf{J})_{no} = 0. \tag{2.100}$$

The matrix element $(\mathbf{J}_{\mathbf{q}}^{+})_{no}$ is thus of order q^{α}, where α is some positive number. On the other hand, the multipair excitation energy, ω_{no}, remains finite when $\mathbf{q} \rightarrow 0$. On making use of Eq. (2.29), we find the result (2.99). Let us emphasize the simplicity of such an argument, which we shall use over and over again in the following chapters.

To the extent that $\mathbf{J}_{\mathbf{q}}$ can be expanded in a power series in $\mathbf{q}$, we expect $(\mathbf{J}_{\mathbf{q}}^{+})_{no}$ to be of order q, corresponding to the choice $\alpha = 1$. We shall assume this to be the case: For a translationally invariant system, $(\rho_{\mathbf{q}}^{+})_{no}$ is then of order q^2 when $\mathbf{q} \rightarrow 0$. If the system is not invariant under translation, $(\mathbf{J})_{no}$ is no longer zero: the matrix element $(\rho_{\mathbf{q}}^{+})_{no}$ is then of order q instead of q^2.

By collecting all these results, we see that the multipair contribution to $S(\mathbf{q}, \omega)$, which extends over a broad range of frequencies, is of order q^4 if the system is invariant under translations (for instance, in liquid ^{3}He), and of order q^2 if it is not (for instance, in real metals). We furthermore proved earlier that it was of order ω^3 when $\omega \rightarrow 0$, because of the exclusion principle.

We finally turn to the zero sound contribution to $S(\mathbf{q}, \omega)$. In the low $\mathbf{q}$ limit, and at zero temperature, the collision damping of collective modes is negligible, as is their decay into multipair configurations (the preceding estimate of the multipair contribution to $S(\mathbf{q}, \omega)$ could in fact be used to evaluate the corresponding decay probability of zero sound). The only damping mechanism which remains efficient is Landau damping, occurring when the collective mode is immersed in the single-pair spectrum.

Let s_0 be the velocity of zero sound (not to be confused with the velocity s of ordinary "acoustic" sound). If $s_0 < v_F$, Landau damping is very large; the collective mode only appears as a broad resonance in the single-pair contribution: there is no need to treat it separately. Let us assume instead that $s_0 > v_F$: there is no Landau damping. The collective mode then contributes a very sharp peak to $S(\mathbf{q}, \omega)$, which for $\mathbf{q} \to 0$ is well represented by the δ-function (2.98).

The excitation energy $\omega_{no} = s_0 q$ is clearly of order q. On the other hand, it is not easy to evaluate the matrix element $(\rho_{\mathbf{q}}{}^{+})_{no}$. We can place an upper bound on it by referring to either the f-sum rule (2.29) or the compressibility sum rule (2.85): we see at once that $(\rho_{\mathbf{q}}{}^{+})_{no}$ cannot be of a lower order than $q^{1/2}$. In fact, detailed calculations show than $(\rho_{\mathbf{q}}{}^{+})_{no}$ is of exactly that order.

The preceding discussion provides us with a satisfactory picture of the general behavior of $S(\mathbf{q}, \omega)$ when $\mathbf{q} \to 0$. We note that the leading terms are those arising from the single-pair and collective excitations. For long wavelengths, such excited states are within reach of the Landau theory of Fermi liquids. It should therefore be possible to calculate $S(\mathbf{q}, \omega)$ using that theory, an approach which we shall follow in Section 2.5.

THE LONG WAVELENGTH LIMIT OF THE SUM RULES

We are now in a position to examine the various contributions to the sum rules when $\mathbf{q} \to 0$. In the long wavelength limit, we found two such rules, namely the f-sum rule (2.29) and the compressibility sum rule (2.85). To those we can add the definition (2.18) of the static form factor $S_{\mathbf{q}}$, which, although not really a "sum rule," has a similar structure. We write again the three relations

$$\sum_n \frac{|(\rho_{\mathbf{q}}{}^{+})_{no}|^2}{\omega_{no}} = \frac{N}{2ms^2}, \tag{2.101a}$$

$$\sum_n |(\rho_{\mathbf{q}}{}^{+})_{no}|^2 = NS_{\mathbf{q}}, \tag{2.101b}$$

$$\sum_n |(\rho_{\mathbf{q}}{}^{+})_{no}|^2 \omega_{no} = \frac{Nq^2}{2m}. \tag{2.101c}$$

Let us first consider the contribution of single-pair excitations, for which ω_{no} is of order q, while the number of states $|n\rangle$ is also of order q. We see at once that such excitations give a contribution of order 1 to Eq. (2.101a), and of order q^2 to (2.101c); this represents a sizable part of the total sum rule when $\mathbf{q} \rightarrow 0$. The corresponding term of Eq. (2.101b) is of order q, and indeed fixes the order of magnitude of $S_{\mathbf{q}}$ when $\mathbf{q} \rightarrow 0$. The zero sound contribution is comparable to that of single pair excitations. Again, it represents an appreciable part of the three sum rules (2.101).

We now consider multipair excitations. Since ω_{no} is finite, the corresponding contributions to the three relations (2.101) are comparable, of order q^4 if the system is translationally invariant, and of order q^2 if it is not. We thus conclude that when $\mathbf{q} \rightarrow 0$, multipair transitions do not contribute to (2.101a) and (2.101b); they contribute to the f-sum rule (2.101c) only if the system is not invariant under translations. The various steps of this derivation are summarized in Table 2.1.

TABLE 2.1. *Matrix Elements, Excitation Energies, and Sum Rule Contributions of the Various Density Fluctuation Excitations in the Long Wavelength Limit*

			Multipair	
	Single pair	Zero sound	Trans. invariance	No trans. invariance
$(\rho_{\mathbf{q}}^{+})_{no}$	1	$q^{\frac{1}{2}}$	q^2	q
ω_{no}	qv_F	qs_0	$\bar{\omega}$	$\bar{\omega}$
Pauli principle restriction on excited states	q/p_F	—	—	—
$\Sigma_n \lvert(\rho_{\mathbf{q}})_{no}\rvert^2/\omega_{no}$	1	1	q^4	q^2
$\Sigma_n \lvert(\rho_{\mathbf{q}}^{+})_{no}\rvert^2$	q	q	q^4	q^2
$\Sigma \lvert(\rho_{\mathbf{q}})_{no}\rvert^2\omega_{no}$	q^2	q^2	q^4	q^2

These results are of great practical importance, as they provide a justification of the Landau theory of Fermi liquids. The latter only takes account of single-pair and collective excitations, and completely neglects multipair excitations. We thus expect the Landau approach to give exact results whenever the contribution of multipair excitations is negligible. We have seen that these excitations do not contribute to $N/2ms^2$ and $NS_{\mathbf{q}}$ (when $\mathbf{q} \rightarrow 0$): These two quantities may therefore be safely calculated in the framework of the Landau theory. That was not *a priori* obvious, especially for $S_{\mathbf{q}}$.

We remark that for a translationally invariant system, the f-sum rule is *exhausted* by single-pair and zero sound excitations (when $\mathbf{q} \rightarrow 0$).

The oscillator strengths corresponding to high-energy transitions are seen to be sizable only for a nonhomogeneous system.

It should further be noted that the above conclusions apply only to neutral systems. The sum rules for charged systems are markedly more complicated than (2.101), because of the occurrence of screening. It is nevertheless possible to use a similar approach in order to discuss the long wavelength limit of the appropriate sum rules, a point which is discussed in detail in Section 4.2.

We mentioned in Section 2.3 that the f-sum rule provided the high-frequency limit of the response function $\chi(\mathbf{q}, \omega)$, given by Eq. (2.57). More precisely, when ω is much larger than the excitation frequencies ω_{no} of importance, we may neglect ω_{no} in the denominator of Eq. (2.57). Then $\chi(\mathbf{q}, \omega)$ has the limiting form (2.60):

$$\chi(\mathbf{q}, \omega) = \frac{Nq^2}{m\omega^2}. \tag{2.60}$$

If the system is not invariant under translation, Eq. (2.60) is only valid if ω is much larger than multipair excitation energies, i.e., if $\omega \gg \mu$. In a translationally invariant system, instead, multipair excitations do not contribute to Eq. (2.57) when $\mathbf{q} \to 0$. In order for Eq. (2.60) to apply in that limit, it suffices that ω be larger than the typical single-pair and zero sound excitation energies, i.e., that

$$qv_F \ll \omega. \tag{2.102}$$

In the next section, we shall calculate $\chi(\mathbf{q}, \omega)$ by means of the Landau theory of Fermi liquids, and we shall verify that for a translationally invariant system, and for frequencies such that

$$qv_F \ll \omega \ll \mu, \tag{2.103}$$

the response function $\chi(\mathbf{q}, \omega)$ is indeed equal to Eq. (2.60).

2.5. CALCULATION OF THE DENSITY RESPONSE FUNCTION BY THE LANDAU THEORY

RESPONSE TO A MACROSCOPIC TEST PARTICLE

When $\mathbf{q} \ll p_F$ and $\omega \ll \mu$, the test particle represents a *macroscopic* probe of system behavior; the response can be treated in the framework of the Landau theory. In physical terms, the test charge exerts on each particle a force $\mathfrak{F}$ given by Eq. (2.64). Since the interaction with the test charge involves only the system *density*, $\mathfrak{F}$ is also the force felt by a *quasiparticle* (see Section 1.4). The response of the system may then be viewed as a flow of quasiparticles in phase space, induced by the force $\mathfrak{F}$. This flow is governed by the general transport equation for quasiparticles, (1.83). Since we are interested only in the *linear* response of the system, we can replace Eq. (1.83) by the transport equation for a given Fourier component $\delta n_{\mathbf{p}}(\mathbf{q}, \omega)$. On replacing $\mathfrak{F}$ by its explicit value (2.64), we find

$$\mathbf{q} \cdot \mathbf{v}_{\mathbf{p}} \delta \bar{n}_{\mathbf{p}} - \omega \delta n_{\mathbf{p}} - \mathbf{q} \cdot \mathbf{v}_{\mathbf{p}} \frac{\partial n^{\circ}}{\partial \epsilon_p} \varphi(\mathbf{q}, \omega) = -iI(\delta n_{\mathbf{p}}). \tag{2.104}$$

The various quantities entering Eq. (2.104) have been defined in Chapter 1. We only remind the reader that $\delta n_{\mathbf{p}}$ measures the departure of the quasiparticle distribution from the *true* equilibrium, while $\delta \bar{n}_{\mathbf{p}}$ corresponds to the departure from *local* equilibrium [cf. Eq. (1.28)].

In the present section, as in the preceding one, we confine our attention to the *collisionless regime*, in which the frequency ω and wave vector $\mathbf{q}$ are such that

$$\nu \ll \omega \text{ or } qv_F, \qquad \omega \text{ and } qv_F \ll \mu, \tag{2.105}$$

where ν is the collision frequency and μ the chemical potential. We can then neglect the collision integral $I(\delta n_{\mathbf{p}})$in Eq. (2.104). The calculation of $\chi(\mathbf{q}, \omega)$ in the opposite limit $(\omega, qv_F \ll \nu)$, corresponding to the hydrodynamic regime, is briefly sketched in Section 2.7.

In Eq. (2.104) let us replace $\delta \bar{n}_{\mathbf{p}}$ by its expression (1.30). We thereby obtain an integral equation satisfied by $\delta n_{\mathbf{p}}$:

$$(\mathbf{q} \cdot \mathbf{v}_{\mathbf{p}} - \omega)\delta n_{\mathbf{p}} - \mathbf{q} \cdot \mathbf{v}_{\mathbf{p}} \frac{\partial n^{\circ}}{\partial \epsilon_p} \sum_{\mathbf{p}'} f_{\mathbf{p}\mathbf{p}'} \delta n_{\mathbf{p}'} - \mathbf{q} \cdot \mathbf{v}_{\mathbf{p}} \frac{\partial n_0}{\partial \epsilon_p} \varphi(\mathbf{q}, \omega) = 0. \tag{2.106}$$

To the extent that Eq. (2.106) can be solved, we may calculate the quasiparticle density fluctuation induced by the test charge,

$$\langle \rho(\mathbf{q}, \omega) \rangle = \sum_{\mathbf{p}} \delta n_{\mathbf{p}}(\mathbf{q}, \omega). \tag{2.107}$$

Since the quasiparticle and particle densities are equal, we thus obtain the density response function $\chi(\mathbf{q}, \omega)$, defined in Eq. (2.49).

We may write down a formal expression for the response function in terms of the scattering amplitudes $A_{\mathbf{p}\mathbf{p}'}(\mathbf{q}, \omega)$ introduced in Chapter 1, and defined by the integral equation (1.112). For that purpose, we set

$$\delta n_{\mathbf{p}} = - \frac{\mathbf{q} \cdot \mathbf{v}_{\mathbf{p}}}{\mathbf{q} \cdot \mathbf{v}_{\mathbf{p}} - \omega - i\eta} \frac{\partial n^{\circ}}{\partial \epsilon_p} x_{\mathbf{p}} \tag{2.108}$$

[in analogy with Eq. (1.110)]. (Remark that we have added a small imaginary term, $i\eta$, to ω in conformity with our adiabatic boundary conditions.) In terms of the new "unknown" $x_{\mathbf{p}}$, the transport equation (2.106) becomes

$$x_{\mathbf{p}} - \sum_{\mathbf{p}'} f_{\mathbf{pp}'} \frac{\mathbf{q} \cdot \mathbf{v}_{\mathbf{p}'}}{\mathbf{q} \cdot \mathbf{v}_{\mathbf{p}'} - \omega - i\eta} \frac{\partial n_o}{\partial \epsilon_{\mathbf{p}'}} x_{\mathbf{p}'} + \varphi(\mathbf{q}, \omega) = 0 \qquad (2.109)$$

[compare with Eq. (1.111)].

Once again, the scattering amplitude $A_{\mathbf{pp}'}(\mathbf{q}, \omega)$, defined by Eq. (1.112), appears as the "resolvent operator" of the integral equation (2.109). [We now see that Eq. (1.112) should also have contained a small imaginary term $+i\eta$, added to the frequency ω: the scattering amplitudes are, in general, complex quantities, and the boundary condition serves to define their imaginary part.] It is straightforward to verify that the solution of Eq. (2.109) may be written as

$$x_{\mathbf{p}} = -\varphi(\mathbf{q}, \omega) \left\{ 1 + \sum_{\mathbf{p}'} A_{\mathbf{pp}'}(\mathbf{q}, \omega) \frac{\mathbf{q} \cdot \mathbf{v}_{\mathbf{p}'}}{\mathbf{q} \cdot \mathbf{v}_{\mathbf{p}'} - \omega - i\eta} \frac{\partial n^o}{\partial \epsilon_{\mathbf{p}'}} \right\}. \qquad (2.110)$$

From $x_{\mathbf{p}}$, we go back to $\delta n_{\mathbf{p}}$ by means of Eq. (2.108), and to $\langle \rho(\mathbf{q}, \omega) \rangle$ by means of Eq. (2.107). We finally obtain the following expression of the density-density linear response function $\chi(\mathbf{q}, \omega)$, defined by Eq. (2.49):

$$\chi(\mathbf{q}, \omega) = \sum_{\mathbf{p}} \frac{\mathbf{q} \cdot \mathbf{v}_{\mathbf{p}}}{\mathbf{q} \cdot \mathbf{v}_{\mathbf{p}} - \omega - i\eta} \frac{\partial n^o}{\partial \epsilon_{\mathbf{p}}} \left[1 + \sum_{\mathbf{p}'} A_{\mathbf{pp}'}(\mathbf{q}, \omega) \frac{\mathbf{q} \cdot \mathbf{v}_{\mathbf{p}'}}{\mathbf{q} \cdot \mathbf{v}_{\mathbf{p}'} - \omega - i\eta} \frac{\partial n^o}{\partial \epsilon_{\mathbf{p}'}} \right] \qquad (2.111)$$

The result (2.111) is correct whenever the collisionless Landau theory is applicable, i.e., when the conditions (2.105) are met.

Let us first consider the limiting form of (2.111) for the case of a noninteracting fermion system. In this case, the second term in brackets of Eq. (2.111) disappears; the corresponding response function, $\chi^o(\mathbf{q}, \omega)$ is just what one would find from the general expression, (2.57), provided the exact matrix elements and excitation frequencies which appear there are replaced by those appropriate to the noninteracting fermion system. This may be readily verified by noting that

$$-\mathbf{q} \cdot \mathbf{v}_{\mathbf{p}} \frac{\partial n^o}{\partial \epsilon_{\mathbf{p}}} = -\mathbf{q} \cdot \frac{\partial n^o}{\partial \mathbf{p}} = \lim_{\mathbf{q} \to 0} [n_{\mathbf{p}}{}^o - n^o_{\mathbf{p}+\mathbf{q}}]. \qquad (2.112)$$

The factor $-\mathbf{q} \cdot \mathbf{v}_{\mathbf{p}}(\partial n^o/\partial \epsilon_{\mathbf{p}})$ may thus be interpreted as limiting the summation over $\mathbf{p}$ to the two crescents of Fig. 2.5, the integrand being multiplied by $+1$ in the shaded crescent, and by -1 in the other one. On changing $\mathbf{p}$ into $-\mathbf{p}$ in the nonshaded crescent, we may write the response function $\chi^o(\mathbf{q}, \omega)$

for a noninteracting system as

$$\chi^{\circ}(\mathbf{q}, \omega) = - \sum_{\substack{p<p_F \\ |\mathbf{p}+\mathbf{q}|>p_F}} \left\{ \frac{1}{\mathbf{q} \cdot \mathbf{v_p} - \omega - i\eta} + \frac{1}{\mathbf{q} \cdot \mathbf{v_p} + \omega + i\eta} \right\}. \quad (2.113)$$

In this form, $\chi^{\circ}(\mathbf{q}, \omega)$ clearly agrees with (2.57).

In an interacting system, χ depends on ω in a much more complicated way, which is determined by the scattering amplitude $A_{\mathbf{pp'}}(\mathbf{q}, \omega)$. We saw in Section 1.7 that the collective modes gave rise to poles of the scattering amplitude: these in turn lead to poles of the response function $\chi(\mathbf{q}, \omega)$. For an isotropic system, by reasons of symmetry, all poles, except that corresponding to zero sound, disappear when we sum over $\mathbf{p}$ and $\mathbf{p'}$ in Eq. (2.111). In the same way, spin antisymmetric waves would appear as poles of the spin-spin response function, transverse symmetric waves as poles of the transverse current-current response function, etc.

The dynamic form factor $S(\mathbf{q}, \omega)$ is obtained from the imaginary part of Eq. (2.111), χ''. Again, a formal solution for $S(\mathbf{q}, \omega)$ can be found by means of the so-called "optical theorem," which provides an explicit expression in terms of $A_{\mathbf{pp'}}$. However, the calculations are lengthy, and not very rewarding, so that we shall confine ourselves to a few qualitative remarks concerning the general case. A first contribution to χ'' arises when the energy denominators of Eq. (2.111) vanish, i.e., when $\omega = \mathbf{q} \cdot \mathbf{v_p}$ or $\mathbf{q} \cdot \mathbf{v_{p'}}$. This part of $S(\mathbf{q}, \omega)$, which extends continuously from 0 to qv_F, corresponds to the single-pair excitation spectrum. The other contribution to χ'' arises from the imaginary part of the scattering amplitude, which acts to modify the continuous single-pair spectrum, and may also provide a discrete peak corresponding to zero sound. This behavior is in accord with our qualitative discussion of Section 2.4.

In order to illustrate in more detail the general character of $\chi(\mathbf{q}, \omega)$, let us consider the simple model of Section 1.7 in which the interaction $f_{\mathbf{pp'}}$ is chosen to be a constant. The scattering amplitude $A_{\mathbf{pp'}}(\mathbf{q}, \omega)$ may then be calculated explicitly; it is given by Eqs. (1.129) and (1.130), in terms of the reduced quantities

$$\lambda = \omega / q v_F,$$

$$F_0 = \frac{m^* p_F}{\pi^2 \hbar^3} f_{\mathbf{pp'}}.$$

We note that $A_{\mathbf{pp'}}(\mathbf{q}, \omega)$ does not depend on $\mathbf{p}$ and $\mathbf{p'}$. The summations of Eq. (2.111) are thus very simple. It is easily verified that

$$\sum_{\mathbf{p}} \frac{\mathbf{q} \cdot \mathbf{v_p}}{\mathbf{q} \cdot \mathbf{v_p} - \omega - i\eta} \frac{\partial n^{\circ}}{\partial \epsilon_p} = - \frac{m^* p_F}{\pi^2 \hbar^3} g(\lambda), \quad (2.114)$$

where we have set

$$g(\lambda) = 1 - \frac{\lambda}{2} \log \left[\frac{\lambda + i\eta + 1}{\lambda + i\eta - 1} \right]. \tag{2.115}$$

Equation (2.111) then reduces to

$$\chi(\mathbf{q}, \omega) = - \frac{m^* p_F}{\pi^2 \hbar^3} \frac{g(\lambda)}{1 + F_o g(\lambda)}. \tag{2.116}$$

Equation (2.116) could in fact have been obtained by solving the transport equation (2.106) directly.

The detailed study of Eq. (2.116) as a function of F_o is left as a problem to the reader. We here remark only on a few important features. First, by taking the imaginary part of Eq. (2.116), we see clearly the two types of contributions to the dynamic form factor $S(\mathbf{q}, \omega)$:

(i) The single-pair terms, which arise from the imaginary part of the logarithm in Eq. (2.115) (it being equal to $i\pi$ in the range $|\lambda| < 1$).

(ii) The zero sound contribution which corresponds to the value λ_o at which the denominator $[1 + F_o g(\lambda)]$ vanishes. Let us assume that there exists such a value $\lambda_o > 1$ (a case which corresponds to $F_o > 0$). In the vicinity of λ_o, we may write

$$\chi(\mathbf{q}, \omega) \sim - \frac{m^* p_F}{\pi^2 \hbar^3} \frac{g(\lambda_o)}{F_o g'(\lambda_o)} \frac{1}{\lambda - \lambda_o + i\eta}. \tag{2.117}$$

If we note that $g(\lambda_o) = -1/F_o$, take the imaginary part of Eq. (2.117), and make use of Eq. (2.63), we find that the zero sound contribution to $S(\mathbf{q}, \omega)$ has the form

$$[S(\mathbf{q}, \omega)]_{\text{zero sound}} = + \frac{m^* p_F}{\pi^2 \hbar^3} \frac{1}{F_o^2 g'(\lambda_o)} \delta(\lambda - \lambda_o). \tag{2.118}$$

Equation (2.118) corresponds to a discrete peak, in agreement with our earlier statements. Let us further note that

$$\delta(\lambda - \lambda_o) = \delta\left(\frac{\omega - s_o q}{q v_F}\right) = q v_F \delta(\omega - s_o q). \tag{2.119}$$

We thereby establish that the matrix element $(\rho_\mathbf{q}^+)_{no}$ for zero sound excitation is of order $q^{1/2}$.

If $-1 < F_o < 0$, the collective mode is subject to substantial Landau damping, and is not sharply defined. It is clear from Eq. (2.116) that $S(\mathbf{q}, \omega)$ will possess a maximum for frequencies in the vicinity of those for which the real part of $[1 + F_o g]$ vanishes. One thus expects to find a broad resonance characteristic of a damped collective mode.

The simple model may be used for studying the strong coupling limit, $F_0 \gg 1$. In that limit, the single pair contribution to $S(\mathbf{q}, \omega)$ may be written as [see Eq. (2.63)]:

$$-\pi[S(\mathbf{q}, \omega)]_{\text{single pairs}} \sim \frac{m^* p_F}{\pi^2 \hbar^3} \frac{1}{F_0^{\,2}} \operatorname{Im}\left(\frac{1}{g(\lambda)}\right). \tag{2.120}$$

The corresponding contribution to the sum rules (2.101a) and (2.101c) is thus of order F_0^{-2}. On the other hand, it is easily verified that the total f-sum rule is of order 1, and the total compressibility sum rule of order F_0^{-1} [see Eq. (1.58)]. We thus conclude that for very strong coupling between quasiparticles, zero sound must exhaust these two sum rules. In other words, the collective behavior is predominant when the quasiparticle interaction is very large.

The predominance of the collective modes may be verified by inspection [using Eq. (2.118) and the limiting form of the zero sound velocity $\lambda_0 \sim \sqrt{F_0/3}$]. The proof we have given is valid only for our simple model. It is, however, not unlikely that the conclusion we have reached is valid in the most general case (even though no simple proof is as yet available).

EXACT FORM OF THE MACROSCOPIC RESPONSE FUNCTION IN SOME LIMITING CASES

The transport equation (2.106) may be solved exactly in the two limiting cases: $\omega \gg qv_F$ or $\omega \ll qv_F$. We shall consider these two limits, and compare the results thus obtained with the sum rules discussed in Section 2.4. For the sake of simplicity, we consider an *isotropic* system.

(*i*) *The static and quasistatic limits* ($\omega \ll qv_F$) Let us first consider a static test particle at zero temperature, so that Eq. (2.105) is satisfied. On making use of Eq. (2.104), we find at once

$$\delta\bar{n}_{\mathbf{p}}(\mathbf{q}, 0) = \frac{\partial n^0}{\partial \epsilon_p} \varphi(\mathbf{q}, 0). \tag{2.121}$$

According to Eq. (2.121), $\delta n_{\mathbf{p}}$ is isotropic. On using Eq. (1.34), we thus obtain

$$\delta n_{\mathbf{p}}(\mathbf{q}, 0) = \frac{\delta\bar{n}_{\mathbf{p}}(\mathbf{q}, 0)}{1 + F_0^{\,s}} = \frac{\varphi(\mathbf{q}, 0)}{1 + F_0^{\,s}} \frac{\partial n^0}{\partial \epsilon_p}. \tag{2.122}$$

It is now straightforward to calculate the induced density fluctuation $\langle\rho(\mathbf{q}, 0)\rangle$ given by Eq. (2.107). From it, one deduces the following value of the response function:

$$\chi(\mathbf{q}, 0) = -\frac{m^* p_F}{\pi^2 \hbar^3} \frac{1}{1 + F_0^s} = \frac{-\nu(0)}{1 + F_0^s}, \tag{2.123}$$

where $\nu(0)$ is the density of quasiparticle states on the Fermi surface. By referring to Eq. (1.58), we may write Eq. (2.123) as

$$\chi(\mathbf{q}, 0) = -\frac{N}{ms^2} \tag{2.124}$$

(where s is the usual sound velocity). Our Landau theory result, (2.124), is seen to be in accord with the compressibility sum rule (2.84); it is, of course, only valid in the limit $\mathbf{q} \to 0$.

We may obtain the response to a slowly varying test particle ($\omega \ll qv_F$) by expressing the solution of Eq. (2.104) in terms of a power series with respect to ω/qv_F. The first term of this series, $\delta n_{\mathbf{p}}^{(0)}$, is given by Eqs. (2.121) and (2.122). By putting this term back in Eq. (2.104), we obtain the next term of the expansion

$$\delta \bar{n}_{\mathbf{p}}^{(1)} = \frac{\omega}{\mathbf{q} \cdot \mathbf{v}_{\mathbf{p}} - i\eta} \frac{\varphi(\mathbf{q}, \omega)}{(1 + F_0^s)} \frac{\partial n^0}{\partial \epsilon_p} = \frac{\omega}{\mathbf{q} \cdot \mathbf{v}_{\mathbf{p}} - i\eta} \delta n_{\mathbf{p}}^{(0)}. \tag{2.125}$$

(We have included the $i\eta$ term which arises from our causal boundary conditions.) Even though we cannot simply calculate the corresponding quantity $\delta n_{\mathbf{p}}^{(1)}$, we may easily obtain the total induced density fluctuation

$$\langle \rho^{(1)}(\mathbf{q}, \omega) \rangle = \sum_{\mathbf{p}} \delta n_{\mathbf{p}}^{(1)} = \frac{\sum_{\mathbf{p}} \delta \bar{n}_{\mathbf{p}}^{(1)}}{1 + F_0^s}. \tag{2.126}$$

(By summing over $\mathbf{p}$, we sample only the $\ell = 0$ part of $\delta n_{\mathbf{p}}^{(1)}$ and $\delta \bar{n}_{\mathbf{p}}^{(1)}$.) As usual, we write

$$\frac{1}{\mathbf{q} \cdot \mathbf{v}_{\mathbf{p}} - i\eta} = P\left(\frac{1}{\mathbf{q} \cdot \mathbf{v}_{\mathbf{p}}}\right) + i\pi\delta(\mathbf{q} \cdot \mathbf{v}_{\mathbf{p}}). \tag{2.127}$$

The contribution of the principal part to the sum (2.126) vanishes for reasons of symmetry. There remains only an imaginary term, arising from the δ function. By performing the summation, we find

$$\chi^{(1)}(\mathbf{q}, \omega) = -\frac{i\pi}{2} \frac{\omega}{qv_F} \frac{\nu(0)}{(1 + F_0^s)^2}. \tag{2.128}$$

From $\chi^{(1)}$, we obtain the leading term of $S(\mathbf{q}, \omega)$ by using Eq. (2.63). We note that the dynamic form factor is of order ω, in agreement with our earlier statement. On comparing with Eq. (2.90), we see that, in that range of frequencies, the many-body effects are all contained in the factor $(1 + F_0^s)^{-2}$.

(*ii*) *The quasihomogeneous limit* We now consider the opposite limit $qv_F \ll \omega$. (We tacitly assume, however, that ω remains small compared to μ.) By retaining only the leading term in Eq. (2.104), we find

$$\delta n_{\mathbf{p}} = -\frac{\mathbf{q}\cdot\mathbf{v}_{\mathbf{p}}}{\omega}\frac{\partial n^0}{\partial \epsilon_p}\varphi(\mathbf{q}, \omega). \tag{2.129}$$

Since this term possesses an odd symmetry, it does not contribute to the net density fluctuation $\langle\rho(\mathbf{q}, \omega)\rangle$. We should therefore push the expansion in powers of qv_F/ω one step further. We shall instead use an artifice; we calculate the induced current

$$\langle\mathbf{J}(\mathbf{q}, \omega)\rangle = \sum_{\mathbf{p}} \delta n_{\mathbf{p}}(\mathbf{q}, \omega)\mathbf{j}_{\mathbf{p}}, \tag{2.130}$$

from which we obtain the induced density by the conservation law

$$\mathbf{q}\cdot\langle\mathbf{J}(\mathbf{q}, \omega)\rangle = \omega\langle\rho(\mathbf{q}, \omega)\rangle. \tag{2.131}$$

On making use of the relations (2.129) to (2.131), we find, after a straightforward calculation,

$$\chi(\mathbf{q}, \omega) = -\sum_{\mathbf{p}} \frac{(\mathbf{q}\cdot\mathbf{v}_{\mathbf{p}})(\mathbf{q}\cdot\mathbf{j}_{\mathbf{p}})}{\omega^2}\frac{\partial n^0}{\partial \epsilon_p} \qquad (qv_F \ll \omega). \tag{2.132}$$

For the most general system, we cannot go further; $\chi(\mathbf{q}, \omega)$ is of order q^2/ω^2, with a coefficient which is not related to simple quantities.

If, however, the system is invariant under translations, we know that $\mathbf{j}_{\mathbf{p}} = \mathbf{p}/m$. The summation (2.132) may then be performed easily; one finds

$$\chi(\mathbf{q}, \omega) = \frac{Nq^2}{m\omega^2}. \tag{2.133}$$

We thus recover the result (2.60) obtained by application of the f-sum rule. These results are in agreement with the qualitative results of Section 2.4: in the range of frequencies $qv_F \ll \omega \ll \mu$, the limiting form (2.133) applies *only* to translationally invariant systems.

In order to bring out more clearly the relation of this discussion to the Landau theory, we note that for the above range of frequencies, the response function (2.57) may be replaced by the approximate form

$$\chi(\mathbf{q}, \omega) \sim \sum_{\substack{\text{single pair,}\\ \text{zero sound}}} |(\rho_{\mathbf{q}}{}^{+})_{no}|^2 \frac{2\omega_{no}}{\omega^2}. \tag{2.134}$$

Let us introduce the oscillator strengths f_{on} defined by Eq. (2.30). We see that the coefficient multiplying $q^2/m\omega^2$ in Eq. (2.134) is simply the *sum* of oscillator strengths for *single pair* and *zero sound* excitations. This partial sum may be identified with the corresponding coefficient of Eq. (2.132). If the system is invariant under translations, the multipair oscillator strengths are negligible; the other two kinds exhaust the sum rule, the sum of oscillator strengths being equal to N; χ is then given by Eq. (2.133). Any departure from Eq. (2.133) gives a measure of the role played by multipair transitions.

Let us emphasize that such an agreement was not *a priori* obvious. As a matter of fact, one may apply the methods of the present chapter to the study of the current-current response function. One then finds that for a nontranslationally invariant system, multipair excitations are quite important. The Landau theory must then be applied with great care, since it cannot take an explicit account of such high-energy excitations.

BACKFLOW AROUND A SLOWLY MOVING PARTICLE

We conclude this section with a brief discussion of the current pattern induced in the system by a slowly moving external particle. Let us consider a point impurity, with velocity $\mathbf{V}_e$. The impurity acts as a scalar probe, of density

$$\rho_e(\mathbf{r}, t) = \delta(\mathbf{r} - \mathbf{V}_e t). \tag{2.135}$$

The Fourier transform of Eq. (2.135) is

$$\rho_e(\mathbf{q}, \omega) = 2\pi\delta(\omega - \mathbf{q}\cdot\mathbf{V}_e). \tag{2.136}$$

We assume that the impurity is much heavier than the system particles, so that its recoil may be neglected: $\mathbf{V}_e$ then may be assumed to be a constant.

Let $\mathcal{V}_{\mathbf{q}}$ be the Fourier component of the interaction between the impurity and system densities. The impurity gives rise to a density fluctuation of the system, $\langle\rho(\mathbf{q}, \omega)\rangle$, given by Eq. (2.49):

$$\langle\rho(\mathbf{q}, \omega)\rangle = \mathcal{V}_{\mathbf{q}}\chi(\mathbf{q}, \omega)\rho_e(\mathbf{q}, \omega). \tag{2.137}$$

From $\langle\rho(\mathbf{q}, \omega)\rangle$, we can deduce the *longitudinal* part of the induced current, $\langle\mathbf{J}(\mathbf{q}, \omega)\rangle$, by means of the conservation law (2.131). Actually, the current $\langle \mathbf{J} \rangle$, being induced by a longitudinal probe, can have no transverse part: $\langle \mathbf{J} \rangle$ is thus completely determined. We find

$$\langle \mathbf{J}(\mathbf{q}, \omega) \rangle = \frac{\omega}{q^2} \mathbf{q} \chi(\mathbf{q}, \omega) \rho_e(\mathbf{q}, \omega) \mathcal{V}_{\mathbf{q}}. \tag{2.138}$$

Let us replace $\rho_e(\mathbf{q}, \omega)$ by its value (2.136); we obtain

$$\langle \mathbf{J}(\mathbf{q}, \omega) \rangle = \frac{\mathbf{q} \cdot \mathbf{V}_e}{q^2} \mathbf{q}\, \chi(\mathbf{q}, \mathbf{q} \cdot \mathbf{V}_e) 2\pi\delta(\omega - \mathbf{q} \cdot \mathbf{V}_e) \mathcal{V}_{\mathbf{q}}. \tag{2.139}$$

Equation (2.139) characterizes the current pattern produced by the impurity; it corresponds to a *backflow* of the system particles around the moving impurity.

Let us first assume that the impurity is very slow, such that

$$V_e \ll v_F. \tag{2.140}$$

We can then replace in Eq. (2.139) the response function χ by its static value $\chi(\mathbf{q}, 0)$. For small values of $\mathbf{q}$, we may use Eq. (2.84) to write the current distribution in the form

$$\langle \mathbf{J}(\mathbf{q}, \omega) \rangle = - \frac{\mathbf{q} \cdot \mathbf{V}_e}{q^2} \mathbf{q} \mathcal{V}_{\mathbf{q}} \frac{N}{ms^2} 2\pi\delta(\omega - \mathbf{q} \cdot \mathbf{V}_e). \tag{2.141}$$

By taking the Fourier transform of Eq. (2.141) (valid for small q), we expect to describe correctly the current distribution *far* from the impurity, at distances large compared with atomic dimensions.

If the interaction between the system and the impurity is short ranged, $\mathcal{V}_{\mathbf{q}}$ is regular when $\mathbf{q} \to 0$; we may replace $\mathcal{V}_{\mathbf{q}}$ by $\mathcal{V}_0$ in Eq. (2.141). It is then straightforward to find the Fourier transform of Eq. (2.141). We thereby obtain the asymptotic form of the current distribution in configuration space,

$$\langle \mathbf{J}(\mathbf{r}, t) \rangle = \frac{\mathcal{V}_0}{4\pi} \frac{N}{ms^2} \nabla(\mathbf{V}_e \cdot \nabla) \frac{1}{R}, \tag{2.142}$$

where we have set

$$\mathbf{R} = \mathbf{r} - \mathbf{V}_e t. \tag{2.143}$$

Equation (2.42) corresponds to a *dipolar* backflow of the system around the impurity.

The fact that the backflow at *large distances* from a *slowly moving* impurity is dipolar in character may be thought of as arising from quite general geometrical considerations. A similar result is found in elementary hydrodynamic theories whenever a spherical obstacle moves slowly in an incompressible fluid. It thus seems that for very low velocities, the backflow current has essentially a hydrodynamic character. The dynamical properties of the system fix the *size* of the backflow, but do not affect its flow pattern at large distances.

We shall return to this question again when studying the properties of a charged electron system. In that case, as a result of perfect screening of a charged impurity, the backflow, which is again dipolar in character, exactly compensates the longitudinal current carried by the impurity. For that reason, backflow plays a far more important role in charged particle systems.

The preceding results are no longer valid if $V_e > v_F$. Under these circumstances, the response function $\chi(\mathbf{q}, \omega)$ depends appreciably on the angle between $\mathbf{q}$ and $\mathbf{V}_e$. The backflow set up by the impurity is no longer dipolar, but has instead a complicated directional dependence. In the limit $V_e \gg v_F$, the backflow becomes a minor effect as compared to the real transitions induced by the impurity (studied in Section 2.1). The response of the system is then in some ways similar to a shock-wave structure, instead of the simple laminar flow that characterizes a slowly moving impurity.

2.6. RESPONSE AND CORRELATION AT FINITE TEMPERATURES

We consider now the formal description of response and correlation in a many-particle system at finite temperatures. We shall see that the analysis of the response to a weak external probe and the definitions of the various correlation functions proceed in much the same way as they did at $T = 0$. There are, however, some important differences, which we comment on briefly.

First, at finite temperatures there is no longer a unique state of well-defined energy which describes the system. A statistical description is necessary. Let us specify the states of the system by the exact eigenstates of the system Hamiltonian, $|m\rangle$, which possess an energy E_m; let us furthermore assume the system is in equilibrium at some temperature, T (an assumption we shall make throughout this section). Under these circumstances, the canonical ensemble of statistical mechanics offers a convenient mathematical formulation. The probability of finding the system in the state $|m\rangle$ is simply

$$W(m) = e^{-\beta E_m}/Z, \tag{2.144}$$

where Z, the sum over states, is given by

$$Z = \sum_m e^{-\beta E_m}. \tag{2.145}$$

Second, in considering the interaction between a time-dependent external probe and the many-particle system we must allow for the possibility that the system will transfer energy to the probe, as well as vice versa. In thermal equilibrium the two probabilities (that the probe heats up the system, and the reverse) are not independent; the relation between them, which we shall establish explicitly, is the famous principle of *detailed balancing.*

Finally, at finite temperatures there is an important physical difference between the dynamic form factor, $S(\mathbf{q}, \omega)$, and the imaginary part of the density-density response function, $\chi''(\mathbf{q}, \omega)$. The former provides a direct measure of the fluctuations in the system (through the static form factor, $S_{\mathbf{q}}$), while the latter describes the dissipative part of the system response. The relation between $S(\mathbf{q}, \omega)$ and $\chi''(\mathbf{q}, \omega)$ may easily be found from the detailed balancing condition; in integrated form it is the well-known *fluctuation-dissipation theorem.*

We begin the section with a consideration of the dynamic form factor at finite temperature, and the principle of detailed balancing. We next consider the density-density response function, as a specific example of a finite temperature linear response function. We establish the fluctuation-dissipation theorem and discuss the various finite temperature sum rules for the density-density correlation functions.

DYNAMIC FORM FACTOR

We consider anew the inelastic scattering of an external beam of particles by the many-particle system, as specified by Eq. (2.6). Within the Born approximation the probability per unit time that a particle transfer momentum $\mathbf{q}$, energy ω, to the system is

$$\mathcal{P}(\mathbf{q}, \omega) = 2\pi|\mathcal{V}_{\mathbf{q}}|^2 Z^{-1} \sum_{mn} e^{-\beta E_m} |(\rho_{\mathbf{q}}{}^+)_{nm}|^2 \delta(\omega - \omega_{nm}). \quad (2.146)$$

The result (2.146) follows from application of second-order perturbation theory to a transition from a given state, $|m\rangle$, to another state, $|n\rangle$, and use of the statistical weighting factor, (2.144); ω_{nm} is the exact energy difference, $E_n - E_m$. The properties of the many-particle system are contained in the dynamic form factor,

$$S(\mathbf{q}, \omega) = Z^{-1} \sum_{mn} e^{-\beta E_m} |(\rho_{\mathbf{q}}{}^+)_{nm}|^2 \delta(\omega - \omega_{nm}). \quad (2.147)$$

We may also write down the probability per unit time that the particle absorbs momentum $-\mathbf{q}$, energy ω, *from* the system. It is

$$\mathcal{P}(\mathbf{q}, -\omega) = 2\pi|\mathcal{V}_{\mathbf{q}}|^2 \sum_{mn} e^{-\beta E_m}|(\rho_{\mathbf{q}}{}^+)_{nm}|^2\delta(\omega + \omega_{nm})$$
$$= 2\pi|\mathcal{V}_{\mathbf{q}}|^2 S(\mathbf{q}, -\omega). \quad (2.148)$$

$S(\mathbf{q}, -\omega)$ thus has a well-defined physical meaning at finite temperature.

There exists a simple thermodynamic relation between $S(\mathbf{q}, -\omega)$ and $S(\mathbf{q}, \omega)$. To obtain it, we interchange indices in Eq. (2.147) to write

$$\begin{aligned} S(\mathbf{q}, \omega) &= Z^{-1} \sum_{mn} e^{-\beta E_n}|(\rho_{\mathbf{q}}{}^+)_{mn}|^2\delta(\omega + \omega_{nm}) \\ &= Z^{-1} \sum_{mn} e^{-\beta(E_n - E_m)}e^{-\beta E_m}|(\rho_{\mathbf{q}})_{nm}|^2\delta(\omega + \omega_{nm}) \quad (2.149) \\ &= e^{\beta\omega}S(-\mathbf{q}, -\omega). \end{aligned}$$

We then make use of time-reversal invariance, which states that

$$S(\mathbf{q}, -\omega) = S(-\mathbf{q}, -\omega). \quad (2.150)$$

We find therefore the important result

$$S(\mathbf{q}, \omega) = e^{\beta\omega}S(\mathbf{q}, -\omega). \quad (2.151)$$

This relation is, in fact, a sophisticated statement of the principle of detailed balancing. To see this we consider the larger system of the many-body system *plus* the probe, and suppose this to be in thermal equilibrium at temperature T. In that case, the principle of detailed balancing states that

$$\frac{\mathcal{P}(\mathbf{q}, \omega)}{\mathcal{P}(-\mathbf{q}, -\omega)} = e^{\beta\omega}, \quad (2.152)$$

which, upon application of Eqs. (2.147) and (2.148), is seen to be simply another statement of Eq. (2.151).

We see that $S(\mathbf{q}, \omega)$ must depend on temperature. In order to illustrate the modifications brought about by the various statistical factors, we consider its evaluation for some simple systems. Consider first a free gas of fermions. We write $S^{\circ}(\mathbf{q}, \omega)$ in the following form:

$$S^{\circ}(\mathbf{q}, \omega) = Z^{-1} \sum_{mn\mathbf{p}\mathbf{p}'} e^{-\beta E_m}\langle m|c^+_{\mathbf{p}'-\mathbf{q}}c_{\mathbf{p}'}|n\rangle\langle n|c^+_{\mathbf{p}+\mathbf{q}}c_{\mathbf{p}}|m\rangle\delta(\omega - \omega_{nm}). \quad (2.153)$$

The initial states, $|m\rangle$, correspond to a given configuration of occupied particle states in momentum space. According to Eq. (2.153), the operator $c^+_{\mathbf{p}+\mathbf{q}}c_{\mathbf{p}}$ acts to couple such a configuration to a new configuration, $|n\rangle$, which differs from $|m\rangle$ in that a particle of momentum $\mathbf{p}$ has been excited into the state $(\mathbf{p} + \mathbf{q})$. It follows at once that for the *free* fermion system, the difference in energy between the two configurations is

$$\omega_{nm} = \omega^{\circ}_{\mathbf{pq}} = \epsilon^{\circ}_{\mathbf{p}+\mathbf{q}} - \epsilon_{\mathbf{p}}{}^{\circ} = (\mathbf{q}\cdot\mathbf{p}/m + q^2/2m). \tag{2.154}$$

We may therefore write

$$S^{\circ}(\mathbf{q},\omega) = Z^{-1}\sum_{m\mathbf{p}\mathbf{p}'} e^{-\beta E_m}\langle m|c^{+}_{\mathbf{p}'-\mathbf{q}}c_{\mathbf{p}'}c^{+}_{\mathbf{p}+\mathbf{q}}c_{\mathbf{p}}|m\rangle\delta(\omega-\omega^{\circ}_{\mathbf{pq}}). \tag{2.155}$$

The matrix element in brackets is nonvanishing only if $\mathbf{p}' = \mathbf{p}+\mathbf{q}$. We have therefore, for $\mathbf{q} \neq 0$,

$$S^{\circ}(\mathbf{q},\omega) = Z^{-1}\sum_{m\mathbf{p}} e^{-\beta E_m}\langle m|c_{\mathbf{p}}{}^{+}c_{\mathbf{p}}(1 - c^{+}_{\mathbf{p}+\mathbf{q}}c_{\mathbf{p}+\mathbf{q}})|m\rangle, \tag{2.156}$$

on making use of the commutation properties of the fermion operators. We now note that by definition the probability of finding a particle $\mathbf{p}$ in the initial state is given by the free fermion distribution function at temperature T, so that

$$Z^{-1}\sum_{m} e^{-\beta E_m}\langle m|c_{\mathbf{p}}{}^{+}c_{\mathbf{p}}|m\rangle = n_{\mathbf{p}}{}^{\circ} = \frac{1}{e^{\beta\epsilon_{\mathbf{p}}{}^{\circ}}+1}\cdot \tag{2.157}$$

Moreover, the states $\mathbf{p}$ and $\mathbf{p}+\mathbf{q}$ are statistically independent. It follows that

$$S_{\circ}(\mathbf{q},\omega) = \sum_{\mathbf{p}} n_{\mathbf{p}}{}^{\circ}(1 - n^{\circ}_{\mathbf{p}+\mathbf{q}})\delta(\omega-\omega^{\circ}_{\mathbf{pq}}). \tag{2.158}$$

The reader may readily verify that Eq. (2.158) satisfies the principle of detailed balancing, (2.151).

As a second example, let us suppose that the operator $\rho_{\mathbf{q}}{}^{+}$ describes a pure collective mode, as might be the case for zero sound at long wavelengths in the strong coupling limit. In that case the states, $|m\rangle$, will correspond to a given number of collective mode quanta, while $\rho_{\mathbf{q}}{}^{+}$ gives rise to absorption or emission processes, according to

$$\rho_{\mathbf{q}}{}^{+} = d_q(A_{\mathbf{q}}{}^{+} + A_{-\mathbf{q}})/\sqrt{2}, \tag{2.159}$$

where d_q is an appropriate constant. Equation (2.159) is the usual transformation from harmonic oscillator coordinates to "phonon" creation and annihilation operators; $A_{\mathbf{q}}{}^{+}$ acts to create a collective mode of momentum $\mathbf{q}$, energy ω_q, $A_{-\mathbf{q}}$ to annihilate one of momentum $-\mathbf{q}$. On substituting Eq. (2.159) into (2.147), one finds at once that

$$S_{\text{coll}}(\mathbf{q},\omega) = Z^{-1}\sum_{m} e^{-\beta E_m}\{\langle m|A_{\mathbf{q}}A_{\mathbf{q}}{}^{+}|m\rangle\delta(\omega-\omega_q) + \langle m|A^{+}_{-\mathbf{q}}A_{-\mathbf{q}}|m\rangle\delta(\omega+\omega_{-q})\}d_q{}^2. \tag{2.160}$$

The phonon statistical factors are defined by

$$n_{\mathbf{q}}^{\circ}(T) = Z^{-1} \sum_{m} e^{-\beta E_m} \langle m | A_{\mathbf{q}}^{+} A_{\mathbf{q}} | m \rangle = \frac{1}{e^{\beta \omega_q} - 1}. \tag{2.161}$$

One thus finds

$$\begin{aligned} S_{\text{coll}}(\mathbf{q}, \omega) &= d_q^{\,2}\{(n_{\mathbf{q}} + 1)\delta(\omega - \omega_q) + n_{-\mathbf{q}}\delta(\omega + \omega_{-q})\} \\ &= d_q^{\,2}\{\delta(\omega - \omega_q) + n_{\mathbf{q}}[\delta(\omega - \omega_q) + \delta(\omega + \omega_q)]\}. \end{aligned} \tag{2.162}$$

The second form follows from time-reversal invariance; the first term on its right-hand side corresponds to spontaneous emission of the collective phonons, while the second represents induced emission, and the third is the induced absorption.

The appropriate generalization to finite temperatures of the various correlation functions considered in Section 2.1 is easily found. For example, the static form factor is clearly

$$NS_{\mathbf{q}} = Z^{-1} \sum_{m} e^{-\beta E_m} \langle m | \rho_{\mathbf{q}}^{+} \rho_{\mathbf{q}} | m \rangle. \tag{2.163}$$

Similar statistical averages are carried out for other quantities of interest.

DENSITY-DENSITY RESPONSE FUNCTION

The density-density response function at finite temperatures may be calculated in a fashion analogous to our second-order perturbation theoretic calculation at $T = 0$. The difference is that one takes an ensemble of initial states $|m\rangle$, weighted by the probability $W(m)$, (2.164), as in the calculation of $\mathcal{P}(\mathbf{q}, \omega)$, (2.146). One finds, on making use of time reversal,

$$\chi(\mathbf{q}, \omega) = Z^{-1} \sum_{mn} e^{-\beta E_m} |(\rho_{\mathbf{q}}^{+})_{nm}|^2 \frac{2\omega_{nm}}{(\omega + i\eta)^2 - \omega_{nm}^2}. \tag{2.164}$$

We again split χ into real and imaginary parts,

$$\chi(\mathbf{q}, \omega) = \chi'(\mathbf{q}, \omega) + i\chi''(\mathbf{q}, \omega),$$

and note that χ' is even in ω, while χ'' is odd.

As at $T = 0$, $\chi''(\mathbf{q}, \omega)$ is the *dissipative* part of the density-density response function, which specifies directly the steady energy transfer from an oscillating probe to the many-particle system (to see this, remark that the derivation leading from Eq. (2.64) to (2.68) is unchanged at finite temperatures). From (2.164), we note that $\chi''(\mathbf{q}, \omega)$

may be written explicitly as

$$\chi''(\mathbf{q}, \omega) = -\pi Z^{-1} \sum_{mn} e^{-\beta E_m} |(\rho_{\mathbf{q}}{}^+)_{nm}|^2 \{\delta(\omega - \omega_{nm}) - \delta(\omega + \omega_{nm})\}. \tag{2.165}$$

It furthermore serves as the spectral density for $\chi(\mathbf{q}, \omega)$, according to

$$\chi(\mathbf{q}, \omega) = \frac{1}{\pi} \int_{-\infty}^{\infty} d\omega' \frac{\chi''(\mathbf{q}, \omega')}{\omega' - \omega - i\eta}. \tag{2.166}$$

The causal properties of $\chi(\mathbf{q}, \omega)$ are specified by Eq. (2.166), from which one may derive directly the Kramers–Kronig relations (2.78).

There continues to exist at finite temperatures a close relationship between the thermal fluctuations of the system [which are readily derived from $S(\mathbf{q}, \omega)$], and the linear response to a density probe [as specified by $\chi''(\mathbf{q}, \omega)$]. According to the definition (2.165), one has

$$\chi''(\mathbf{q}, \omega) = -\pi[S(\mathbf{q}, \omega) - S(\mathbf{q}, -\omega)]. \tag{2.167}$$

If we now make use of the detailed balancing condition (2.151), we obtain the following relation:

$$\chi''(\mathbf{q}, \omega) = -\pi[1 - e^{-\beta\omega}]S(\mathbf{q}, \omega), \tag{2.168}$$

known as the fluctuation dissipation theorem [Nyquist (1928); Callen and Welton (1951)].

At finite temperatures, $\chi''(\mathbf{q}, \omega)$ is a somewhat simpler quantity to use to specify the density fluctuation spectrum than $S(\mathbf{q}, \omega)$, essentially because it need not depend on temperature. For example, one finds directly from Eq. (2.162) that for the pure collective mode in $S(\mathbf{q}, \omega)$ considered above,

$$\chi''(\mathbf{q}, \omega) = -\pi d_q{}^2 \{\delta(\omega - \omega_q) - \delta(\omega + \omega_q)\}. \tag{2.169}$$

We shall accordingly use χ'' whenever possible, when we require only a specification of the *character* of the density fluctuations at finite temperatures.

The relation (2.168) may be written in somewhat more symmetric form if we define a "symmetrized" dynamic form factor,

$$\tilde{S}(\mathbf{q}, \omega) = \frac{S(\mathbf{q}, \omega) + S(\mathbf{q}, -\omega)}{2} = S(\mathbf{q}, \omega) \frac{[1 + e^{-\beta\omega}]}{2}. \tag{2.170}$$

On combining Eqs. (2.168) and (2.170) we find

$$\tilde{S}(\mathbf{q}, \omega) = -\frac{1}{2\pi} \chi''(\mathbf{q}, \omega) \coth \frac{\beta\omega}{2}. \tag{2.171}$$

We may therefore write, for the mean square thermal fluctuations,

$$\langle \rho_{\mathbf{q}}^{+} \rho_{\mathbf{q}} \rangle = NS_{\mathbf{q}} = -\int_{-\infty}^{\infty} \frac{d\omega}{2\pi} \chi''(\mathbf{q}, \omega) \coth \frac{\beta\omega}{2}, \tag{2.172}$$

which is one of the standard forms of the fluctuation-dissipation theorem.

If one goes over to a space-time description, one sees readily that the response of the system is specified (in the Heisenberg representation) by the commutator of the density fluctuations at two different space-time points, while the anticommutator furnishes a measure of the system fluctuations. (We leave the proof of this statement as an exercise for the formally inclined reader.) In equilibrium the thermal fluctuations in the density are determined completely once the dissipative behavior of the system is known; this is no longer the case when the system is in a nonequilibrium state.

At finite temperatures χ'' and S obey essentially the same sum rules that they obey at $T = 0$. We leave it as an exercise for the reader to show that the f-sum rule may be written as

$$2 \int_{-\infty}^{\infty} d\omega\, \omega S(\mathbf{q}, \omega) = -\frac{1}{\pi} \int_{-\infty}^{\infty} d\omega\, \omega \chi''(\mathbf{q}, \omega) = \frac{Nq^2}{m}. \tag{2.173}$$

The compressibility sum rule follows from the definition

$$\lim_{q \to 0} \chi(\mathbf{q}, 0) = -N^2 \kappa_{\mathrm{iso}} = -\frac{N}{ms_{\mathrm{iso}}^2}, \tag{2.174}$$

where κ_{iso} is the *isothermal* compressibility, and s_{iso} is the *isothermal* sound velocity, defined through Eq. (2.174). On making use of Eqs. (2.166) and (2.167), we see that

$$\lim_{q \to 0} \left[2 \int_{-\infty}^{\infty} d\omega \frac{S(\mathbf{q}, \omega)}{\omega} \right] = \lim_{q \to 0} \left[-\frac{1}{\pi} \int_{-\infty}^{\infty} d\omega \frac{\chi''(\mathbf{q}, \omega)}{\omega} \right] = \frac{N}{ms_{\mathrm{iso}}^2}. \tag{2.175}$$

The relations (2.173) and (2.175) provide useful information on the response function $\chi''(\mathbf{q}, \omega)$. We shall use them in the following section to describe the structure of χ'' in the hydrodynamic regime.

2.7. THE HYDRODYNAMIC LIMIT

HYDRODYNAMIC VS. COLLISIONLESS REGIMES

We now consider the macroscopic limit ($\mathbf{q}, \omega \to 0$) of the response functions at finite temperatures. Just as at $T = 0$, this limit is of special interest, as it provides direct information on the nature of the low-lying excitation spectrum. At first sight, one might think that for low enough temperatures, the limit of $\chi(\mathbf{q}, \omega)$ is similar to that obtained at $T = 0$. In fact, such a guess is incorrect: however small the temperature, as soon as it is finite there appears a new physical effect, namely the collisions between thermally excited quasiparticles. The corresponding collision frequency ν is of order T^2. At any finite T, the collisions will always dominate the system's behavior for small enough $\mathbf{q}$ and ω. One must thus exert considerable care in passing to the limit $(\mathbf{q}, \omega) \to 0$.

We already encountered an example of this collision-dominated regime in Section 1.8; there we found that when ω was decreased below the collision frequency ν, the "zero sound" mode of a neutral Fermi liquid was replaced by the first sound mode. A similar transition occurs for any system in which there exist random collisions. One must then clearly distinguish between

(i) a collisionless regime, in which the system behavior is essentially unaffected by collisions;
(ii) a "hydrodynamic" regime, which may be described in terms of the usual macroscopic laws of hydrodynamics and thermodynamics.

We describe briefly the main physical features of response functions in these two regimes.

Qualitatively, random collisions act to disrupt the coherent response of the system to an external field. Their influence will be dominant when the collision frequency is much lower than the frequency of the exciting field seen by a given particle. The latter frequency is modified by the Doppler effect; for a particle with velocity $\mathbf{v}_\mathbf{p}$, it is equal to $(\omega - \mathbf{q} \cdot \mathbf{v}_\mathbf{p})$, rather than to ω. (Mathematically, this feature is reflected in the form of the energy denominators involved in the perturbation expansion, which always contain a factor $(\omega - \mathbf{q} \cdot \mathbf{v}_\mathbf{p} + i/\tau)$.) Let $\gamma = \nu^{-1}$ be the collision time, and $\ell = \bar{v}\tau$ the mean free path (where $\bar{v}$ is the average quasiparticle velocity). If

$$\omega\tau \gg 1 \qquad \text{and/or} \qquad q\ell \gg 1, \tag{2.176}$$

then $1/\tau$ may be neglected as compared to $(\omega - \mathbf{q} \cdot \mathbf{v}_\mathbf{p})$ for most values of $\mathbf{p}$: the collisions play a negligible role. On the other hand, collisions are most important when

$$\omega\tau \gg 1 \qquad \text{and} \qquad q\ell \gg 1. \tag{2.177}$$

The conditions (2.177) define the hydrodynamic limit.

In the collisionless regime, a quasiparticle may build up its response for a large number of periods (or wavelengths) without suffering a collision. In the expression (2.164) for the response function, the relevant excitations are essentially the same as at zero temperature: collisions act only to broaden slightly the energy levels. The matrix elements or excitation energies may be temperature dependent, yet the structure of $\chi''(\mathbf{q}, \omega)$ is qualitatively the same. For instance, the response of neutral Fermi liquids may still be described in terms of single-pair or collective mode excitations; the energy of the excitations is somewhat blurred. Put another way, if we are in the collisionless regime, we may pass smoothly to the limit $T \equiv 0$.

The situation is completely different in the *hydrodynamic regime*, characterized by Eq. (2.177). In this case, a quasiparticle is subject to many collisions during one period (or one wavelength) of the exciting field. It is thus completely "thermalized" well before it can sample the periodicity of the field. Clearly, it is no longer possible to describe the system's response in terms of elementary excitations. Instead, we must assume that the collisions act to bring about everywhere a state of *local thermodynamic equilibrium*, taking into account the action of the external force. The state of the system is completely described by macroscopic "local" quantities (density, current, pressure, etc.); the response to the external field may be obtained by using the usual laws of thermodynamics and hydrodynamics. In this regime, the structure of response functions displays information on the average "equilibrium" properties. The singularities arising from the details of the microscopic excitation spectrum have been washed out by the collisions.

The intermediate region corresponds to

$$\omega\tau + q\ell \sim 1.$$

It is obviously very complicated, as one has lost the simplicity of both limits. The response to an external field may be viewed either as a highly damped coherent response, or as a thermodynamic response which departs appreciably from equilibrium. We have seen in Section 1.9 that this transition regime corresponds to a maximum damping of sound oscillations. By analogy, we expect that the dissipative part of the response function will be most important in that region.

At a given temperature T, however small, one always passes into the hydrodynamic regime when $\mathbf{q}$ and ω go to zero. The extreme "macroscopic limit" is thus basically different from that found at $T \equiv 0$. One must be very careful *not* to interchange the limits $T \to 0$ and $(\mathbf{q}, \omega) \to 0$. If T goes to zero first, the collision frequency $\nu \sim T^2$ vanishes: the limit of $\chi(\mathbf{q}, \omega)$ corresponds to the collisionless regime. If instead $(\mathbf{q}, \omega)$ go to zero first, one obtains the hydrodynamic value of $\chi(\mathbf{q}, \omega)$. Mathematically, we may say that $\chi(\mathbf{q}, \omega, T)$ has an essential singularity at the origin.

In this discussion, we have considered only the collisions between thermally excited quasiparticles. Actually, a hydrodynamic regime may be achieved by any sort of random collisions. In Chapter 3, for instance, we shall discuss in detail the properties of a charged electron gas interacting with random fixed impurities. At low frequency and wave vectors, the system displays characteristic "diffusive" behavior, which may be described by macroscopic equations. In a metal, the electrons may also collide with phonons, etc. In all these cases, what is really important are the collision frequency ν and the nature of the quantities which are *conserved* in the collisions (particle number, and possibly momentum).

Strictly speaking, there is no "macroscopic" limit for response functions in the collisionless regime. However, at low enough temperatures, such that $1/\tau \ll \mu$ (where μ is the chemical potential), one may achieve the following conditions:

$$\frac{1}{\tau} \ll \omega,\ qv_F \ll \mu. \tag{2.178}$$

One is then in the collisionless region; yet the frequency and wave vector are still much smaller than the corresponding atomic quantities. The response therefore occurs on a macroscopic scale. The corresponding response functions will be very similar to those found at $T = 0$ (see, for instance, Sections 2.4 and 2.5 in the case of neutral Fermi liquids); this limit needs no further discussion here.

We now focus our attention on the hydrodynamic limit. We shall show how the response functions may be calculated either by microscopic arguments or from hydrodynamic equations; we shall further compare our results with the value of $\chi(\mathbf{q}, \omega)$ found in the collisionless regime.

HYDRODYNAMIC LIMIT OF THE DENSITY-DENSITY RESPONSE FUNCTION FOR A NEUTRAL FERMI LIQUID

We consider a neutral Fermi liquid at very low temperature ($\kappa T \ll \mu$), such that the Landau theory is applicable. The collision frequency, $\nu \sim (\kappa T)^2/\mu$, is then very small. The system is subject to an external density probe, characterized by a scalar potential $\varphi(\mathbf{q}, \omega)$; the corresponding transport equation for quasiparticles is given by Eq. (2.104). In the hydrodynamic limit, the collision integral plays an essential role, in contrast to the collisionless regime described in Section 2.5.

In view of the complexity of the collision integral (see Section 1.8), the solution of Eq. (2.104) appears hopeless. Actually, we may bypass the difficulty in the *extreme* hydrodynamic limit ($\omega\tau$, $ql \ll 1$) by using the artifice introduced in Section 1.9 to describe first sound. Since the collisions act to restore local thermodynamic equilibrium, they will damp any fluctuation $\delta n_{\mathbf{p}}$ in the quasiparticle distribution. Such damping must, however, remain consistent with the fundamental laws of particle number and momentum conservation. Thus, if we expand $\delta n_{\mathbf{p}}$ in a series of spherical harmonics,

$$\delta n_{\mathbf{p}} = \sum_{\ell=0}^{\infty} \delta n_\ell P_\ell(\cos\theta), \tag{2.179}$$

the $\ell = 0$ and $\ell = 1$ components cannot be affected by collisions. In the extreme hydrodynamic limit, we expect $\delta n_{\mathbf{p}}$ to contain only these first two components; the other terms, corresponding to $\ell \geq 2$, are "damped out" by the collisions. To put it another way, the main effect of collisions is to *truncate* $\delta n_{\mathbf{p}}$, by killing all the nonconserved components.

The solution of Eq. (2.104) is now straightforward. We expand the equation in a series of spherical harmonics (taking $\mathbf{q}$ as the polar axis), and we keep only the $\ell = 0$ and $\ell = 1$ components. We thus obtain

$$\begin{aligned} \frac{qv_F}{3}\,\delta\bar{n}_1 - \omega\delta n_0 &= 0, \\ qv_F\delta\bar{n}_0 - \omega\delta n_1 - qv_F\frac{\partial n^0}{\partial\epsilon_p}\varphi(\mathbf{q},\omega) &= 0. \end{aligned} \tag{2.180}$$

In these equations the collision integral has disappeared, because of number and momentum conservation [the first of Eqs. (2.180) is nothing but the continuity equation (1.88)]. On making use of Eq. (1.34) to relate δn_ℓ and $\delta\bar{n}_\ell$, we find

$$\delta n_0 = \frac{(\partial n^0/\partial\epsilon_p)\varphi(\mathbf{q},\omega)}{1 + F_0{}^s - 3\omega^2/q^2v_F{}^2(1 + F_1{}^s/3)}. \tag{2.181}$$

From Eq. (2.181), we obtain the induced particle density by means of Eq. (2.107), and thus the density response function. On referring to Eqs. (1.61) and (1.100), we may write the result as

$$\chi(\mathbf{q},\omega) = -\frac{Nq^2/m}{s^2q^2 - \omega^2}. \tag{2.182}$$

The form (2.182), corresponding to the hydrodynamic limit, is quite different from the result (2.111) found in the collisionless regime. Note that it is only valid at very low temperatures, such that $\kappa T \ll \mu$.

Let us examine more closely the result (2.182); we note several important features:

(i) $\chi(\mathbf{q}, \omega)$ has a pole at the first sound frequency. This was to be expected, since first sound is after all a collective density mode. What is perhaps more surprising is that $\chi(\mathbf{q}, \omega)$ has no other singularity. The spectral density, defined by Eq. (2.166), reduces to a single discrete peak:

$$\chi''(\mathbf{q}, \omega) = \frac{\pi}{2} \frac{Nq}{ms} [\delta(\omega + sq) - \delta(\omega - sq)]. \tag{2.183}$$

Equation (2.183) is to be contrasted with the result found in the collisionless regime: there $\chi''(\mathbf{q}, \omega)$ had a discrete peak at the zero sound frequency and a *continuum* corresponding to single-pair excitations; the latter has been washed out by collisions. Let us emphasize that Eq. (2.183) is only valid at very low T; at higher temperatures, the structure of χ'' is complicated by the appearance of thermal diffusion. This point is briefly discussed in the following paragraph.

(ii) $\chi''(\mathbf{q}, \omega)$, given by Eq. (2.183), clearly obeys the f-sum rule (2.173). If we further notice that at low temperatures $s = s_{\text{iso}}$, we see that $\chi''(\mathbf{q}, \omega)$ also satisfies the compressibility sum rule (2.174) (as it should). Indeed, if we guess at the beginning that $\chi''(\mathbf{q}, \omega)$ reduces to a single discrete peak, the result (2.183) may be obtained directly by application of the two sum rules (the calculation is left as an exercise to the reader). We note that the "homogeneous" limit of $\chi(\mathbf{q}, \omega)$, corresponding to $\omega \gg qv_F$, is the same in the collisionless and hydrodynamic regimes:

$$\chi(\mathbf{q}, \omega) \sim \frac{Nq^2}{m\omega^2} \quad \text{if} \quad \omega \gg qv_F \tag{2.184}$$

[compare with Eq. (2.133)]. This result is a direct consequence of the f-sum rule.

(iii) Finally, we remark that the *static* response ($\omega = 0$) is the same as in the collisionless regime

$$\chi(\mathbf{q}, 0) = -\frac{N}{ms^2} \tag{2.185}$$

[compare with Eq. (2.124)]. This result is in fact very important, as it corresponds to a general feature of all interacting systems: the response to a static probe is unaffected by collisions (as long as $1/\tau \ll \mu$). Such a property is easily understood if we realize that a static probe leads to a new *equilibrium state*, distorted by the probe potential. Since the system remains at equilibrium, collisions are not expected to play any major role.

We have made implicit use of result (iii) in our derivation of the compressibility sum rule, (2.84). There we calculated the collisionless response to a static probe by means of hydrostatic arguments. This *a priori* questionable procedure is actually correct, since $\chi(\mathbf{q}, 0)$ is the same in both the collisionless and hydrodynamic regimes. Similarly, we are now in a position to answer a question which the alert reader may have raised in connection with Chapter 1. There, we discussed the stability of the ground state, on the one hand by studying unstable collective modes, on the other by requiring that the free energy be minimum. The latter argument is typically of a thermodynamic nature, while the former refers to the collisionless regime. It is not immediately obvious that these two different approaches should lead to the same stability criterion. In order to answer that paradox, we note that a collective mode becomes unstable when its frequency goes to zero; thus the instability threshold for zero sound corresponds to $\chi(\mathbf{q}, 0) = \infty$ in the collisionless regime. On the other hand, the free energy is minimum with respect to density fluctuations if the compressibility $\kappa > 0$; the instability threshold corresponds to $1/\kappa = -(1/\Omega)(\partial P/\partial\Omega) = 0$, and thus again to $\chi(\mathbf{q}, 0) = \infty$ but here in the hydrodynamic regime [compare with Eq. (2.83)]. Since $\chi(\mathbf{q}, 0)$ is independent of $q\ell$, the two criteria do in fact agree. This conclusion is easily extended to any other instability (spin waves, etc.).

It should be emphasized that it is not necessary to go through a solution of the transport equation in order to find the hydrodynamic limit of response functions. The latter may be obtained more simply by solving the appropriate hydrodynamic equations. For instance, the linear response to a scalar potential φ may be derived by application of the following macroscopic laws:

$$
\begin{aligned}
N \operatorname{div} \mathbf{v} + \frac{\partial \rho}{\partial t} &= 0, \\
Nm \frac{\partial \mathbf{v}}{\partial t} &= -\mathbf{grad}\, P - N\, \mathbf{grad}\, \varphi, \qquad (2.186)\\
\frac{\partial P}{\partial \rho} = \frac{1}{\kappa \rho} &= ms^2,
\end{aligned}
$$

where $\mathbf{v}(\mathbf{r}, t)$ is the average *local velocity* of the fluid, ρ the density, and P the pressure [cf. Eq. (1.49)]. By taking the Fourier transform of Eqs. (2.186), we recover the earlier result (2.182).

The system's behavior in the hydrodynamic limit is therefore fully described by the usual macroscopic equations, which in fact are obtained by taking the appropriate *moments* of the full transport equation. Our major purpose in solving Eq. (2.104) in the hydrodynamic limit was to stress the differences with the (more complicated) solution in the collisionless case.

THE INFLUENCE OF THERMAL DIFFUSION

The preceding results are only valid in the limit of very low temperatures, when the thermal and mechanical degrees of freedom are decoupled. In this limit, isothermal and adiabatic processes are essentially equivalent. It is for that reason that our result, (2.182), is so simple.

As the temperature is raised, the adiabatic sound velocity, s, departs from its isothermal counterpart s_{iso}. More specifically, one has

$$\frac{s^2}{s_{\text{iso}}^2} = \gamma, \tag{2.187}$$

where $\gamma = C_p/C_v$ is the specific heat ratio. Furthermore, density fluctuations are now influenced by thermal diffusion processes, as a result of the coupling between mechanical and thermal degrees of freedom. This coupling is of order $(\gamma - 1)$; to that order, thermal diffusion should affect the density-density response function $\chi(\mathbf{q}, \omega)$. A detailed calculation of the correlation functions in this case has been carried out by Kadanoff and Martin (1963); it is based on the usual laws of thermodynamics. Because the algebra is cumbersome, we quote only a few salient features of their results.

As expected, the spectral density $\chi''(\mathbf{q}, \omega)$ contains a peak at the first sound frequency $\omega = sq$. In addition, $\chi''(\mathbf{q}, \omega)$ possesses a continuous background arising from thermal diffusion; this contribution extends over all frequencies and is proportional to

$$\operatorname{Im}\left[\frac{D_T q^2}{i\omega + D_T q^2}\right] = \frac{D_T q^2 \omega}{\omega^2 + [D_T q^2]^2}, \tag{2.188}$$

where D_T is the thermal diffusion coefficient. The strength of the contribution (2.188) is of order $(\gamma - 1)$, as expected. We note that the first sound peak is immersed in the above continuum. Thermal diffusion thus gives rise to a *damping* of first sound, and to a broadening of the corresponding peak in $\chi''(\mathbf{q}, \omega)$. This broadening is of order $(\gamma - 1)$, and vanishes when $T \to 0$.

By using the sum rules, (2.173) and (2.174), one may estimate the strength of the thermal diffusion pole in $\chi''(\mathbf{q}, \omega)$ (and the damping of first sound). On combining the two sum rules, we may write

$$\begin{aligned} -\frac{1}{\pi}\int_{-\infty}^{+\infty} \chi''(\mathbf{q}, \omega)\left[\frac{s^2 q^2}{\omega} - \omega\right] d\omega &= \frac{Nq^2}{m}\left[\frac{s^2}{s_{\text{iso}}^2} - 1\right] \\ &= \frac{Nq^2}{m}(\gamma - 1). \end{aligned} \tag{2.189}$$

Let us write $\chi''(\mathbf{q}. \omega)$ in the form

$$\chi''(\mathbf{q}, \omega) = z_q\{\delta(\omega + sq) - \delta(\omega - sq)\} + \chi_{\text{th}}, \tag{2.190}$$

where the continuous term χ_{th} arises from thermal diffusion. Clearly, the integral (2.189) only involves χ_{th}. Since χ_{th} is a smooth function of ω, it follows that the strength of χ_{th} is of order $(\gamma - 1)$, as predicted. (This discussion provides a good example of how sum rules can be used to obtain simple qualitative results.)

In conclusion, let us emphasize that the response function formalism is equally valid in the collisionless and hydrodynamic regimes. All the general theorems, dealing with sum rules, causality, etc., can be applied in either limit. What is affected by the transition from one regime to the other is not the validity of the method, but only the structure of the results. We thus appreciate the usefulness of response function methods, which provide a unified description of the system properties under widely different conditions.

2.8. CONCLUSION

Our main purpose in this chapter has been to introduce the reader to the proper *language* in which to express with precision results of both theory and experiment concerning the response of a many-particle system to macroscopic and microscopic external probes. In this response-function language, the connections between macroscopic correlations and microscopic excitations are made evident. One may use the language both to analyze the behavior of a given quantum liquid, and to compare the behavior of different liquids. We have considered a particular example, that of a scalar probe coupled to the density, and have analyzed the behavior of the density fluctuation of a neutral Fermi liquid. The approach is, however, completely general; it applies equally well to a vector probe and current fluctuations, a spin-dependent probe and spin-density fluctuations, etc.

We have seen that the structure of $S(\mathbf{q}, \omega)$ in the long wavelength limit for a normal Fermi liquid at zero temperature may be understood qualitatively on the basis of general arguments. A rough evaluation of the matrix elements and excitation energies provides information on the relative importance of the different types of excitations created by a density fluctuation, namely single-pair, multipair, and collective excitations. Such an analysis is possible only when $\mathbf{q} \to 0$ (in fact, it is only in that region that the various contributions to $S(\mathbf{q}, \omega)$ can be sharply distinguished). An equivalent approach can be used for systems other than a normal Fermi liquid: it will, of course, yield different results, depending on the structure of the corresponding excitation spectrum. The case of a superfluid Bose liquid will be discussed in Chapter 6.

To illustrate the formal theory of linear response, we have used the Landau theory to calculate the macroscopic density-density response function of a neutral Fermi liquid. The formal approach of the present chapter shows clearly why the Landau theory is rigorous in the long wavelength limit: as we pointed out earlier, the Landau theory is equivalent to *neglecting multipair excitations* in the calculation of the response function. (The latter are difficult to describe, and always require a microscopic theory.) In the light of the general discussion presented in this chapter, we see that the Landau theory succeeds because in the limit of small $\mathbf{q}$ and ω, the contribution of multipair excitations is usually negligible. When this is not true (for instance, in the f-sum rule for a nonhomogeneous system), we expect the Landau theory answer to be erroneous. In such cases, one may combine the two approaches in order to obtain exact results. More generally, a study of the interrelation between the two methods is always very fruitful.

Finally, by studying the structure of the linear response function, one may describe *explicitly* the transition from the *collisionless* high-frequency regime (where the system response is essentially microscopic), to the low-frequency *hydrodynamic* regime (where the response is governed by macroscopic average laws). Such a transition is an essential feature of quantum liquids at finite temperature, which we shall meet again when describing the properties of a Bose liquid.

PROBLEMS

2.1. Calculate the system's response to a space- and time-dependent magnetic field, $\mathcal{H}(\mathbf{r}, t)$. Show that the Fourier transform of the induced magnetization, $\mathfrak{M}(\mathbf{q}, \omega)$, may be written as

$$\mathfrak{M}(\mathbf{q}, \omega) = \chi_\sigma(\mathbf{q}, \omega)(g\beta\sigma)^2\mathcal{H}(\mathbf{q}, \omega),$$

where $\chi_\sigma(\mathbf{q}, \omega)$ is the *spin-spin correlation function*. Calculate χ_σ in terms of the exact system's eigenstates.

Show that $\lim_{\mathbf{q}\to 0} [\chi_\sigma(\mathbf{q}, 0)]$ is related to the spin susceptibility χ_P. Calculate $\chi_\sigma''(\mathbf{q}, \omega)$, and show that it satisfies an f-sum rule in the absence of a magnetic field.

Calculate the various limits of $\chi_\sigma(\mathbf{q}, \omega)$ when $(\mathbf{q}, \omega) \to 0$ by using the Landau theory of Fermi liquids.

2.2. Calculate the current-current response tensor, $\chi_{\alpha\beta}(\mathbf{q}, \omega)$, in terms of the exact system's eigenstates. By using particle conservation, (2.27), together with the f-sum rule, show that

$$\sum_{\alpha\beta} q_\alpha q_\beta \chi_{\alpha\beta}(\mathbf{q}, \omega) + \frac{Nq^2}{m} = \omega^2\chi(\mathbf{q}, \omega).$$

2.3. Evaluate $\chi''(\mathbf{q}, \omega)$ for a system of noninteracting fermions at finite temperature.

REFERENCES

Brockhouse, B. N. (1960), *Proc. Symp. Inelastic Scattering of Neutrons in Solids and Liquids*, Intern. At. Energy Agency, Vienna, 1960.

Callen, H. B. and Welton, T. R. (1951), *Phys. Rev.* **83**, 34.

Kadanoff, L. P. and Martin, P. C. (1962), *Ann. Phys.* **24**, 419.

Kramers, H. C. (1927), *Atti. Congr. Intern. Fisica, Como* **2**, 545. See *Collected Scientific Papers*, North-Holland, Amsterdam.

Kronig, R. (1926), *J. Opt. Soc. Am.* **12**, 547.

Kubo, R. (1956), *Can. J. Phys.* **34**, 1274.

Kubo, R. (1957), *J. Phys. Soc. Japan* **12**, 570.

Nozières, P. and Pines, D. (1958), *Phys. Rev.* **109**, 741.

Nyquist, H. (1928), *Phys. Rev.* **32**, 110.

Placzek, G. (1952), *Phys. Rev.* **86**, 377.

CHAPTER 3

CHARGED FERMI LIQUIDS

We now begin our study of systems of interacting electrons. For simplicity we shall initially confine our attention to a homogeneous system. In order to guarantee its stability, we assume that the electrons are immersed in a uniform background of positive charge, of density equal to the average electron density. For the high electron densities and low temperatures in which we shall be interested, the noninteracting electrons obey Fermi–Dirac statistics. The system may be regarded as a *quantum plasma*, in analogy to the usual classical plasmas, for which the noninteracting electrons are described by a Maxwell–Boltzmann distribution. The quantum plasma is likewise the natural analog, for charged particles, of the neutral Fermi liquid we have thus far considered; it may equally well be regarded as a charged Fermi liquid.

Quantum plasmas are interesting in themselves. They also serve as a useful model for the behavior of electrons in the conduction band of simple metals. For many purposes, the effect of the periodic array of ions in the latter systems may be well approximated by the uniform charge background.

The Coulomb interaction between a pair of electrons falls off slowly with distance; it is a *long-range* interaction. As a result a charged Fermi liquid differs appreciably from its neutral counterpart. The present chapter is primarily devoted to an elucidation of the new physical features introduced by the Coulomb interaction, and their description in terms of Silin's generalization of the Landau theory of the Fermi liquid [Silin (1957)].

In Section 3.1 we consider briefly the divergences (present in elementary calculations) that originate in the long range of the Coulomb interaction; we introduce the important physical concepts of screening and plasma oscillation. We show in the following section, 3.2, that the dielectric response function offers a natural and precise description of these concepts. In Section 3.3 we derive the macroscopic transport equation for the quantum plasma; we contrast this transport equation with the Landau equation of Chapter 1, and discuss the relationship between plasma oscillations and zero sound.

The transport equation is applied to a calculation of the macroscopic dielectric response of a charged Fermi liquid in Section 3.4. Following a discussion of backflow in Section 3.5, the transport equation is generalized to describe the response to an external electromagnetic field in Section 3.6. There exist certain important differences between the longitudinal and transverse response of a quantum plasma, both as regards the physical processes which occur, and (to a lesser extent) their mathematical description. These are likewise considered in this section. Impurity scattering is studied in Section 3.7, while the extent to which the Landau theory is applicable to a description of electrons in metals is analyzed once more in the concluding section, 3.8.

The present chapter may be considered to be the analog of Chapter 1, in that important physical concepts are introduced and described in the macroscopic limit, while microscopic considerations are not pursued in any detail. Such microscopic considerations for the charged Fermi liquid, together with the more formal description of the dielectric functions, and their application to a number of other problems of physical interest, are reserved for Chapter 4.

3.1. SCREENING AND PLASMA OSCILLATION: AN ELEMENTARY INTRODUCTION

DIVERGENCES

The early theoretical treatments of quantum plasmas were haunted by the appearance of divergences. For example, one finds a logarithmic divergence if one carries out a second-order perturbation-theoretic calculation of the ground state energy of a quantum plasma. A similar divergence appears in the Hartree–Fock approximation calculation of the group velocity of a particle on the Fermi surface. Again, one finds that the interaction energy, $f_{\mathbf{pp}'}$, of quasiparticles in an inhomogeneous charged Fermi liquid is divergent; the transport equation developed for a system with short-range particle interactions is clearly inapplicable. Such a divergence must be "cured" before one can hope to develop an analog of the Landau theory for a charged Fermi liquid.

The origin of the aforementioned divergences is the long range of the Coulomb interaction between electrons. A given electron interacts not just with a few near neighbors, but with a *very* large number of other electrons, so that its motion cannot easily be decoupled from that of its neighbors. Looked at in momentum space, the difficulties arise in the treatment of the low momentum transfer part of the interaction. The Fourier transform of the Coulomb interaction is

$$V_{\mathbf{q}} = \int d^3\mathbf{r} \exp\left[-i\mathbf{q}\cdot\mathbf{r}\right]\frac{e^2}{r} = \frac{4\pi e^2}{q^2}.$$

One finds, in any low-order perturbation-theoretic calculation, a piling up of factors $1/q^2$, which gives rise to logarithmic, or higher-order, divergences.

How one might, in principle, "cure" such divergences has also been known for some time. Physically, the *effective* interaction between electrons in a plasma (quantum or classical) is not a particularly long-range one. When an electron moves, it tends to push other electrons out of its way, as a result of their mutual Coulomb repulsion. We may say that the electron moves surrounded by a "screening hole" (the region in which one is not likely to find another electron). The screening hole corresponds to a distribution of positive charge, which, taken as a whole, tends to compensate the negative charge of the electron in question. The electrostatic field of the electron is thus screened at large distances.

As the electron moves, it tends to carry along the screening hole. Note that the *dynamic* screening around a moving charge will certainly be different from the *static* screening about a fixed charge. (In other words, the screening hole possesses inertia.) Indeed, the central problem in developing a divergence-free theory of electron systems is that of introducing the concept of dynamic screening in consistent fashion. It is, nonetheless, illuminating to study briefly the screening of a fixed charge, a question to which we now turn our attention.

STATIC SCREENING

The first theoretical treatments of screening dealt with the response of classical and quantum plasmas to a fixed external charge [Debye and Hückel (1923), Mott and Jones (1936)]. Let us consider a charge z introduced at the origin of a quantum plasma. The electrostatic potential $\varphi(\mathbf{r})$ felt by an electron far from the origin will not be simply that due to z alone. The external charge acts to polarize the electrons in its immediate vicinity; a distant electron therefore responds both to the external charge and the induced polarization charge, $e\langle\rho(\mathbf{r})\rangle$. The effective potential and the polarization charge are related by Poisson's equation:

$$\nabla^2\varphi(\mathbf{r}) = -4\pi\{z\delta(\mathbf{r}) + e\langle\rho(\mathbf{r})\rangle\}. \tag{3.1}$$

The induced charge density, $e\langle\rho(\mathbf{r})\rangle$, may be calculated with the aid of the *Fermi–Thomas* approximation [Mott and Jones (1936)]. In that approximation, the chemical potential of the electrons is regarded as the sum of a potential energy $e\varphi(\mathbf{r})$ and of a kinetic energy $\epsilon_F{}^o$ given by the relation appropriate to a system of noninteracting electrons:

$$N = \frac{(2m\epsilon_F{}^o)^{3/2}}{3\pi^2}. \tag{3.2a}$$

At equilibrium, the chemical potential must be constant: hence $\epsilon_F{}^o$ is a function of position. At a given point $\mathbf{r}$, we have

$$\epsilon_F{}^o(\mathbf{r}) = \epsilon_F{}^o - e\varphi(\mathbf{r}). \tag{3.2b}$$

The fluctuation in $\epsilon_F{}^o(\mathbf{r})$ implies a fluctuation in the particle density N. Up to first order in this fluctuation, we have

$$\langle\rho(\mathbf{r})\rangle = \frac{3}{2}\frac{Ne\varphi(\mathbf{r})}{\epsilon_F{}^o}. \tag{3.3}$$

By introducing Eq. (3.3) into (3.1), and taking the Fourier transform of the resulting equation, one then obtains

$$(q^2 + q_{FT}^2)\varphi_{\mathbf{q}} = 4\pi z \tag{3.4}$$

and

$$\varphi(\mathbf{r}) = \frac{z}{r}\exp(-q_{FT}r), \tag{3.5}$$

where the so-called Fermi–Thomas screening wave vector is given by

$$q_{FT} = \left(\frac{6\pi Ne^2}{\epsilon_F{}^o}\right)^{1/2} \tag{3.6}$$

We see in Eq. (3.5) that the field of the external charge is effectively screened within a distance of the order of $\lambda_{FT} = q_{FT}^{-1}$. We further remark that the screening length is determined by the competition between the influence of the potential energy and kinetic energy on the motion of the electrons. The interaction between the electrons acts to bring about screening; the kinetic energy of the electrons, which represents their essentially random motion, opposes it. One finds easily that the screening length, λ_{FT}, measured in units of the interparticle spacing,

r_o, is proportional to the square root of the ratio of the average kinetic and potential energies:

$$\frac{\lambda_{FT}}{r_o} = \left[\left(\frac{8\pi}{9}\right)\frac{\epsilon_F{}^{o}}{e^2/r_o}\right]^{1/2}, \tag{3.7}$$

where r_o is given by

$$N = \frac{3}{4\pi r_o{}^3}.$$

For an electron gas at metallic densities, the potential energy per particle, e^2/r_o, is essentially comparable to the kinetic energy per particle, so that λ_{FT} is comparable to r_o. Screening is thus seen to be quite effective in quantum plasmas of physical interest.

It should be emphasized that the Fermi–Thomas approximation is not exact. First of all, it is justified only when the inhomogeneities in the problem correspond to distances which are long compared to the interparticle spacing; otherwise the assumption of a local relationship, (3.2a), between particle density and Fermi energy is not justified. Thus the Fermi–Thomas approximation can properly be applied only to the calculation of the response of the electrons to that part of the external potential that represents a slow spatial variation; one should therefore state the results of the Fermi–Thomas analysis as

$$\varphi_{\mathbf{q}} = \frac{4\pi z}{q^2 + q_{FT}^2} \qquad (qr_o \ll 1). \tag{3.8}$$

Second, even in this limit the analysis is not exact. The relation, (3.2a), is that appropriate to noninteracting electrons; we therefore expect that it will be valid only in the limit of a very high-density electron system, for which the interaction between the electrons may properly be regarded as weak (see Chapter 5).

DYNAMIC SCREENING

The above calculations, while suggestive, cannot be applied directly to the screening of electron-electron interactions in plasmas because a moving electron does not, in general, give rise to a nearly static charge disturbance in the plasma. What is required is a time-dependent version of the self-consistent calculation we have described. To be specific, suppose the charge z which is introduced into the plasma moves with some velocity $\mathbf{V}_e$. In that case, the charge density it produces is given by

$$z\rho_e(\mathbf{r}, t) = z\delta(\mathbf{r} - \mathbf{V}_e t) \tag{3.9}$$

(if the charge is assumed to be at the origin at time $t = 0$). Again we might expect that the external charge will polarize the electrons in its neighborhood. The resultant electronic polarization charge, $e\langle\rho(\mathbf{r},t)\rangle$, is now time dependent. The effective potential and polarization charge are related by a time-dependent Poisson equation:

$$\nabla^2\varphi(\mathbf{r}, t) = -4\pi[z\delta(\mathbf{r} - \mathbf{V}_e t) + e\langle\rho(\mathbf{r}, t)\rangle]. \tag{3.10}$$

Let us take the Fourier transform in space and time of Eq. (3.10); we find

$$\varphi(\mathbf{q}, \omega) = \frac{4\pi}{q^2}\{z2\pi\delta(\omega - \mathbf{q}\cdot\mathbf{V}_e) + e\langle\rho(\mathbf{q}, \omega)\rangle\}. \tag{3.11}$$

The problem reduces to that of determining the frequency and wave vector–dependent polarization density, $\langle\rho(\mathbf{q}, \omega)\rangle$. This quantity offers a direct measure of the *dynamic screening* in the electron gas.

The above problem serves as a prototype for a consideration of the screening of electron-electron interactions. Qualitatively, each electron in the system behaves like a moving test charge. It acts to polarize its surroundings. Another electron sees the electron *plus* its accompanying time-dependent polarization cloud; the effective interaction between the electrons is thus dynamically screened. By using such a concept of dynamic screening, it is not difficult to reformulate the transport equation of the Landau theory of Fermi liquids, so that it can be applied to quantum plasmas without difficulty. In this fashion, one can obtain a rigorous expression for the response of a quantum plasma to an external electric field that varies slowly in space and time (such being the limitations of the Landau theory). Moreover, one finds quite generally that inclusion of dynamic screening serves to remove the divergences in the ground state energy and quasiparticle velocity calculations discussed earlier.

PLASMA OSCILLATION

A second important physical phenomenon which is characteristic of homogeneous electron systems is a collective oscillation of the electrons as a whole, the plasma oscillation.* The existence of organized oscillations in the plasma is complementary to the existence of screening. When the electrons move to screen a charge disturbance in the plasma, they will, in general, tend to overshoot the mark somewhat. They are consequently pulled back toward that region, overshoot again, etc., in

* Organized oscillations in classical plasmas were first studied by Langmuir (1928) and Tonks and Langmuir (1929); the pioneering investigations of their quantum counterpart in metals were carried out by Pines (1950) and Pines and Bohm (1952).

such a way that an oscillation is set up about the state of charge neutrality. The restoring force responsible for the oscillation is simply the average self-consistent field of all the electrons, exactly the same mechanism as was operative for zero sound. Because of the long range of the Coulomb force, the frequency of the oscillations is very high. For long wavelengths it is very nearly constant, and equal to

$$\omega_p = \left(\frac{4\pi N e^2}{m}\right)^{1/2}. \tag{3.12}$$

An elementary derivation of Eq. (3.12) may be given along the following lines. Let us suppose that a charge imbalance is established in the plasma by the displacement of a slab of charge of thickness d by a small distance x ($x \ll d$), as shown in Fig. 3.1. In that case, the slab behaves

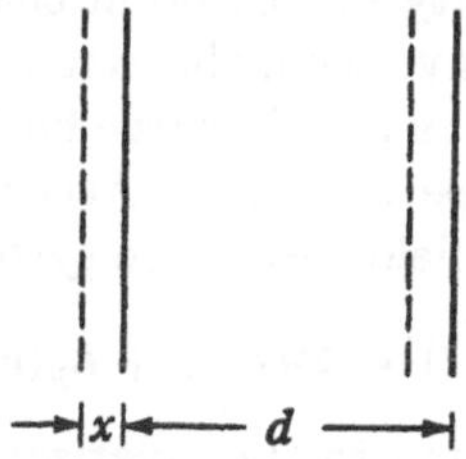

FIGURE 3.1. *Displacement of a slab of charge.*

like a condenser; a constant electric field, $\mathcal{E}$, is set up which acts to restore charge neutrality. The magnitude of the field is

$$\mathcal{E} = -4\pi N e x, \tag{3.13}$$

since the surface charge on either end of the condenser is simply Nex. An electron inside the slab obeys the equation of motion

$$m\ddot{x} = e\mathcal{E} = -4\pi N e^2 x \tag{3.14}$$

with the result that the slab will oscillate at frequency ω_p. Such a simple derivation neglects altogether the random motion of the electrons, an approximation which turns out to be justified in the limit of a very long wavelength oscillation.

Plasma oscillations, like screening, represent a typical polarization phenomenon in the plasma. Indeed both effects can be easily described within the framework of a general treatment of polarization, or dielectric processes, in an electron system. We therefore proceed to develop such a theory, by following in detail the response of the plasma to an external longitudinal field. The formalism so developed will, in addition, enable us to extend, in straightforward fashion, the Landau theory to the case of a charged Fermi liquid.

3.2. DIELECTRIC RESPONSE FUNCTION

RESPONSE TO A LONGITUDINAL FIELD

Let us suppose an external longitudinal electric field, $\mathfrak{D}(\mathbf{r}, t)$, is applied to an electron system. The vector $\mathfrak{D}$ is the "dielectric displacement" defined in elementary electrostatics. It satisfies Poisson's equation

$$\operatorname{div} \mathfrak{D}(\mathbf{r}, t) = 4\pi z \rho_e(\mathbf{r}, t), \tag{3.15}$$

where $z\rho_e(\mathbf{r}, t)$ is the density of "external charge" introduced into the gas at a point $\mathbf{r}$. As we have seen, the external field will act to polarize the electron system. The induced charge fluctuations may be regarded as producing a *space charge* field, $\mathcal{E}_p$. According to the usual laws of electrostatics, the electric field inside the system is equal to

$$\mathcal{E}(\mathbf{r}, t) = \mathfrak{D}(\mathbf{r}, t) + \mathcal{E}_p(\mathbf{r}, t), \tag{3.16}$$

where $\mathcal{E}_p(\mathbf{r}, t)$ may be related to the polarization charge density by the equation

$$\operatorname{div} \mathcal{E}_p(\mathbf{r}, t) = 4\pi e \langle \rho(\mathbf{r}, t) \rangle. \tag{3.17}$$

When dealing with dielectrics, one must modify Eq. (3.17) in order to take into account a surface term, due to the accumulation of charge at the boundary of the sample. Such a term does not enter in the present case, because we assume the system to be closed on itself (for instance, by means of an external "perfectly conducting" wire which connects the ends perpendicularly to $\mathcal{E}$).

We may combine Eqs. (3.15)–(3.17) to write

$$\operatorname{div} \mathcal{E}(\mathbf{r}, t) = 4\pi \{ z\rho_e(\mathbf{r}, t) + e\langle \rho(\mathbf{r}, t) \rangle \} \tag{3.18}$$

as the equation that relates the electric field to the external and induced charge fluctuations. Let us take the Fourier transform in space and time of Eqs. (3.15) and (3.18):

$$i\mathbf{q} \cdot \mathfrak{D}(\mathbf{q}, \omega) = 4\pi z \rho_e(\mathbf{q}, \omega), \tag{3.19}$$
$$i\mathbf{q} \cdot \mathcal{E}(\mathbf{q}, \omega) = 4\pi \{ z\rho_e(\mathbf{q}, \omega) + e\langle \rho(\mathbf{q}, \omega) \rangle \}. \tag{3.20}$$

For a macroscopic external field, these equations express the usual laws of electrostatics for a dielectric material; we now extend them to a *microscopic* level by considering them as applicable for all wave-vectors $\mathbf{q}$

and frequency ω, corresponding to fields which vary arbitrarily rapidly in both space and time. (3.20) thus serves as a definition of the electric field $\mathcal{E}(\mathbf{q}, \omega)$ on a scale for which the usual macroscopic definition does not hold.

We now make a key assumption: that the dielectric response of the electrons, $\langle\rho\rangle$, is proportional to the applied field $\mathfrak{D}$. This will be the case if the external field is sufficiently weak: we thus assume that in computing the system response one can neglect coupling terms proportional to $\mathfrak{D}^2$, etc. It follows at once that $\mathcal{E}$ will be proportional to $\mathfrak{D}$. Both $\mathfrak{D}$ and $\mathcal{E}$ are purely longitudinal fields. We may therefore write, again in analogy to the usual laws of electrostatics,

$$\mathcal{E}(\mathbf{q}, \omega) = \frac{\mathfrak{D}(\mathbf{q}, \omega)}{\epsilon(\mathbf{q}, \omega)}. \tag{3.21}$$

Here $\epsilon(\mathbf{q}, \omega)$, the *frequency and wave vector–dependent dielectric constant*, is the natural generalization (to fields which vary in space and time) of the static homogeneous dielectric constant of electrostatics. We see in Eq. (3.16) that $\epsilon(\mathbf{q}, \omega)$ furnishes a direct measure of the dielectric response of the electron system; it tells us the extent to which the external field, $\mathfrak{D}$, is screened by the electronic polarization that it induces.

Two useful expressions for $\epsilon(\mathbf{q}, \omega)$ may be obtained directly from Eqs. (3.19)–(3.21):

$$\epsilon(\mathbf{q}, \omega) = 1 + \frac{4\pi ie\langle\rho(\mathbf{q}, \omega)\rangle}{\mathbf{q}\cdot\mathcal{E}(\mathbf{q}, \omega)}, \tag{3.22}$$

$$\frac{1}{\epsilon(\mathbf{q}, \omega)} - 1 = -\frac{4\pi ie\langle\rho(\mathbf{q}, \omega)\rangle}{\mathbf{q}\cdot\mathfrak{D}(\mathbf{q}, \omega)}. \tag{3.23}$$

These serve to determine ϵ once the appropriate relationship between $\langle\rho\rangle$ and $\mathcal{E}$ or $\mathfrak{D}$ is established.

SCREENING AND PLASMA OSCILLATION

We see that $\epsilon(\mathbf{q}, \omega)$ furnishes a natural description of the linear response of the plasma to an external field. As an example, consider the case, discussed in the preceding section, of a static charge of strength z located at the origin. The Fourier transform, in space, of the net potential, $\varphi(\mathbf{r})$, is given by

$$\varphi(\mathbf{q}) = \frac{4\pi z}{q^2\epsilon(\mathbf{q}, 0)}. \tag{3.24}$$

The static wave vector–dependent dielectric constant, $\epsilon(\mathbf{q}, 0)$, describes the screening action of the electron system. In the Fermi–Thomas approximation one finds, on comparing Eqs. (3.24) and (3.8),

$$\epsilon_{FT}(\mathbf{q}, 0) = 1 + (q_{FT}^2/q^2) \qquad (qr_o \ll 1). \tag{3.25}$$

If we now suppose the charge z to move at some velocity, $\mathbf{V}_e$, we would write, in place of Eq. (3.11),

$$\varphi(\mathbf{q}, \omega) = \frac{4\pi z 2\pi\delta(\omega - \mathbf{q}\cdot\mathbf{V}_e)}{q^2\epsilon(\mathbf{q}, \omega)} \tag{3.26}$$

as the Fourier transform in space and time of the net potential, $\varphi(\mathbf{r}, t)$. The dynamic screening action of the electron gas is described by $\epsilon(\mathbf{q}, \omega)$.

A knowledge of $\epsilon(\mathbf{q}, \omega)$ also permits us to write down the dispersion relation for the plasma oscillations. Consider Eqs. (3.19) and (3.20) in the absence of an external charge. We may write these as

$$\epsilon(\mathbf{q}, \omega)\mathbf{q}\cdot\boldsymbol{\mathcal{E}}(\mathbf{q}, \omega) = 0,$$
$$i\mathbf{q}\cdot\boldsymbol{\mathcal{E}}(\mathbf{q}, \omega) = 4\pi e\langle\rho(\mathbf{q}, \omega)\rangle.$$

The usual solution of these equations is

$$\langle\rho(\mathbf{q}, \omega)\rangle = \langle\mathbf{q}\cdot\boldsymbol{\mathcal{E}}(\mathbf{q}, \omega)\rangle = 0,$$

which corresponds to no net charge density or electric field present in the plasma. However, for frequencies, ω_q, such that

$$\epsilon(\mathbf{q}, \omega_q) = 0 \tag{3.27}$$

we see that one may have a nonvanishing value of $\boldsymbol{\mathcal{E}}(\mathbf{q}, \omega)$ and $\langle\rho(\mathbf{q}, \omega)\rangle$. In other words, one has a *free* oscillation of the charge density (there being no external field). The condition (3.27) is thus the condition for existence of plasma oscillations at frequency ω_q; such oscillations correspond, of course, to a net longitudinal electric field in the plasma. We shall see that in the limit of very long wavelengths ($\mathbf{q} \to 0$), the dielectric constant is equal to

$$\epsilon(0, \omega) = 1 - (\omega_p{}^2/\omega^2). \tag{3.28}$$

The corresponding frequency of plasma oscillation is thus seen to be ω_p, in accordance with the simple calculation of the preceding section.

LONGITUDINAL CONDUCTIVITY

The response of an electron system to an applied electric field is frequently specified in terms of the conductivity, which is the ratio of the current induced in the system, $e\langle\mathbf{J}\rangle$, to the effective electric field, $\boldsymbol{\mathcal{E}}$. For the present case of fields which vary in space and time, we may write

$$e\langle\mathbf{J}(\mathbf{q}, \omega)\rangle = \sigma(\mathbf{q}, \omega)\boldsymbol{\mathcal{E}}(\mathbf{q}, \omega). \tag{3.29}$$

Here $e\langle \mathbf{J}(\mathbf{q}, \omega)\rangle$ is the Fourier transform of the induced particle current, while $\sigma(\mathbf{q}, \omega)$ is the scalar longitudinal conductivity. (We are considering only a longitudinal external field, and have made our customary assumption that the system is isotropic.) There is a simple relation between σ and ϵ. To obtain it, we make use of the particle conservation equation for the charge and current induced in the electron gas:

$$\operatorname{div} \langle \mathbf{J}\rangle + \frac{\partial\langle\rho\rangle}{\partial t} = 0. \tag{3.30}$$

Let us Fourier transform Eq. (3.30):

$$\mathbf{q} \cdot \langle \mathbf{J}(\mathbf{q}, \omega)\rangle = \omega\langle\rho(\mathbf{q}, \omega)\rangle. \tag{3.31}$$

On taking the divergence of Eq. (3.29), and making use of (3.22), we find at once

$$\epsilon(\mathbf{q}, \omega) = 1 + \frac{4\pi i\sigma(\mathbf{q}, \omega)}{\omega}. \tag{3.32}$$

3.3. MACROSCOPIC TRANSPORT EQUATION

LANDAU-SILIN EQUATION

We now show how the concept of dynamic screening enables one to extend the Landau transport equation to electron systems. We consider, as in Chapter 1, a state characterized by a distribution function

$$n_{\mathbf{p}}(\mathbf{r}, t) = n_{\mathbf{p}}{}^{0} + \delta n_{\mathbf{p}}(\mathbf{r}, t). \tag{3.33}$$

The departure, $\delta n_{\mathbf{p}}$, from the ground state is assumed to be small, to contain only long wavelength fluctuations, and to be restricted to the vicinity of the Fermi surface, where quasiparticles are well defined. Let us again regard the total energy as a functional, $E[n_{\mathbf{p}}(\mathbf{r}, t)]$, of the distribution function. If we then attempt a straightforward expansion of this functional in powers of $\delta n_{\mathbf{p}}$, we run into difficulties with the expression for the interaction energy of two quasiparticles, $f(\mathbf{pr}, \mathbf{p'r'})$; as we have remarked, an expression of the form (1.76) is divergent, as a direct consequence of the long range of the Coulomb interaction between the electrons.

As Silin (1957) has first shown, these difficulties are removed if we allow for the dynamic screening of the particle motion in self-consistent fashion. An expression of the form (1.76) is meaningful if it refers to the interaction between *screened* quasiparticles, that is, the quasiparticle plus its associated screening cloud. What is necessary in the construction of the theory is that one take into account *ab initio* the formation of the screening cloud. This one may do by means of the following procedure.

(i) Consider first the electrostatic interaction between the *averaged* charge distributions of the excited quasiparticles. The departure from equilibrium gives rise to an average density fluctuation,

$$\langle \rho(\mathbf{r}, t) \rangle = \sum_{\mathbf{p}} \delta n_{\mathbf{p}}(\mathbf{r}, t), \tag{3.34}$$

and hence to a space charge electrostatic field, $\mathbf{\mathcal{E}}_p(\mathbf{r}, t)$, which is given by

$$\operatorname{div} \mathbf{\mathcal{E}}_p(\mathbf{r}, t) = 4\pi e \sum_{\mathbf{p}} \delta n_{\mathbf{p}}(\mathbf{r}, t). \tag{3.35}$$

This part of the interaction between quasiparticles can be accounted for in the transport equation by regarding $\mathbf{\mathcal{E}}_p(\mathbf{r}, t)$ as an additional applied field, which acts to screen the field produced by any given quasiparticle.

(ii) As a result of step (i), each excited quasiparticle is surrounded by a polarization cloud of other quasiparticles. The interaction between any given pair of quasiparticles is thus screened; the range of the effective interaction is relatively short (of atomic size). Associated with this screened interaction (and *not* accounted for by the *average* polarization field of the dielectric screening), are short-range correlations between the two quasiparticles. In addition, when a given quasiparticle approaches another, it acts to alter the polarization cloud of that particle, giving rise to a departure from the plain Coulomb interaction. Both effects appear as a *fluctuating* field associated with each quasiparticle.

Once the average polarization field has been singled out, what we have is a system of quasiparticles with a short-range interaction, which may be treated *à la* Landau by the introduction of an interaction energy, $f_{\mathbf{pp}'}$. We are therefore led to express the local excitation energy of a quasiparticle in the form

$$\bar{\epsilon}_{\mathbf{p}}(\mathbf{r}) = \epsilon_p + \sum_{\mathbf{p}'} f_{\mathbf{pp}'} \delta n_{\mathbf{p}'}(\mathbf{r}). \tag{3.36}$$

The derivation of a transport equation then proceeds exactly as in Chapter 1. In the absence of an external field, and with collisions neglected, one has for the transport equation,

$$\frac{\partial}{\partial t}\,\delta n_{\mathbf{p}} + \mathbf{v}_{\mathbf{p}}\cdot\nabla_{\mathbf{r}}\delta n_{\mathbf{p}} + \mathbf{v}_{\mathbf{p}}\delta(\epsilon_p - \mu)\cdot\sum_{\mathbf{p}} f_{\mathbf{pp}'}\nabla_{\mathbf{r}}\delta n_{\mathbf{p}'} - e\mathcal{E}_p\cdot\mathbf{v}_{\mathbf{p}}\delta(\epsilon_p - \mu) = 0, \tag{3.37}$$

where the polarization field $\mathcal{E}_p$ is obtained from Eq. (3.35). If we consider a periodic disturbance, with wave vector $\mathbf{q}$ and frequency ω, Eq. (3.37) takes the simpler form

$$i(\mathbf{q}\cdot\mathbf{v}_{\mathbf{p}} - \omega)\delta n_{\mathbf{p}} + i\mathbf{q}\cdot\mathbf{v}_{\mathbf{p}}\delta(\epsilon_p - \mu)\sum_{\mathbf{p}'} f_{\mathbf{pp}'}\delta n_{\mathbf{p}'} - e\mathcal{E}_p\cdot\mathbf{v}_{\mathbf{p}}\delta(\epsilon_p - \mu) = 0, \tag{3.38}$$

where the polarization field $\mathcal{E}_p$ is equal to

$$\mathcal{E}_p = -i\mathbf{q}\,\frac{4\pi e}{q^2}\sum_{\mathbf{p}}\delta n_{\mathbf{p}}. \tag{3.39}$$

Using Eqs. (3.38) and (3.39), we may write the transport equation in the absence of an external field as an integral equation homogeneous in $\delta n_{\mathbf{p}}$:

$$(\mathbf{q}\cdot\mathbf{v}_{\mathbf{p}} - \omega)\delta n_{\mathbf{p}} + \mathbf{q}\cdot\mathbf{v}_{\mathbf{p}}\delta(\epsilon_p - \mu)\sum_{\mathbf{p}'}\left\{f_{\mathbf{pp}'} + \frac{4\pi e^2}{q^2}\right\}\delta n_{\mathbf{p}'} = 0. \tag{3.40}$$

This result is quite striking, and shows that the *total* interaction between charged quasiparticles is characterized by Fourier coefficients

$$\frac{4\pi e^2}{q^2} + f_{\mathbf{pp}'} \tag{3.41}$$

(at least when $\mathbf{q} \to 0$).

On comparing Eq. (3.40) with the result (1.103) obtained for a neutral system, we see clearly the peculiar behavior of an electron gas. For a neutral system, $f_{\mathbf{pp}'}$ represented the *total* interaction between quasiparticles, which remained regular in the limit $\mathbf{q} \to 0$. For a charged system, on the other hand, the total interaction (3.41) is singular, as a consequence of the long range of Coulomb interactions. In order to avoid this difficulty, we split away the singular term $4\pi e^2/q^2$, and treat it in terms of an average polarization field acting on the electrons. What remains is the *screened* interaction $f_{\mathbf{pp}'}$, which is regular when $\mathbf{q} \to 0$.

The transport equation (3.37) is easily extended to cover the case of a longitudinal external electric field $\mathfrak{D}_{\text{ext}}(\mathbf{r}, t)$. Because the charge on a quasiparticle is e (charge being conserved when the interaction between particles is switched on), the force produced by $\mathfrak{D}_{\text{ext}}(\mathbf{r}, t)$ on a quasiparticle is simply

$$\mathfrak{F} = e\mathfrak{D}_{\text{ext}}(\mathbf{r}, t). \tag{3.42}$$

One thus adds to Eq. (3.37) an additional forcing term,

$$-e\mathfrak{D}_{\text{ext}}(\mathbf{r}, t) \cdot \mathbf{v}_{\mathbf{p}}\delta(\epsilon_p - \mu). \tag{3.43}$$

We may distinguish two cases of interest, depending on the character of the external field.

(i) Inhomogeneous longitudinal field In this case, $\mathfrak{D}_{\text{ext}}(\mathbf{r}, t) = \mathfrak{D}(\mathbf{r}, t)$. On combining Eqs. (3.43) and (3.37) we see that it is the effective electric field, $\mathcal{E}(\mathbf{r}, t)$, to which the electrons respond. We obtain the electron response with the aid of Eq. (3.37) (with $\mathcal{E}_p$ replaced by $\mathcal{E}$), together with the relation (3.21),

$$\mathfrak{D}(\mathbf{q}, \omega) = \epsilon(\mathbf{q}, \omega)\mathcal{E}(\mathbf{q}, \omega),$$

where $\epsilon(\mathbf{q}, \omega)$ is, in turn, determined from Eq. (3.22) or (3.23). The dynamic screening action of the electrons plays an essential role in determining the response of the electrons, and, indeed, in making possible a consistent calculation of that response.

(ii) Homogeneous external field ($q \equiv 0$) In this case there is no screening. A polarization field, $\mathcal{E}_p$, could only result from surface charges located at the boundary of the system; such an effect has been ruled out by our boundary conditions. We therefore have:

$$\mathcal{E}(0, \omega) = \mathfrak{D}_{\text{ext}}(0, \omega). \tag{3.44}$$

In both cases it is appropriate to use as our transport equation,

$$\frac{\partial}{\partial t}\,\delta n_{\mathbf{p}} + \mathbf{v}_{\mathbf{p}} \cdot \nabla_{\mathbf{r}}\delta n_{\mathbf{p}} + \mathbf{v}_{\mathbf{p}} \cdot \left\{\sum_{\mathbf{p}'} f_{\mathbf{p}\mathbf{p}'}\nabla_{\mathbf{r}}\delta n_{\mathbf{p}'} - e\mathcal{E}\right\}\delta(\epsilon_p - \mu) = 0. \tag{3.45}$$

The distinction between homogeneous and inhomogeneous fields appears only when one considers the relation between $\mathcal{E}$ and $\mathfrak{D}$.

CHARGED VS. NEUTRAL SYSTEMS

We see on the basis of the above discussion that the essentially different character of the electron system manifests itself in the response to an inhomogeneous longitudinal field. It is only in this case that charge fluctuations occur; the long-range character of the particle interactions then requires that one introduce *at the outset* the space charge field responsible for the dynamic screening of the particle interactions.

Most of the equilibrium properties of the electron gas, involving *homogeneous* disturbances, can be treated along the same lines used for neutral systems in Chapter 1. There is, however, one important exception, namely the compressibility of the electron system. When defining the compressibility, one assumes that the uniform background of positive charge is compressed along with the electrons, in order to maintain neutrality everywhere. This point of view is just opposite to that underlying the previous discussion, where we tacitly assumed that the positive background always remained uniform. We must therefore revise the calculation of the compressibility given in Chapter 1.

For a neutral system we have argued that the force per unit volume, $\mathfrak{F}(\mathbf{r})$, produced by a longitudinal external field is balanced by a pressure distribution, $\delta P(\mathbf{r})$, such that

$$\mathbf{grad}\ \delta P = \mathfrak{F}(\mathbf{r}). \tag{3.46}$$

The additional pressure δP will in turn change the density, by an amount

$$\langle \delta \rho \rangle = \kappa N \delta P, \tag{3.47}$$

where κ is the compressibility. The results (3.46) and (3.47) apply equally well to a charged particle system, provided one takes appropriate care in the definition of $\mathfrak{F}(\mathbf{r})$.

Let $e\rho(\mathbf{r})$ be the charge density fluctuation of the electrons, produced by a force per unit volume

$$\mathfrak{F}(\mathbf{r}) = Ne\mathfrak{D}(\mathbf{r}).$$

The electrons give rise to a polarization field $\mathcal{E}_p$. Since we wish to maintain overall charge neutrality, we must distort the positive background by an equal amount $\rho(\mathbf{r})$; this distortion gives rise to an additional space charge $-e\rho(\mathbf{r})$, which just compensates the electronic space charge. The positive background thus creates a space charge field $-\mathcal{E}_p$. The total field $\mathfrak{D}(\mathbf{r})$ felt by the electron system is due partly to its interaction with the positive background (which contributes a term $-\mathcal{E}_p(\mathbf{r})$), and partly to "external" charges, the field of which is $\mathfrak{D}({}_{\text{ext}}\mathbf{r})$:

$$\mathfrak{D}(\mathbf{r}) = \mathfrak{D}_{\text{ext}}(\mathbf{r}) + [-\mathcal{E}_p(\mathbf{r})]. \tag{3.48}$$

When defining the compressibility, we are only interested in the response of the system to those forces that are "external" to the complete system "electrons + background," i.e., to the field $\mathfrak{D}_{\text{ext}}$. According to Eq. (3.48), the compressibility κ thus refers to the response to the field $(\mathfrak{D} + \mathcal{E}_p)$, i.e., to the screened electric field $\mathcal{E}(\mathbf{r})$. We may therefore calculate κ from the Landau–Silin equation, (3.45), with the proviso that the relation between the induced density and the electric field is given by [cf. Eqs. (3.46) and (3.47)]

$$\mathbf{grad}\,\langle\delta\rho\rangle = N^2 e\kappa\boldsymbol{\mathcal{E}}. \tag{3.49}$$

One finds easily that κ is related to the *screened* quasiparticle energy, $f_{\mathbf{pp}'}$, by the same expression as that found for the neutral system:

$$\kappa = \frac{1}{N^2}\frac{p_F m^*}{\pi^2\hbar^3}\frac{1}{1+F_0{}^s}. \tag{3.50}$$

PLASMA OSCILLATIONS VS. ZERO SOUND

A collective mode always corresponds to a possible oscillation of the system in the absence of an *external* field. For a neutral system, the only force *explicitly* applied to the particles is the external force $\mathfrak{F}$: the collective modes are then governed by the transport equation (1.105), in which we have set $\mathfrak{F} = 0$. Such an equation does *not* apply to the longitudinal oscillations of a charge Fermi liquid. In that case, the density fluctuations give rise to a space charge field $\boldsymbol{\mathcal{E}}_p$, which in turn acts to drive the electrons. Put another way, there is a "feedback" of the electron response into the electric field $\boldsymbol{\mathcal{E}}$ felt by the system. In order to describe the longitudinal oscillations of the electron gas, one must solve *jointly* the transport equation (3.38) and Poisson's equation (3.39); it is this feature which is responsible for the difference between the longitudinal oscillations of charged and neutral Fermi liquids.

The special behavior of charged systems is displayed very clearly when we combine Eqs. (3.38) and (3.39) into the single equation (3.40): It is obvious that the long wavelength longitudinal oscillations are considerably affected by the singular interaction term $4\pi e^2/q^2$. Indeed, when $\mathbf{q} \to 0$, we can neglect $f_{\mathbf{pp}'}$ as compared to $4\pi e^2/q^2$. The solution of Eq. (3.40) is then straightforward. By using the results of Chapter 1, one may show that for a translationally invariant system, the longitudinal collective mode has a frequency $\omega = \omega_p$: it is nothing but the plasma oscillation studied in the preceding section. (See Problem 3.1.) More generally, it is straightforward to prove by the methods of the following section (see Problem 3.2) that the dispersion relation for longitudinal collective modes based on Eq. (3.40) is identical to the relation (3.27),

$$\epsilon(\mathbf{q}, \omega_q) = 0,$$

which served to define the plasma oscillations.

The plasma oscillations therefore appear as the "zero sound mode" of a charged liquid. In the limit of long wavelengths, the frequency of that mode is enormously shifted by the long range of Coulomb interactions: the linear spectrum $\omega = s_0 q$, characteristic of a neutral Fermi liquid, is replaced by a constant frequency $\omega = \omega_p$ when $\mathbf{q} \to 0$. Mathematically, this difference in behavior arises as a consequence of the singular interaction term $4\pi e^2/q^2$ in Eq. (3.40).

As $\mathbf{q}$ increases, one can no longer neglect $f_{\mathbf{pp}'}$ in Eq. (3.40). Indeed, we may expect that for a high enough value of $\mathbf{q}$, $4\pi e^2/q^2$ will become negligible as compared to $f_{\mathbf{pp}'}$; one should therefore recover the characteristic behavior of a neutral system. To the extent that $f_{\mathbf{pp}'}$ is repulsive, the plasma oscillation would thus go over to a zero sound mode. (See Problem 3.3.) In practice, it is not possible to observe such a transition. We shall see that $f_{\mathbf{pp}'}$ is always of order $4\pi e^2/p_F^2$: the transition from plasma oscillations to zero sound only occurs for $q \sim p_F$, in a region where the Landau theory is not applicable and where collective modes are highly damped.

In fact the alert reader will have noticed that even the plasma frequency when $\mathbf{q} \to 0$ should not be calculated by the Landau theory, since the latter only applies to macroscopic frequencies, much smaller than ω_p. Actually, we shall see in Chapter 4 that the plasma frequency $\omega = \omega_p$ derived by the Landau theory happens to be correct because of simple sum rule arguments. On the other hand, use of the Landau theory to calculate corrections of order q^2 to the plasma frequency is not justified, and will generally lead to an incorrect result.

The complications arising from the space charge field $\mathcal{E}_p$ do not occur for the other collective modes (either spin symmetric waves or antisymmetric waves with a quantum number $m \geq 1$); for such waves, there are no density fluctuations, and $\mathcal{E}_p = 0$. Nonetheless, one encounters new features in the case of *transverse symmetric waves.* The latter involve transverse current fluctuations, which give rise to an electromagnetic field. The transverse modes are thus coupled to the electromagnetic field, in much the same way as the longitudinal modes were coupled to the electrostatic field. In order to describe the transverse symmetric modes, one must solve jointly the transport equation (3.38) and Maxwell's equations. We shall discuss this problem in Section 3.6.

All the other collective modes (symmetric modes with $m \geq 2$ and antisymmetric modes) are not affected by the fact that the particles are charged. In such cases, the question of whether a collective mode exists reduces to the question of the sign, and size, of the relevant components of $f_{\mathbf{pp}'}$. For example, one does not expect to find longitudinal spin waves for conduction electrons in metals. All such systems exhibit a spin susceptibility which is somewhat greater than the free-electron, Pauli value; thus they display negative values of F_0^a, which correspond to the case of an effective weak attraction in the singlet spin state. As a result the longitudinal spin wave mode is strongly damped, just as is the case for ^{3}He.

In concluding this section, let us emphasize once more the conditions of applicability of the Landau transport equation (3.45). First, it is necessary that the wave vectors and frequencies associated with any inhomogeneities have a "macroscopic" scale such that

$$\omega \ll \mu, \qquad qv_F \ll \mu, \tag{3.51}$$

in order that the theory be applicable. Second, in order that collisions may be neglected, at least one of the following conditions must be satisfied:

$$\nu \ll \omega, \qquad \nu \ll qv_F, \tag{3.52}$$

where ν is a collision frequency.

3.4. MACROSCOPIC DIELECTRIC RESPONSE OF A QUANTUM PLASMA

We now calculate the *macroscopic* dielectric response with the aid of the Landau–Silin equation, (3.45). Because the response of the system is linear, it suffices to consider the external field, $\mathfrak{D}$, as taking the form of a propagating plane wave, of wave vector $\mathbf{q}$ and frequency ω,

$$\mathfrak{D}(\mathbf{r}, t) = \lim_{\eta \to 0} \mathfrak{D} \exp i(\mathbf{q} \cdot \mathbf{r} - \omega t) \exp \eta t. \tag{3.53}$$

All other quantities of interest will have the same periodic behavior. (We shall again denote the system quantities of interest by their amplitude, $\mathcal{E}$, $\langle \rho \rangle$, etc. The amplitudes will be complex, with the arguments characterizing the phase of the quantity.) We shall also find it convenient to re-express Eq. (3.45) by making use of the "local" distribution function, $\delta \bar{n}_\mathbf{p}$, as defined in Chapter 1. There we saw that the departure of the distribution function from the *local* ground state, $\bar{n}_\mathbf{p}{}^\circ$, was given by

$$\delta \bar{n}_\mathbf{p} = n_\mathbf{p} - \bar{n}_\mathbf{p}{}^\circ = \delta n_\mathbf{p} + \sum_{\mathbf{p}'} f_{\mathbf{p}\mathbf{p}'} \delta(\epsilon_p - \mu) \delta n_{\mathbf{p}'}. \tag{3.54}$$

In terms of $\delta \bar{n}_\mathbf{p}$, our basic transport equation takes the form:

$$\mathbf{q} \cdot \mathbf{v}_\mathbf{p} \delta \bar{n}_\mathbf{p} - \omega \delta n_\mathbf{p} + ie\mathcal{E} \cdot \mathbf{v}_\mathbf{p} \delta(\epsilon_p - \mu) = 0. \tag{3.55}$$

Taken together, Eqs. (3.54) and (3.55) correspond to an integral equation for $\delta n_\mathbf{p}$; the solution of that equation is, of course, far more complicated than the corresponding equation for a weakly interacting system (for which $f_{\mathbf{p}\mathbf{p}'} = 0$ and $\delta \bar{n}_\mathbf{p} = \delta n_\mathbf{p}$). Nonetheless it is possible to obtain explicit solutions of the Landau–Silin equations for several cases of physical interest.

The calculation of the dielectric response from (3.55) is formally equivalent to that for the density-density correlation function in a neutral system. We denote the latter quantity by $\chi^n(\mathbf{q}, \omega)$, to avoid confusion with the corresponding correlation function for the electron gas (which is markedly affected by screening, see Chapter 4). In the macroscopic collisionless regime, $\chi^n(\mathbf{q}, \omega)$ was obtained by solving the transport equation (2.104). To make the correspondence explicit, we write (2.104) as

$$\mathbf{q} \cdot \mathbf{v_p} \delta \bar{n}_\mathbf{p} - \omega \delta n_\mathbf{p} + \mathbf{q} \cdot \mathbf{v_p} \varphi^n \delta(\epsilon_p - \mu) = 0, \qquad (3.55a)$$

where $\varphi^n(\mathbf{q}, \omega)$ is the Fourier transform of the scalar potential applied to the neutral system. On comparing Eqs. (3.55) and (3.55a), we see that for the electron system, $e\mathcal{E}$ plays a role directly analogous to that of $(-\mathbf{grad}\, \varphi^n)$. We may construct a pseudo response function in the electron system, $\chi^n(\mathbf{q}, \omega)$, which would be that of a neutral system with the same N, ϵ_p, and $f_{\mathbf{pp}'}$. On making use of the definition of ϵ, (3.22), it is straightforward to show that

$$\epsilon(\mathbf{q}, \omega) = 1 - \frac{4\pi e^2}{q^2} \chi^n(\mathbf{q}, \omega). \qquad (3.56)$$

With the aid of Eq. (3.56), and the results obtained in Section 2.5, one can readily obtain various asymptotic expressions for $\epsilon(\mathbf{q}, \omega)$ in the long wavelength limit. We shall not, in fact, adopt that procedure, but rather go once more through the details of the calculations in the limits we consider. It is hoped that the reader may thereby acquire an increased familiarity with the solutions of the basic Landau transport equation.

SCREENING OF A STATIC LONGITUDINAL FIELD ($\omega = 0$)

In this case, $\mathcal{E}$ is in the direction of $\mathbf{q}$; Eq. (3.55) at once reduces to

$$\delta \bar{n}_\mathbf{p} = -\frac{ie}{q} \mathcal{E} \delta(\epsilon_p - \mu). \qquad (3.57)$$

Moreover, we have, from Eq. (1.34), the relation

$$\delta n_\mathbf{p} = \frac{\delta \bar{n}_\mathbf{p}}{1 + F_0^s}, \qquad (3.58)$$

since $\delta \bar{n}_\mathbf{p}$ is isotropic and spin independent. From $\delta n_\mathbf{p}$, we at once calculate

$$\langle\rho\rangle = \sum_{\mathbf{p}} \delta n_{\mathbf{p}} = -\frac{3iNem^*}{qp_F^2(1+F_0^s)}\,\mathcal{E}. \tag{3.59a}$$

Using the value (1.61) of the macroscopic sound velocity s, we can cast the above result in the form

$$\langle\rho\rangle = -\frac{iNe}{qms^2}\,\mathcal{E}. \tag{3.59b}$$

On substituting Eq. (3.59) into (3.22) we find

$$\epsilon(\mathbf{q}, 0) = 1 + \frac{4\pi Ne^2}{ms^2q^2} \tag{3.60}$$

a result which is, it must be emphasized, valid only for small values of q.

It is instructive to compare the result (3.60) with that one obtains in the Fermi–Thomas approximation. We may write the Fermi–Thomas result for $\epsilon(\mathbf{q}, 0)$, (3.25), in the following way:

$$\epsilon_{FT}(\mathbf{q}, 0) = 1 + \frac{q_{FT}^2}{q^2} = 1 + \frac{4\pi Ne^2}{mq^2(v_F^2/3)}. \tag{3.61}$$

We see that the exact result of the Landau theory differs from the approximate Fermi–Thomas result only in that the macroscopic sound velocity for an interacting particle system replaces the value, $v_F/\sqrt{3}$, appropriate to a noninteracting system. Comparison of the Fermi–Thomas approach and that considered here makes it clear that in both cases one is calculating the self-consistent response of the electrons to the effective field, $\mathcal{E}$. However, in the Landau theory, one takes account of particle interaction through the additional forcing term in Eq. (3.45) proportional to $f_{\mathbf{pp}'}$ (that is, through the difference in $\delta\bar{n}_{\mathbf{p}}$ and $\delta n_{\mathbf{p}}$).

By means of Eq. (1.49), we may express the static dielectric constant in terms of the compressibility κ (remember that now $\Omega = 1$):

$$\epsilon(\mathbf{q}, 0) = 1 + \frac{4\pi N^2e^2}{q^2}\,\kappa. \tag{3.62}$$

The result, (3.62), could have been obtained directly by a macroscopic argument based on Eq. (3.49) and Poisson's equation. It is illuminating to write Eq. (3.49) in terms of the net electrostatic potential, $\varphi(\mathbf{r})$:

$$\langle\rho(\mathbf{r})\rangle = -\kappa N^2 e\varphi(\mathbf{r}). \tag{3.63}$$

Equation (3.63) replaces the Fermi–Thomas approximation result, (3.3); it is exact provided the spatial variation in φ is slow.

The result (3.60) is a very important one. It shows that screening of a long wavelength static charge disturbance is an *exact* property of a quantum plasma, and that the screening length, λ_s, is simply related to the macroscopic sound velocity by

$$\lambda_s = \frac{s}{\omega_p},$$

where ω_p is the classical plasma frequency, (3.12).

EFFECTIVE ELECTRON–ION INTERACTION

Another static problem which may be usefully studied with the Landau theory is the matrix element for the scattering of a dressed electron by an ion (either an impurity introduced into the system, or one of the ions which make up a given metal). Assume that a point impurity, with charge Ze, is introduced at the origin; it produces an "external" electrostatic potential Ze/r, whose Fourier transform is

$$\varphi_{\mathbf{q}}^{\text{ext}} = \frac{4\pi Ze}{q^2}.$$

As a result, a quasiparticle with momentum $\mathbf{p}$ acquires an additional *effective* potential energy, $\mathcal{V}_{\mathbf{p}}^{\text{eff}}(\mathbf{r})$ with Fourier transform $\mathcal{V}_{\mathbf{qp}}^{\text{eff}}$. (Strictly speaking $\mathcal{V}_{\mathbf{qp}}^{\text{eff}}$ is only defined in the long wavelength limit, $q \ll p_F$, where the concept of a *local* quasiparticle energy is meaningful.) The quantity $\mathcal{V}_{\mathbf{qp}}^{\text{eff}}$ may be viewed as the *matrix element* for scattering of a quasiparticle by the impurity from the state of momentum $\mathbf{p}$ to that of momentum $(\mathbf{p} + \mathbf{q})$.

The above interpretation of $\mathcal{V}_{\mathbf{qp}}^{\text{eff}}$ is obviously correct if one neglects the Coulomb interaction between electrons. In that case, quasiparticles reduce to bare electrons, while the effective potential energy is given by

$$\mathcal{V}_{\mathbf{qp}} = e\varphi_{\mathbf{q}}^{\text{ext}} = \frac{4\pi Ze^2}{q^2}.$$

On the other hand, the Hamiltonian which describes the electron-impurity coupling may be written as

$$\sum_i \frac{Ze^2}{r_i} = \sum_{\mathbf{q}} \frac{4\pi Ze^2}{q^2}\, \rho_{\mathbf{q}}^{+} = \sum_{\mathbf{qp}} \frac{4\pi Ze^2}{q^2}\, c_{\mathbf{p+q}}^{+} c_{\mathbf{p}}$$

(on going to the second quantized representation, see the Appendix). The matrix element for the transition $(\mathbf{p} \to \mathbf{p} + \mathbf{q})$ is clearly equal to $4\pi Ze^2/q^2$, and thus to $\mathcal{V}_{\mathbf{qp}}$, as predicted.

As a result of the Coulomb interaction between electrons, the impurity is surrounded by a screening cloud; the latter involves a distortion

$\delta n_{\mathbf{p}'}(\mathbf{r})$ of the quasiparticle distribution, with Fourier transform $\delta n_{\mathbf{p}'}(\mathbf{q})$. The effective potential energy, $\mathcal{V}_{\mathbf{qp}}^{\text{eff}}$, is then modified for two reasons:

(i) The electrostatic potential produced by the impurity is screened, $\varphi_{\mathbf{q}}^{\text{ext}}$ being divided by the dielectric function $\epsilon(\mathbf{q}, 0)$.

(ii) The scattering quasiparticle, with momentum $\mathbf{p}$, interacts with the screening cloud via the screened quasiparticle interaction $f_{\mathbf{pp}'}$.

We assume that Z is sufficiently small, such that the response of the electrons to the impurity is linear; the net effective potential is then equal to

$$\mathcal{V}_{\mathbf{qp}}^{\text{eff}} = \frac{4\pi Ze^2}{q^2\epsilon(\mathbf{q}, 0)} + \sum_{\mathbf{p}'} f_{\mathbf{pp}'}\delta n_{\mathbf{p}'}(\mathbf{q}). \tag{3.64a}$$

Let us set $\omega = 0$ in the transport equation, (3.55), and further remark that the screened electric field produced by the impurity is given by

$$\mathcal{E}_{\mathbf{q}} = -i\mathbf{q}\,\frac{4\pi Ze}{q^2\epsilon(\mathbf{q}, 0)}.$$

On using (3.54) and (3.64a), we may write (3.55) in the form

$$\delta n_{\mathbf{p}}(\mathbf{q}) = -\mathcal{V}_{\mathbf{qp}}^{\text{eff}}\delta(\epsilon_p - \mu). \tag{3.64b}$$

Equation (3.64b) may be considered as a definition of the effective potential energy $\mathcal{V}_{\mathbf{qp}}^{\text{eff}}$, exerted by the impurity on the quasiparticle $\mathbf{p}$. Starting from that relation, one may show explicitly that $\mathcal{V}_{\mathbf{qp}}^{\text{eff}}$ is indeed the matrix element for scattering of a *quasiparticle* from the state with momentum $\mathbf{p}$ to that with momentum $(\mathbf{p} + \mathbf{q})$.

Since the electric field produced by the ion is a static longitudinal field, we may obtain $\mathcal{V}_{\mathbf{qp}}^{\text{eff}}$ at once from the considerations of the previous subsection. The departure from equilibrium, $\delta n_{\mathbf{p}}(\mathbf{q})$, is given by Eqs. (3.57) and (3.58). On inserting these results into (3.64a), we find

$$\mathcal{V}_{\mathbf{qp}}^{\text{eff}} = \frac{4\pi Ze^2}{q^2}\,\frac{1}{\epsilon(\mathbf{q}, 0)}\,\frac{1}{1 + F_0{}^s}. \tag{3.65a}$$

[Note that Eq. (3.65a) is only exact in the long wavelength limit.] On using Eqs. (3.60) and (1.61), we may also write (3.65a) in the form

$$\mathcal{V}_{\mathbf{qp}}^{\text{eff}} = Z\,\frac{p_F{}^2}{3m^*N} = \frac{Z}{\nu(0)}, \tag{3.65b}$$

where $\nu(0)$ is the density of states per unit energy at the Fermi surface. The result, (3.65b), is due to Heine (unpublished).

The right-hand side of Eq. (3.64a) displays clearly the physical nature of the two corrections to $\mathcal{V}_{qp}^{eff}$ arising from electron-electron interaction. The first term involves Coulomb interaction between the colliding electron and the average charge density of the screening cloud (the latter giving rise to dielectric screening of the impurity field). To lowest order in the electron interaction, such a term would correspond to a *direct* process, characteristic of a "Hartree" approximation (see Section 5.1). In contrast, the second term represents a screened interaction with specific electrons of the screening cloud; the lowest-order contribution to this term involves *exchange* scattering of the colliding electron, of momentum **p**, with an electron of the screening cloud. The two corrections, dielectric screening and exchange scattering, tend to cancel one another; as a result, the coefficient F_o^s disappears from the final result, (3.65b); the only effect of electron interaction which remains is the modification in the density of states per unit energy, $\nu(0)$.

Such a cancellation does not occur if we consider the mutual scattering of two external charges, say Ze and $Z'e$. In that case, there can be no exchange process. The matrix element for scattering of the two charges with a momentum transfer **q** is simply

$$\mathcal{V}_{\mathbf{q}}^{eff} = \frac{4\pi ZZ'e^2}{q^2}\frac{1}{\epsilon(\mathbf{q},0)} = \frac{ZZ'}{\nu(0)}(1 + F_o^s) \tag{3.66}$$

[on making use of Eqs. (3.60) and (1.61)]. We see that $\mathcal{V}_{\mathbf{q}}^{eff}$ depends on the interaction coefficient F_o^s, in contrast to the matrix element for electron-ion scattering. Once more, we note that Eq. (3.66) is only valid in the "linear" regime, when Z and Z' are small enough.*

As such, the result (3.65b) is only valid in the case of scattering by a static point impurity. Actually, any ion in a real metal possesses inner electronic shells, with which the scattering electron may interact (via a polarization of the ion core, or even through an exchange coupling with the deep lying electrons). As a result, the effective electron-ion interaction is no longer given by Eq. (3.65b). However, the latter expression

* If one introduces an impurity Ze into the electron liquid, the total charge displaced is

$$\sum_{\mathbf{p}} e\delta n_{\mathbf{p}} = -e\sum_{\mathbf{p}} [Z/\nu(0)]\delta(\epsilon_p - \mu) = -Ze$$

[see Eqs. (3.64a) and (3.64b)]. One may actually work backward, and obtain $\mathcal{V}_{qp}^{eff}$ from the assumption of perfect screening. Such a procedure is closely related to the Friedel sum rule governing the scattering of electrons by a screened impurity potential [Friedel (1954)]. Equation (3.65b) is easily derived by using the expression of the pseudo potential in terms of scattering phase shifts, and the Friedel sum rule [see, for example, Greene and Kohn (1965)].

provides a useful approximation, which is of importance in the pseudopotential theory of metals [Harrison (1966)].

QUASISTATIC RESPONSE ($0 < \omega \ll qv_F$)

We next consider the response of the electrons to an almost static longitudinal field, of frequency ω small compared to a typical one-electron frequency qv_F. Energy will be absorbed by electrons that move in phase with the wave, with velocities $\mathbf{v}_\mathbf{p}$ such that

$$\mathbf{q} \cdot \mathbf{v}_\mathbf{p} = \omega.$$

Since $\omega \ll qv_F$ and $v_p = v_F$, those electrons move in a direction nearly perpendicular to $\mathbf{q}$.

We search for a solution to Eq. (3.55) which is valid to first order in ω/qv_F. Let us write

$$\delta\bar{n}_\mathbf{p} = \delta\bar{n}_\mathbf{p}^{(0)} + \delta\bar{n}_\mathbf{p}^{(1)} + \cdots.$$

The zeroth-order solution, $\delta\bar{n}_\mathbf{p}^{(0)}$, is specified by (3.56). The first-order term in $\delta\bar{n}_\mathbf{p}$ is given by

$$\delta\bar{n}_\mathbf{p}^{(1)} = \frac{\omega\,\delta n_\mathbf{p}^{(0)}}{\mathbf{q} \cdot \mathbf{v}_\mathbf{p} - i\eta}, \tag{3.67}$$

where the term $i\eta$ has been included to account for our adiabatic causal boundary condition. On making use of the relations (3.57) and (3.58), which are valid to lowest order in ω/qv_F, we may write

$$\delta\bar{n}_\mathbf{p}^{(1)} = -\frac{ie\omega\mathcal{E}\,\delta(\epsilon_p - \mu)}{q(\mathbf{q} \cdot \mathbf{v}_\mathbf{p} - i\eta)(1 + F_0{}^s)}. \tag{3.68}$$

We then calculate easily the change in the *local* density brought about by the external field; it is, on keeping terms through first order,

$$\langle\bar{\rho}\rangle = \sum_\mathbf{p} \delta\bar{n}_\mathbf{p} = -\frac{ie\mathcal{E}}{q}\left\{\frac{N(1 + F_0{}^s)}{ms^2} + \sum_\mathbf{p} \frac{\omega}{(\mathbf{q} \cdot \mathbf{v}_\mathbf{p} - i\eta)}\,\frac{\delta(\epsilon_p - \mu)}{1 + F_0{}^s}\right\}. \tag{3.69}$$

We now seek to relate $\langle\bar{\rho}\rangle$ to the real change of density $\langle\rho\rangle$. Because we are calculating a scalar quantity, we can show that

$$\langle\rho\rangle = \frac{\langle\bar{\rho}\rangle}{1 + F_0{}^s} \tag{3.70}$$

[once we carry out the summation over $\mathbf{p}$, the only contribution comes from the $\ell = 0$ terms of $\delta\bar{n}_\mathbf{p}$ and $\delta n_\mathbf{p}$, the ratio of which is $(1 + F_0{}^s)$]. Remark that the relation (3.70) is an *integral* relation; one does not have, for example, $\delta\bar{n}_\mathbf{p}^{(1)} = (1 + F_0{}^s)\,\delta n_\mathbf{p}^{(1)}$. It is left as a problem to the reader to show that with the aid of Eq. (3.70), and the relation

$$\sum_\mathbf{p} \frac{\delta(\epsilon_p - \mu)}{\mathbf{q} \cdot \mathbf{v}_\mathbf{p} - i\eta} = \pi i \sum_\mathbf{p} \delta(\epsilon_p - \mu)\,\delta(\mathbf{q} \cdot \mathbf{v}_\mathbf{p}), \tag{3.71}$$

one obtains, in place of Eq. (3.60),

$$\epsilon(\mathbf{q}, \omega) = 1 + \frac{\omega_p^2}{s^2q^2}\left\{1 + i\left(\frac{\pi}{6}\right)\frac{\omega}{qp_F}\left(\frac{p_F^2}{ms^2}\right)\right\} + \cdots \qquad (0 < \omega \ll qv_F). \quad (3.72)$$

The new, dissipative term in the dielectric constant vanishes in the limit of $\omega \rightarrow 0$, as it should. In view of its dependence on the actual sound velocity, s, the low-frequency dissipative part of the conductivity differs from its counterpart for the free electron gas.

We note that one cannot obtain in any simple fashion the reactive term of order ω^2 in $\epsilon(\mathbf{q}, \omega)$. For that purpose, one requires a knowledge of

$$\delta\bar{n}_\mathbf{p}^{(2)} = \frac{\omega\delta n_\mathbf{p}^{(1)}}{\mathbf{q}\cdot\mathbf{v}_\mathbf{p} - i\eta}$$

and hence of $\delta n_\mathbf{p}^{(1)}$. There is not, however, any simple relationship between $\delta n_\mathbf{p}^{(1)}$ and $\delta\bar{n}_\mathbf{p}^{(1)}$, as the reader may easily verify with the aid of Eqs. (3.54) and (3.68).

SLIGHTLY INHOMOGENEOUS TIME-VARYING FIELD $(0 < qv_F \ll \omega)$

In this case the applied field is nearly uniform, its phase velocity being large compared to v_F. We may write at once, from Eq. (3.55),

$$\delta n_\mathbf{p}(0, \omega) = \frac{ie\boldsymbol{\mathcal{E}}\cdot\mathbf{v}_\mathbf{p}}{\omega}\,\delta(\epsilon_p - \mu). \quad (3.73)$$

We may then calculate the conductivity with the aid of the equivalent expression for the current density,

$$e\mathbf{J} = e\sum_\mathbf{p} \delta n_\mathbf{p}\mathbf{j}_\mathbf{p}. \quad (3.74)$$

In a translationally invariant system, for which $\mathbf{j}_\mathbf{p} = \mathbf{p}/m$, a straightforward integration leads to

$$\sigma(0, \omega) = \frac{iNe^2}{m\omega}. \quad (3.75)$$

Equation (3.75) is a well-known result, which can be derived directly. Let us emphasize that σ depends on the *bare* mass m, and not on the effective mass m^*.

With the aid of Eq. (3.32), we find that

$$\epsilon(0, \omega) = 1 - \frac{\omega_p^2}{\omega^2} \qquad (0 < qv_F \ll \omega). \quad (3.76)$$

Strictly speaking this result has been proved only for small ω $(\omega \ll E_F)$.

However, as we shall see in Chapter 4, by virtue of two longitudinal sum rules, Eq. (3.76) is valid for all frequencies. As we have seen, the zeros of $\epsilon(\mathbf{q}, \omega)$ yield the plasma oscillation frequencies. Equation (3.76) therefore tells us that

$$\lim_{\mathbf{q}\to 0} \omega_q = \omega_p,$$

a result which is in accord with our simple calculation at the beginning of the chapter, and which we shall see is in fact exact. Let us emphasize once more that the Landau theory *cannot* be used to obtain the corrections of order q^2 in the plasma frequency.

POLES OF $\epsilon(\mathbf{q}, \omega)$

We have seen in Chapter 2 that the response function of neutral systems, $\chi^n(\mathbf{q}, \omega)$, possesses a pole at the zero sound frequency. This pole exists when zero sound is a well-defined collective mode, i.e., for F_0^s sufficiently large. In view of Eq. (3.56), we expect that under similar circumstances in the electron liquid there will be a pole in $\epsilon(\mathbf{q}, \omega)$. For example, in the strong coupling limit, $F_0^s \gg 1$, $\epsilon(\mathbf{q}, \omega)$ takes the following form:

$$\epsilon(\mathbf{q}, \omega) = 1 - \frac{\omega_p^2}{(\omega + i\eta)^2 - s^2q^2}, \tag{3.77}$$

where s is the first and zero sound velocity (identical in this limit). (This expression is valid *both* in the strong-coupling collisionless regime and in the "extreme" hydrodynamic regime, its applicability to the latter regime following directly from Eqs. (2.182) and (3.56).) More generally, one expects a "zero sound–like" pole in $\epsilon(\mathbf{q}, \omega)$, plus additional structure associated with the single-pair excitation contributions. Such a pole in $\epsilon(\mathbf{q}, \omega)$ is not readily observable; as we shall see in the following chapter, it corresponds to a *zero* in the dissipative part of the density-density response function, $\chi''(\mathbf{q}, \omega)$. In other words, an external scalar probe cannot measure structure in $\epsilon(\mathbf{q}, \omega)$; it measures, in fact, structure in $\epsilon^{-1}(\mathbf{q}, \omega)$, a result which is apparent from the defining equations, (3.19) and (3.23). It is possible that for anisotropic systems, or as a result of boundary effects, a coupling between a scalar probe and $\epsilon(\mathbf{q}, \omega)$ may take place; such a possibility deserves exploration. In general the appearance of a pole in $\epsilon(\mathbf{q}, \omega)$ is primarily of academic interest.

3.5. BACKFLOW

We have earlier studied the screening of a fixed charged impurity in the quantum plasma. We now extend that study to the response of the electron system to a moving impurity. We shall confine our attention to the case of a slowly moving massive impurity; in this case dynamic screening produces a current flow around the impurity of just such a kind as to screen the long wavelength current fluctuations associated with the impurity motion. This *backflow* of electrons corresponds to an induced polarization cloud which moves along with the impurity; it is a purely reactive effect. There will be dissipative effects as well, which correspond to transfer of energy from the impurity to the plasma. However, for a slowly moving impurity, of velocity $V_e \ll v_F$, such dissipative effects are negligible.

Let us assume that the impurity possesses a charge z, and that it is at the origin at time $t = 0$. To the extent that we may neglect the recoil of the impurity (and that is why we assume it to be very massive), we may regard it as an external test charge, of density

$$\rho_e(\mathbf{r}, t) = \delta(\mathbf{r} - \mathbf{V}_e t) \tag{3.78a}$$

with Fourier transform

$$\rho_e(\mathbf{q}, \omega) = 2\pi\delta(\omega - \mathbf{q} \cdot \mathbf{V}_e). \tag{3.78b}$$

Associated with the impurity motion is a current fluctuation, of which the Fourier transform is

$$\mathbf{J}_e(\mathbf{q}, \omega) = \mathbf{V}_e 2\pi\delta(\omega - \mathbf{q} \cdot \mathbf{V}_e). \tag{3.79}$$

If the charge z is sufficiently weak, the response of the electron gas will be linear and is easily calculated from Eq. (3.23). The induced electron density fluctuation is such that

$$e\langle\rho(\mathbf{q}, \omega)\rangle = \left\{\frac{1}{\epsilon(\mathbf{q}, \omega)} - 1\right\} z 2\pi\delta(\omega - \mathbf{q} \cdot \mathbf{V}_e). \tag{3.80}$$

In order to obtain the induced current, we remark that it is purely longitudinal (since it results from the *longitudinal* field created by the impurity). Thus $\langle \mathbf{J}(\mathbf{r}, t)\rangle$ may be obtained from the particle conservation equation, (3.31); for our isotropic system it possesses the Fourier transform

$$\langle \mathbf{J}(\mathbf{q}, \omega)\rangle = \frac{\mathbf{q}\omega}{q^2}\langle\rho(\mathbf{q}, \omega)\rangle = \mathbf{q}\frac{\mathbf{q} \cdot \mathbf{V}_e}{q^2}\left\{\frac{1}{\epsilon(\mathbf{q}, \omega)} - 1\right\}\frac{z}{e} 2\pi\delta(\omega - \mathbf{q} \cdot \mathbf{V}_e). \tag{3.81}$$

So far our results have been exact (within the framework of the linear treatment of the electron response). We now assume that the impurity is moving very slowly, such that $V_e \ll v_F$. We are then in the quasi-static limit. We may neglect, to a first approximation, the transfer of energy from the impurity to the plasma, and write

$$\langle \mathbf{J}(\mathbf{q}, \omega)\rangle = \mathbf{q}\frac{\mathbf{q} \cdot \mathbf{V}_e}{q^2}\left\{\frac{1}{\epsilon(\mathbf{q}, 0)} - 1\right\}\frac{z}{e} 2\pi\delta(\omega - \mathbf{q} \cdot \mathbf{V}_e). \tag{3.82}$$

For long wavelengths ($q \ll p_F$), the dielectric constant $\epsilon(\mathbf{q}, 0)$, given by Eq. (3.60), is of order $1/q^2$. Within corrections of order q^2, we can neglect $1/\epsilon$ as compared to 1, and write

$$\langle \mathbf{J}(\mathbf{q}, \omega) \rangle = -\mathbf{q} \frac{\mathbf{q} \cdot \mathbf{V}_e}{q^2} \frac{z}{e} 2\pi\delta(\omega - \mathbf{q} \cdot \mathbf{V}_e) \qquad (q \ll p_F). \quad (3.83)$$

The result (3.83) is quite a striking one. We see, on comparing Eqs. (3.83) and (3.79), that the *net* electric current fluctuation, $z\mathbf{J}_e(\mathbf{q}, \omega) + e\langle \mathbf{J}(\mathbf{q}, \omega) \rangle$, satisfies the relation

$$\mathbf{q} \cdot (z\mathbf{J}_e + e\langle \mathbf{J}(\mathbf{q}, \omega) \rangle) = 0 \qquad (q \ll p_F). \quad (3.84)$$

In other words, the long wavelength current fluctuation induced in the electron gas is just such as to compensate the longitudinal portion of the

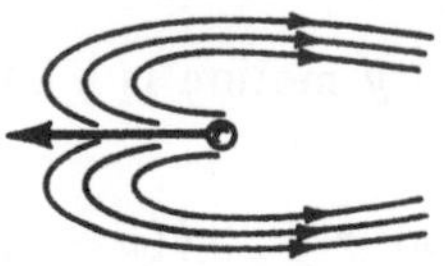

FIGURE 3.2. *Dipolar backflow around a charged impurity atom.*

current fluctuation carried by the impurity. We may speak of the induced electron current as a *backflow* of electrons around the impurity, which cancels out the long wavelength part of the longitudinal impurity current.

In order to achieve a qualitative understanding of the spatial character of the induced current, let us assume the relation (3.83) to hold for all $\mathbf{q}$. By taking its Fourier transform, we find the current distribution in $\mathbf{r}$-space,

$$\langle \mathbf{J}(\mathbf{r}, t) \rangle = \frac{z}{4\pi e} \nabla(\mathbf{V}_e \cdot \nabla) \frac{1}{R} = \frac{z}{4\pi e} \left\{ \frac{3(\mathbf{V}_e \cdot \mathbf{R})\mathbf{R}}{R^5} - \frac{\mathbf{V}_e}{R^3} \right\}, \quad (3.85)$$

where we have set

$$\mathbf{R} = \mathbf{r} - \mathbf{V}_e t. \quad (3.86)$$

Equation (3.85) describes a *dipolar flow* around the impurity, of the type shown on Fig. 3.2; the flow pattern extends to large distances from the moving impurity.

At first sight, this result is rather surprising, since the induced charge is localized in the immediate vicinity of the impurity. If we assume perfect screening, the induced electron density is

$$\langle \rho(\mathbf{r}, t)\rangle = -\frac{z}{e}\,\delta(\mathbf{R}). \tag{3.87}$$

Since this disturbance moves at a velocity $\mathbf{V}_e$, one would expect to find an electron current

$$\langle \mathbf{J}(\mathbf{r}, t)\rangle = -\frac{z}{e}\,\mathbf{V}_e\delta(\mathbf{R}) \tag{3.88}$$

instead of the dipolar flow (3.85). The reason for the difference is that the charged impurity acts as a longitudinal disturbance, and hence can only induce longitudinal currents in the electron system. The impurity current may be split into a transverse and a longitudinal part. Each of these extends over the whole volume, even though the total impurity current is localized at $\mathbf{R} = 0$. The induced current (3.85) cancels exactly the longitudinal part of the impurity current, leaving the transverse part untouched. In fact, the dipolar form of $\langle \mathbf{J}(\mathbf{R})\rangle$ could have been predicted by a purely geometrical argument: $\langle \mathbf{J}\rangle$ is curl free (being longitudinal), and divergence free wherever the induced density is negligible. The only such function with vector symmetry is a dipole of the form (3.85).

The existence of a "perfect" dipolar backflow around the impurity arises directly as a consequence of perfect "dielectric" screening. It is an obvious consequence of the long range of the Coulomb interaction. For the case of a slowly moving impurity considered here, the backflow is the dominant effect. This is no longer true if the impurity moves rapidly as compared to the electrons ($V_e \gg v_F$): the major effect is then the transfer of energy from the impurity to the plasma (see Section 4.4).

The dipolar flow (3.85) has been obtained under the assumption that $1/\epsilon$ could be neglected as compared to 1 in Eq. (3.81). Obviously this is an approximation which is only valid for small $\mathbf{q}$. We might expect it to provide an accurate description of the flow pattern at large distances from the impurity (R much larger than the screening radius). Actually, this is not quite true: we shall see in Chapter 5 that the dielectric constant $\epsilon(\mathbf{q}, 0)$ displays a singularity at $q = 2p_F$, where p_F is the Fermi momentum. This singularity gives rise to oscillations of the electron density, decreasing as $\cos(2p_FR)/R^3$. The longitudinal current associated with these oscillations varies as $1/R^3$, and is thus comparable with the dipolar backflow (3.85); however, it oscillates very rapidly, over distances of atomic size. The dipolar flow (3.85) therefore represents the electron current far from the impurity, *averaged* over a volume large as compared to the atomic dimensions.

Let us emphasize that the backflow for the charged Fermi liquid differs from that characteristic of the neutral Fermi liquid. On comparing the results of the present section with those of Section 2.5, [for instance, Eq. (3.85) with (2.162)], we see that the longitudinal response has the same structure in the two cases, but a different strength; for a neutral system, the backflow is not perfect. The special behavior of the charged Fermi liquid occurs essentially because the screening action of the electrons dominates their response to any quasistatic longitudinal field.

3.6. RESPONSE TO ELECTROMAGNETIC FIELDS

SEMICLASSICAL RESPONSE THEORY

We consider now the response of an electron system to a transverse electromagnetic field, which may vary in both space and time. We shall adopt the usual semiclassical approach to this problem, in which the electric and magnetic fields, $\mathcal{E}$ and $\mathcal{H}$, are determined in self-consistent fashion from Maxwell's equations:

$$\operatorname{curl} \mathcal{E} = -\frac{1}{c}\frac{\partial \mathcal{H}}{\partial t}, \tag{3.89}$$

$$\operatorname{curl} \mathcal{H} = \frac{1}{c}\frac{\partial \mathcal{E}}{\partial t} + \frac{4\pi e}{c}\langle \mathbf{J} \rangle. \tag{3.90}$$

In Eq. (3.90), $\langle \mathbf{J} \rangle$ is the particle current induced in the electron system by the electromagnetic field. We assume the fields are sufficiently weak that there exists a linear relation between the induced current and the electric field strength (a condition well satisfied for all but laser beams). We therefore write

$$e\langle J_\mu \rangle = \sigma_{\mu\nu}\mathcal{E}_\nu, \tag{3.91}$$

an equation which serves to define the conductivity tensor, $\sigma_{\mu\nu}$. The equations must then be solved in self-consistent fashion, since the currents that the field induces in turn serve to specify the field.

We will again specialize to the case of an isotropic electron system. In that case, with a transverse electric field $\mathcal{E}$, such that

$$\mathbf{q} \cdot \mathcal{E}(\mathbf{q}, \omega) = 0, \tag{3.92}$$

there can be no induced charge density: there is no electrostatic polarization field. The relation of $\mathbf{J}$ and $\mathcal{E}$ is a scalar one:

$$e\langle \mathbf{J}_\perp(\mathbf{q}, \omega)\rangle = \sigma_\perp(\mathbf{q}, \omega)\mathcal{E}_\perp(\mathbf{q}, \omega). \tag{3.93}$$

Equation (3.93) is directly analogous to the relation (3.29) used to define the longitudinal conductivity.

It is often useful to specify the system's response in terms of a *transverse* frequency and wave vector–dependent dielectric constant, $\epsilon_\perp(\mathbf{q}, \omega)$. We define $\epsilon_\perp(\mathbf{q}, \omega)$ by introducing the displacement field strength, $\mathfrak{D}$, into Maxwell's equation, (3.90):

$$\text{curl } \mathfrak{K}(\mathbf{r}, t) = \frac{1}{c}\frac{\partial \mathfrak{D}_\perp(\mathbf{r}, t)}{\partial t} = \frac{1}{c}\frac{\partial \mathcal{E}_\perp(\mathbf{r}, t)}{\partial t} + \frac{4\pi e}{c}\langle \mathbf{J}(\mathbf{r}, t)\rangle. \tag{3.94}$$

For an isotropic system, $\mathfrak{D}$ will be proportional to $\mathcal{E}$, so that we may write

$$\mathfrak{D}_\perp(\mathbf{q}, \omega) = \epsilon_\perp(\mathbf{q}, \omega)\mathcal{E}_\perp(\mathbf{q}, \omega) \tag{3.95}$$

as the equation that defines $\epsilon_\perp(\mathbf{q}, \omega)$. On taking the Fourier transform of Eq. (3.94), and making use of Eq. (3.93), we find

$$\epsilon_\perp(\mathbf{q}, \omega) = 1 + \frac{4\pi i}{\omega}\sigma_\perp(\mathbf{q}, \omega), \tag{3.96}$$

a relation quite analogous to Eq. (3.32).

We have thus arrived at the usual symmetric definition of longitudinal and transverse conductivities and dielectric constants. In the case of a homogeneous time-varying external field, the two response functions for an isotropic system are equal:

$$\epsilon(0, \omega) = \epsilon_\perp(0, \omega) \tag{3.97}$$

[as are $\sigma(0, \omega)$ and $\sigma_\perp(0, \omega)$]. This important result follows directly from the defining equations (3.29) and (3.93): in the limit $\mathbf{q} \equiv 0$, one obviously cannot distinguish between a longitudinal and a transverse electric field.

TRANSPORT EQUATION

We now proceed to write down the transport equation for quasiparticles in the presence of a transverse field. We expect it to be very similar to Eq. (3.45) derived in the case of a longitudinal perturbation. At first sight, the only change should be the replacement of the driving force $e\mathcal{E}$ by the Lorentz force

$$\mathfrak{F} = e\mathcal{E} + \frac{e}{c}\mathbf{v}_\mathbf{p} \times \mathfrak{K}. \tag{3.98}$$

Actually, Eq. (3.98) is not completely obvious. There is no *a priori* reason that the magnetic force should involve the velocity $\mathbf{v}_\mathbf{p}$ rather than the current $\mathbf{j}_\mathbf{p}$. Furthermore, the transport equation (3.45) is based on the assumption that the force $\mathfrak{F}$ is small, of the same order as $\delta n_\mathbf{p}$. In practice, one is often interested in the response of an electron gas exposed to a large uniform magnetic field (as in cyclotron resonance or magnetoconductance experiments). The magnetic force is then of "zeroth order"; the transport equation (3.45) must be revised accordingly.

For all these reasons, we shall construct the transport equation in the presence of an electromagnetic field from first principles. As usual, we express the fields $\mathcal{E}$ and $\mathcal{H}$ in terms of a vector potential $\mathcal{A}(\mathbf{r}, t)$:

$$\begin{aligned} \mathcal{E} &= -\frac{1}{c}\frac{\partial \mathcal{A}}{\partial t}, \\ \mathcal{H} &= \operatorname{curl} \mathcal{A}, \end{aligned} \tag{3.99}$$

$\mathcal{A}(\mathbf{r}, t)$ is determined up to a gauge transformation (we choose the gauge where the scalar potential vanishes).

We first review the motion of a single particle in the vector potential $\mathcal{A}(\mathbf{r}, t)$. Let $\mathbf{p}$ be the momentum of the particle *in the absence* of an electromagnetic field ($\mathcal{A} = 0$). In the presence of the field, the quantity that is canonically conjugate to the position $\mathbf{r}$ is no longer $\mathbf{p}$, but rather

$$\mathsf{p} = \mathbf{p} + \frac{e\mathcal{A}(\mathbf{r}, t)}{c}. \tag{3.100}$$

The Hamiltonian of the particle may be written as

$$H = \frac{1}{2m}\left[\mathsf{p} - \frac{e\mathcal{A}(\mathbf{r}, t)}{c}\right]^2. \tag{3.101}$$

The motion of the particle is then governed by Hamilton's equations

$$\begin{aligned} \frac{dr_\alpha}{dt} &= \frac{\partial H}{\partial \mathsf{p}_\alpha} = \frac{1}{m}\left[\mathsf{p}_\alpha - \frac{e\mathcal{A}_\alpha(\mathbf{r}, t)}{c}\right], \\ \frac{d\mathsf{p}_\alpha}{dt} &= -\frac{\partial H}{\partial r_\alpha} = -\frac{e}{mc}\frac{\partial \mathcal{A}_\beta}{\partial r_\alpha}\left[\mathsf{p}_\beta - \frac{e\mathcal{A}_\beta(\mathbf{r}, t)}{c}\right]. \end{aligned} \tag{3.102}$$

We note that the particle velocity is given by

$$\mathbf{v} = \frac{1}{m}\left[\mathsf{p} - \frac{e\mathcal{A}}{c}\right]. \tag{3.103}$$

On combining Eqs. (3.102) with the definitions (3.99), we obtain easily the usual equation of motion

$$m\frac{d^2\mathbf{r}}{dt^2} = e\mathcal{E} + \frac{e}{c}\mathbf{v} \times \mathcal{H}. \tag{3.104}$$

The Hamiltonian (3.101) [as well as the velocity (3.103)] is simply obtained by replacing the momentum $\mathbf{p}$ by $(\mathsf{p} - e\mathcal{A}/c)$ in the corresponding expression in the absence of a field. The only effect of the vector potential $\mathcal{A}$ is thus to shift the origin in p-space by an amount $e\mathcal{A}(\mathbf{r}, t)/c$. The quantity p canonically conjugate to $\mathbf{r}$ is measured from this shifted origin, instead of the real one; the Hamiltonian is unchanged when expressed in terms of $\mathbf{p}$ and $\mathbf{r}$.

The above conclusion still holds for a *system* of particles, the Hamiltonian of which may be written as

$$H = \frac{1}{2m}\sum_i \left[\mathsf{p}_i - \frac{e\mathcal{A}(\mathbf{r}_i, t)}{c}\right]^2 + V \tag{3.105}$$

(where V is the Coulomb interaction). It furnishes the appropriate clue for the construction of the Hamiltonian of excited quasiparticles in a Fermi liquid, from which we can derive the transport equation.

Consider a quasiparticle at point $\mathbf{r}$, and let p be the momentum conjugate to $\mathbf{r}$ *in the presence of the field*. In order to obtain the quasiparticle Hamiltonian, we introduce the quantity

$$\mathbf{p} = \mathsf{p} - \frac{e\mathcal{A}(\mathbf{r}, t)}{c}. \tag{3.106}$$

$\mathbf{p}$ is the usual momentum. According to the preceding discussion, the Hamiltonian *expressed in terms of* $\mathbf{p}$ *and* $\mathbf{r}$ is the same as when $\mathcal{A} = 0$: it is thus equal to the local energy $\bar{\epsilon}_\mathbf{p}$, itself a functional of the quasiparticle distribution $n_\mathbf{p}(\mathbf{r}, t)$ [see Chapter 1, Eq. (1.77)]. Upon differentiation, we may write

$$d\bar{\epsilon}_\mathbf{p} = \mathbf{v}_\mathbf{p} \cdot d\mathbf{p} + \sum_{\mathbf{p}'} f_{\mathbf{p}\mathbf{p}'}[\nabla_\mathbf{r} n_{\mathbf{p}'}]_{\mathbf{p}'} \cdot d\mathbf{r}. \tag{3.107}$$

Making use of Eq. (3.106), we can relate $d\mathbf{p}$ to $d\mathsf{p}$:

$$dp_\alpha = d\mathsf{p}_\alpha - \frac{e}{c}\frac{\partial \mathcal{A}_\alpha}{\partial r_\beta}dr_\beta. \tag{3.108}$$

By combining Eqs. (3.107) and (3.108), we finally obtain

$$\begin{aligned} \left(\frac{\partial \bar{\epsilon}_\mathbf{p}}{\partial \mathsf{p}_\alpha}\right)_\mathbf{r} &= v_{p\alpha}, \\ \left(\frac{\partial \bar{\epsilon}_\mathbf{p}}{\partial r_\alpha}\right)_\mathsf{p} &= \sum_{\mathbf{p}'} f_{\mathbf{p}\mathbf{p}'}\left(\frac{\partial n_{\mathbf{p}'}}{\partial r_\alpha}\right)_{\mathbf{p}'} - \frac{e}{c}v_{p\beta}\frac{\partial \mathcal{A}_\beta}{\partial r_\alpha}. \end{aligned} \tag{3.109}$$

These two equations, together with Hamilton's equation, describe completely the motion of a quasiparticle.

Let us now study the flow of quasiparticles in phase space. Let $\mathsf{n}_\mathsf{p}(\mathbf{r}, t)$ be the distribution function as measured from the shifted origin in $\mathbf{p}$-space; it is equal to the distribution n_p measured from the true origin $\mathbf{p} = 0$. In the absence of collisions, the quasiparticle flow is governed by Liouville's equation

$$\left(\frac{\partial \mathsf{n}_\mathbf{p}}{\partial t}\right)_{\mathbf{p},\mathbf{r}} + \left(\frac{\partial \mathsf{n}_\mathbf{p}}{\partial r_\alpha}\right)_\mathbf{p} \left(\frac{\partial \bar{\epsilon}_\mathbf{p}}{\partial \mathsf{p}_\alpha}\right)_\mathbf{r} - \left(\frac{\partial \mathsf{n}_\mathbf{p}}{\partial \mathsf{p}_\alpha}\right)_\mathbf{r} \left(\frac{\partial \bar{\epsilon}_\mathbf{p}}{\partial r_\alpha}\right)_\mathbf{p} = 0. \tag{3.110}$$

Equation (3.110) is a direct consequence of the Hamilton equations satisfied by the canonically conjugate variables $\mathbf{p}$ and $\mathbf{r}$; it provides the quasiparticle transport equation.

In practice, Eq. (3.110) is not very manageable, as the momentum $\mathbf{p}$ depends on the vector potential $\mathcal{A}(\mathbf{r}, t)$ applied to the system. It is much more convenient to express the transport equation in terms of the variable $\mathbf{p}$, which is characteristic of the system itself in the absence of an electromagnetic field. For this purpose, we note that

$$\begin{aligned}
\left(\frac{\partial \mathsf{n}_\mathbf{p}}{\partial t}\right)_{\mathbf{p},\mathbf{r}} &= \left(\frac{\partial n_\mathbf{p}}{\partial t}\right)_{\mathbf{p},\mathbf{r}} - \frac{e}{c}\frac{\partial n_\mathbf{p}}{\partial p_\alpha}\frac{\partial \mathcal{A}_\alpha}{\partial t}, \\
\left(\frac{\partial \mathsf{n}_\mathbf{p}}{\partial r_\alpha}\right)_\mathbf{p} &= \left(\frac{\partial n_\mathbf{p}}{\partial r_\alpha}\right)_\mathbf{p} - \frac{e}{c}\frac{\partial n_\mathbf{p}}{\partial p_\beta}\frac{\partial \mathcal{A}_\beta}{\partial r_\alpha}, \\
\left(\frac{\partial \mathsf{n}_\mathbf{p}}{\partial \mathsf{p}_\alpha}\right)_\mathbf{r} &= \left(\frac{\partial n_\mathbf{p}}{\partial p_\alpha}\right)_\mathbf{r}.
\end{aligned} \tag{3.111}$$

By inserting Eqs. (3.111) into (3.110), we obtain our final version of the transport equation, expressed in terms of the variables $(\mathbf{p}, \mathbf{r}, t)$:

$$\frac{\partial n_\mathbf{p}}{\partial t} + \mathbf{v}_\mathbf{p} \cdot \nabla_\mathbf{r} n_\mathbf{p} - \sum_{\mathbf{p}'} \nabla_\mathbf{p} n_\mathbf{p} \cdot \nabla_\mathbf{r} n_{\mathbf{p}'} f_{\mathbf{p}\mathbf{p}'} + e\mathcal{E} \cdot \nabla_\mathbf{p} n_\mathbf{p} + \frac{e}{c}[\mathbf{v}_\mathbf{p} \times \mathcal{H}] \cdot \nabla_\mathbf{p} n_\mathbf{p} = 0 \tag{3.112}$$

[where we have used the definitions (3.99) of $\mathcal{E}$ and $\mathcal{H}$]. We thus substantiate the guess formulated at the beginning of this section: the quasiparticles move as if they were subjected to the Lorentz force (3.98).

The current density carried by the quasiparticles may be analyzed in a similar way. In terms of the momentum $\mathbf{p}$, the current is given by the usual expression

$$\mathbf{J}(\mathbf{r}, t) = \sum_\mathbf{p} \mathbf{j}_\mathbf{p} n_\mathbf{p}(\mathbf{r}, t). \tag{3.113}$$

Let us now consider a quasiparticle with a canonical momentum $\mathbf{p}$ given by Eq. (3.106). It carries a current $\mathbf{j}_{\mathbf{p}-e\mathcal{A}/c}$. The total current expressed in terms of the variables $(\mathbf{p}, \mathbf{r}, t)$ is thus given by

$$\mathbf{J}(\mathbf{r}, t) = \sum_\mathbf{p} \mathsf{n}_\mathbf{p}(\mathbf{r}, t)\mathbf{j}_{\mathbf{p}-e\mathcal{A}/c}. \tag{3.114}$$

For the special case of a translationally invariant system, we recover the well-known result

$$\mathbf{J}(\mathbf{r}, t) = \sum_{\mathbf{p}} n_{\mathbf{p}}(\mathbf{r}, t) \left[\frac{\mathbf{p}}{m} - \frac{e\mathcal{A}(\mathbf{r}, t)}{mc}\right]. \qquad (3.115)$$

Equations (3.113) and (3.115) are two equivalent expressions of the current density, which differ by the *definition* of the momentum variable.

It now remains to linearize the transport equation (3.112). Let us first assume that there is no applied dc magnetic field. $\mathcal{H}$ then reduces to the ac magnetic field associated with the propagating electromagnetic wave. $\mathcal{H}$ and $\mathcal{E}$ are both first-order perturbations, of the same order as $\delta n_{\mathbf{p}}$. Making use of the definition (3.54), we can write the transport equation (3.112) in the form

$$\frac{\partial \delta n_{\mathbf{p}}}{\partial t} + \mathbf{v}_{\mathbf{p}} \cdot \nabla_{\mathbf{r}} \delta \bar{n}_{\mathbf{p}} + e\left\{\mathcal{E} + \frac{\mathbf{v}_{\mathbf{p}}}{c} \times \mathcal{H}\right\} \cdot \nabla_{\mathbf{p}} n_{\mathbf{p}}{}^{\circ} = 0. \qquad (3.116)$$

Equation (3.116) is similar to Eq. (3.55) derived in the longitudinal case, except for the additional term describing the driving action of the ac magnetic field. In most cases of interest, $n_{\mathbf{p}}{}^{\circ}$ is the ground state distribution $n^{\circ}(\epsilon_{\mathbf{p}})$. Then $\nabla_{\mathbf{p}} n_{\mathbf{p}}{}^{\circ}$ is parallel to $\mathbf{v}_{\mathbf{p}}$: the magnetic driving term of (3.116) vanishes (it is only of importance if the unperturbed distribution carries a dc current). For a system at equilibrium, Eq. (3.116) reduces exactly to (3.55).

Let us now assume that a large dc magnetic field $\mathcal{H}_0$ is applied to the system, and that, furthermore, $n_{\mathbf{p}}{}^{\circ}$ is the ground state distribution. Since $\mathcal{H}_0$ is of zeroth order with respect to $\delta n_{\mathbf{p}}$, we must look for the first-order contribution to the other two factors, $\mathbf{v}_{\mathbf{p}}$ and $\nabla_{\mathbf{p}} n_{\mathbf{p}}$ (the velocity $\mathbf{v}_{\mathbf{p}}$ is here defined as the gradient with respect to $\mathbf{p}$ of $\tilde{\epsilon}_{\mathbf{p}}$, and thus contains first-order corrections). The first-order contributions to Eq. (3.112) are found to be

$$\frac{e}{c}(\mathbf{v}_{\mathbf{p}} \times \mathcal{H}_0) \cdot \nabla_{\mathbf{p}} \delta n_{\mathbf{p}} + \frac{e}{c}\left\{\nabla_{\mathbf{p}}\left[\sum_{\mathbf{p}'} f_{\mathbf{p}\mathbf{p}'} \delta n_{\mathbf{p}'}\right] \times \mathcal{H}_0\right\} \cdot \nabla_{\mathbf{p}} n_{\mathbf{p}}{}^{\circ}. \qquad (3.117)$$

We can then, after some algebra, write this transport equation in the form

$$\frac{\partial\, \delta n_{\mathbf{p}}}{\partial t} + \mathbf{v}_{\mathbf{p}} \cdot \nabla_{\mathbf{r}} \delta \bar{n}_{\mathbf{p}} + \frac{e}{c}(\mathbf{v}_{\mathbf{p}} \times \mathcal{H}_0) \cdot \nabla_{\mathbf{p}} \delta \bar{n}_{\mathbf{p}} + e\mathcal{E} \cdot \nabla_{\mathbf{p}} n_{\mathbf{p}}{}^{\circ} = 0 \qquad (3.118)$$

(according to the previous discussion, the ac magnetic field does not contribute to the transport equation). We see that the applied dc magnetic field $\mathcal{H}_0$ acts on the departure $\delta \bar{n}_{\mathbf{p}}$ from *local equilibrium*, instead of $\delta n_{\mathbf{p}}$, which was not immediately obvious. Equation (3.118) is extremely useful, as it forms the starting point for a study of cyclotron resonance, magnetoresistance, etc. (see Problem 3.8); we wish therefore to stress its importance. In fact, we shall not have further occasion to consider it, as we shall henceforth confine our attention to the case of small ac transverse perturbations; the transport equation is then identical to that used for a longitudinal perturbation, with $\mathcal{E}$ representing the transverse electric field.

MACROSCOPIC TRANSVERSE RESPONSE

We now apply the transport equation (3.55) to the evaluation of the macroscopic conductivity tensor $\sigma_{\alpha\beta}(\mathbf{q}, \omega)$, in the limit in which $\mathbf{q}$ and ω are small. The calculations are practically identical to those carried out for a longitudinal field.

Let us first consider the limit $\mathbf{q} = 0$. The departure from equilibrium $\delta n_\mathbf{p}$ is then given by Eq. (3.73), which we can write as

$$\delta n_\mathbf{p}(0, \omega) = -\frac{ie}{\omega}\, \boldsymbol{\mathcal{E}} \cdot \nabla_\mathbf{p} n_\mathbf{p}^{\,0}. \tag{3.119}$$

We see that $\delta n_\mathbf{p}$ does not depend on the direction of $\mathbf{q}$. The current is the same, whether the field $\mathcal{E}$ is transverse or longitudinal. The transverse and longitudinal conductivities are thus seen to be equal, given by Eq. (3.75), in agreement with our earlier statement.

Let us now turn to the opposite limit $\omega = 0$. We wish to calculate the response to a quasistatic transverse field. Using the transport equation (3.55), we obtain

$$\delta \bar{n}_\mathbf{p} = -\frac{ie\boldsymbol{\mathcal{E}} \cdot \mathbf{v}_\mathbf{p}}{\mathbf{q} \cdot \mathbf{v}_\mathbf{p} - i\eta}\, \delta(\epsilon_p - \mu) \tag{3.120}$$

(where the term $i\eta$ expresses the adiabatic switching-on of the field). In the present case, $\mathcal{E}$ is perpendicular to $\mathbf{q}$, so that Eq. (3.120) does not reduce to the simple form (3.57). The current $\mathbf{J}$ is given by

$$\mathbf{J} = \sum_\mathbf{p} \delta \bar{n}_\mathbf{p} \mathbf{v}_\mathbf{p} = -ie \sum_\mathbf{p} \delta(\epsilon_p - \mu)(\boldsymbol{\mathcal{E}} \cdot \mathbf{v}_\mathbf{p})\mathbf{v}_\mathbf{p} \left\{ P\left(\frac{1}{\mathbf{q} \cdot \mathbf{v}_\mathbf{p}}\right) + i\pi\delta(\mathbf{q} \cdot \mathbf{v}_\mathbf{p}) \right\}. \tag{3.121}$$

For reasons of symmetry, the only contribution arises from the δ-function. One thus finds the following transverse conductivity:

$$\sigma_\perp(\mathbf{q}, 0) = \frac{3\pi N e^2}{4 q p_F}. \tag{3.122}$$

We note that $\sigma_\perp$ is real, corresponding to a dissipative process; the electromagnetic wave transfers energy to those electrons that move in a direction perpendicular to $\mathbf{q}$.

The result (3.122) depends only on the Fermi momentum p_F. Neither the effective mass m^* nor the interaction $f_{pp'}$ enters into $\sigma_\perp$. The static transverse conductivity is thus the same as that for a noninteracting system. This conclusion remains valid for a nonisotropic system, such as electrons in a real metal: a measurement of $\sigma_\perp(\mathbf{q}, 0)$ may then be used to determine the shape of the real Fermi surface.

Experimentally, the conductivity $\sigma_\perp(\mathbf{q}, 0)$ enters in the anomalous skin effect. The latter occurs whenever the wavelength q^{-1} of the electromagnetic field is much shorter than both the mean free path and the distance v_F/ω which an electron travels during a period. The ratio ω/qv_F is then vanishingly small; the conductivity of the electron gas is equal to $\sigma_\perp(\mathbf{q}, 0)$. Under such conditions, the relation between the electric field and the current density is nonlocal; there exists a q-dependent conductivity, hence the name "anomalous."

When ω is finite, the calculation of $\sigma_\perp$ is much more difficult. We might attempt to expand $\sigma_\perp$ in powers of ω/qv_F, as we did for the longitudinal case; even that is rather complicated. The first term of the expansion (which is linear in ω) is imaginary, corresponding to a reactive conductivity. It bears no simple relation to the corresponding longitudinal σ. The explicit calculation of this term is left as a problem to the reader.

Let us stress the essential difference between the longitudinal and transverse conductivities in the limit $\omega \ll qv_F$. The main term of σ is imaginary, of order ω, while the main term of $\sigma_\perp$ is real, of order 1. In this region, the behaviors of σ and $\sigma_\perp$ are thus completely different, in contrast with the opposite limit, $\omega \gg qv_F$, where they are both given by Eq. (3.75).

TRANSVERSE COLLECTIVE MODES

As we have remarked in Section 3.3, a transverse motion of the system gives rise to a current, which in turn produces an electromagnetic field. There exists a "feedback" of the system's response into the exciting field, similar to that found for longitudinal oscillations. As a consequence of this feedback, the transverse collective modes of a charged system are coupled to the elecromagnetic waves. They are obtained by a joint solution of the transport equation (3.55) and Maxwell's equations (3.89) and (3.90).

In fact, the dispersion relation for the coupled "electromagnetic-collective modes" can be obtained directly once we know the transverse conductivity $\sigma_\perp(\mathbf{q}, \omega)$. By combining Eqs. (3.89) and (3.90), we find that the frequency of a *transverse* wave satisfies the equation

$$c^2q^2 = \omega^2\epsilon_\perp(\mathbf{q}, \omega) = \omega^2 + 4\pi i\omega\sigma_\perp(\mathbf{q}, \omega). \tag{3.123}$$

The various roots of Eq. (3.123) provide *all* the transverse modes that can propagate in the system.

Let us first consider the high-frequency roots of Eq. (3.123), such that $\omega \gg qv_F$. We can then replace $\epsilon_\perp(\mathbf{q}, \omega)$ by the expression (3.76). In that case, the solution of Eq. (3.123) is straightforward, and yields

$$\omega^2 = \omega_p^2 + c^2q^2 \tag{3.124}$$

(we verify *a posteriori* that $\omega > cq \gg qv_F$). The solution (3.124) corresponds to the usual electromagnetic wave propagation. There are clearly two regions of interest.

(i) The short wavelength regime ($cq \gg \omega_p$), in which the frequency ω is very close to cq: the electromagnetic field is then *weakly coupled* to the system; its frequency of oscillation is only slightly modified by the presence of the electrons.

(ii) The long wavelength regime ($cq \ll \omega_p$), in which $\omega \cong \omega_p$: The electromagnetic field is then strongly coupled to the electrons, as witnessed by the enormous modification in its natural frequency, cq.

We notice that the frequency (3.124) is finite and large. In principle, the system's motion cannot be obtained from the Landau theory, which only applies to very low frequencies. Actually as was the case with longitudinal plasma oscillations Eq. (3.124) turns out to be correct because of simple sum rule arguments. (See Chapter 4.)

In addition to the solution (3.124), corresponding to photon propagation, Eq. (3.123) may exhibit other, low-frequency roots, such that $\omega \sim qv_F$. Such roots correspond to genuine transverse collective modes. Their *existence* is best studied by considering a simple model, in which the interaction $f_{\mathbf{pp}'}$ only has $\ell = 0$ and $\ell = 1$ components. The transport equation may then be solved explicitly, and provides the dispersion equation for transverse collective modes. We leave the detailed calculation as a problem to the reader (see Problem 3.7), and describe here only a few relevant features of the result.

(i) In the weak coupling limit ($cq \gg \omega_p$), the admixture of electromagnetic field (the "pure" photon field) into the transverse collective mode is negligible. In the dispersion equation (3.123), we can let cq go to infinity; the collective mode is the same as in a neutral system. The frequency of the collective mode appears as a pole of $\sigma_\perp(\mathbf{q}, \omega)$ (i.e., as a pole of the transverse current-current response function, in analogy with the results of Chapter 2). It is easily verified that the transverse collective mode only exists for strong repulsion between quasiparticles, such that $F_1^s > 6$.

(ii) In the strong coupling region, ($cq < \omega_p$) the transverse collective mode is repelled by the nearby photon root. It disappears in the continuum of single-pair excitations, as shown in Fig. 3.3. In that region, the effect of electromagnetic coupling is seen to be very important.

The above considerations are somewhat academic as the interaction F_1^s in an electron gas is usually too small to allow for a transverse collective mode. Nevertheless, such a problem is important as it clearly

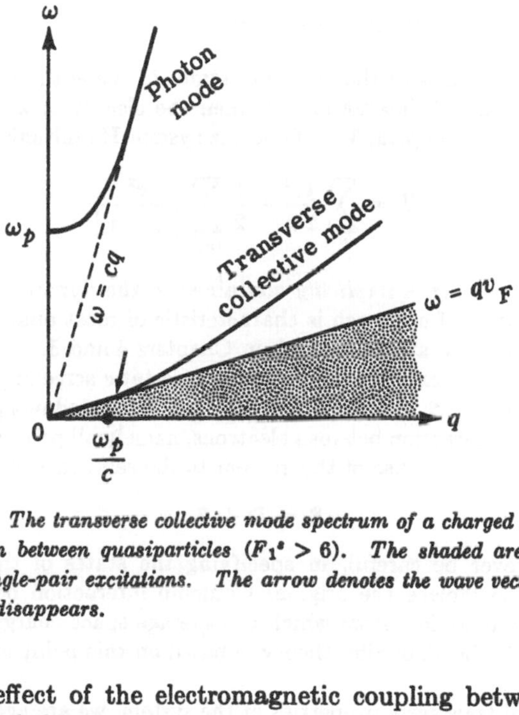

FIGURE 3.3. *The transverse collective mode spectrum of a charged Fermi liquid for strong interaction between quasiparticles ($F_1^s > 6$). The shaded area represents the continuum of single-pair excitations. The arrow denotes the wave vector for which the transverse mode disappears.*

displays the effect of the electromagnetic coupling between electrons as a function of the wave vector $\mathbf{q}$. Moreover, the same methods can be used in order to describe the transverse collective modes in the presence of a dc magnetic field, a problem which has many experimental applications. (See Problem 3.8.)

TRANSVERSE VS. LONGITUDINAL RESPONSE

We now compare in more detail the longitudinal and transverse properties of an electron gas. When dealing with the longitudinal response, we first considered the response to an *external* field, measured by the displacement vector $\mathfrak{D}$. It was only later that we found it convenient to work with the electric field $\mathcal{E}$, which embodies the average screening action of the electron gas. On the other hand, we formulated the transverse properties directly in terms of the electric field (the displacement defined in Eq. (3.94) being only a convenient mathematical step). We wish to explain the physical origin of this difference.

Let us first consider the response to an external test charge. The latter gives rise to a longitudinal "applied" displacement $\mathfrak{D}$. In the presence of $\mathfrak{D}$, the electron gas is polarized; the corresponding induced density fluctuations give rise to a space charge field $\mathcal{E}_p$. Each electron interacts with the space charge field as well as with $\mathfrak{D}$. However, it should be noted that the interaction with $\mathcal{E}_p$ is a direct consequence of the Coulomb interaction between electrons; we must be careful not to count it twice. We are therefore free to choose either of the following two points of view:

(a) We may work with the quantum states of the entire system, including of course the Coulomb interaction between the electrons; we study then the response to the *external field*, $\mathfrak{D}$. Thus the system Hamiltonian is taken to be

$$H = \sum_i \frac{p_i^2}{2m} + \frac{1}{2} \sum_{i \neq j} \frac{e^2}{|\mathbf{r}_i - \mathbf{r}_j|}. \tag{3.125a}$$

Electrostatic screening is *implicitly* contained in the dynamic response of the system. This type of approach is characteristic of most quantum treatments, and will be considered at some length in Chapters 4 and 5.

(b) We may take *explicit* account of electrostatic screening by introducing the polarization field, $\mathcal{E}_p$. In this way, we have separated out part of the effect of the Coulomb interaction between electrons, namely all polarization processes. We then study the response of the system to the self-consistent screened field,

$$\mathcal{E} = \mathfrak{D} + \mathcal{E}_p.$$

We must however be careful, in specifying the states of the system which respond to $\mathcal{E}$, to replace the original Coulomb interaction between electrons by a screened interaction, from which the average space charge effect has been eliminated. The Landau–Silin theory is based on this point of view.

In the case of transverse properties of the system, we are again free to adopt either of the two attitudes in the treatment of the system response:

(a) We may take *explicit* account of the transverse *magnetic* interactions between the electrons, and study the response to an *external* field, $\mathfrak{D}_{\text{ext}}$. We work therefore with the quantum states of the entire system, specified by the Hamiltonian,

$$H = \sum_i \left[\mathbf{p}_i - \frac{e}{c}\,\mathfrak{A}(\mathbf{x}_i)\right]^2 + \frac{1}{2} \sum_{i \neq j} \frac{e^2}{|\mathbf{r}_i - \mathbf{r}_j|}, \tag{3.125b}$$

where $\mathfrak{A}$ is the transverse vector potential produced by the electron currents within the system. Magnetic screening is contained implicitly in the dynamic response of the system. This approach is generally necessary in dealing with a relativistic electron gas, for which the magnetic interactions between electrons play a role equal to that of the Coulomb interactions.

(b) We may study directly the response to the local transverse field, $\mathcal{E}$, which arises both from external sources and from the electron currents:

$$\mathcal{E} = \mathfrak{D}_{\text{ext}} + \mathcal{E}_p.$$

Magnetic screening is thereby explicitly taken into account through the introduction of the appropriate transverse polarization field, $\mathcal{E}_p$. There remains an effective screened magnetic interaction between electrons which gives rise, at best, to corrections of order v^2/c^2, which are negligible for the nonrelativistic electron system. We can then calculate the response to $\mathcal{E}$ using the states appropriate to a system with *no* magnetic interactions, that is, with the Hamiltonian, (3.125a). In other words, for a nonrelativistic system, the only important consequences of the magnetic interaction between electrons are the polarization processes, which may be treated explicitly by introducing the self-consistent field, $\mathcal{E}_p$. All other consequences of magnetic interactions may be neglected.

Clearly one might treat transverse and longitudinal response on an equivalent basis. For the nonrelativistic sytems with which we deal, it is, however, by no means useful to do so. For the transverse response, approach (b), based on calculating the response to a screened field, $\mathcal{E}$, is to be preferred. For the longitudinal response, both approaches possess advantages and disadvantages; which one uses depends on the problem at hand.

3.7. IMPURITY SCATTERING

Thus far we have discussed transport phenomena within the framework of the "collisionless" Landau–Silin equation. Such an equation is, of course, only an approximation. In fact, excited quasiparticles collide both with "external" defects (phonons or impurities in the case of electrons in metals) and with one another. Such collisions modify the response of electrons to an external field and must, in general, be taken into account in developing a general theory of transport phenomena for a charged Fermi liquid.

We have already discussed in Chapter 1 the collisions between quasiparticles, which are of considerable importance in liquid ^{3}He at not-too-low temperatures ($T \gtrsim 0.01°$K). For electrons in metals, all three collision mechanisms (electron-phonon, electron-impurity, and electron-electron) are potentially of importance. We quote here some order-of-magnitude estimates of the collision times of interest, in order to obtain some perspective on which mechanisms play a role in a given temperature regime. One finds the following approximate results [Pines (1963)] for collision mechanisms in sodium.

Electron-phonon collisions:

$$(1/\tau)_{e^- - ph} \cong \kappa T \qquad \text{(at room temperature)},$$
$$(1/\tau)_{e^- - ph} \cong T^5 \times 10^4 \qquad \text{(at liquid helium temperatures)}.$$

Electron-impurity collisions:

$$(1/\tau)_{e^- - imp} \cong (N_i/N) \times 10^{15} \qquad \text{(independent of temperature)}.$$

Electron-electron collisions:

$$(1/\tau)_{e^- - e^-} \cong (\kappa T/\epsilon_F)^2 \times 10^{15} \qquad (\kappa T \ll \epsilon_F).$$

At room temperatures, collisions with phonons provide the dominant mechanism for electron scattering. At liquid helium temperatures (4°K) electron-phonon collisions and electron-electron collisions are roughly of the same importance; they are, however, much less frequent than electron-impurity collisions, with which they compete only for an impurity concentration (N_i/N) of the order of 10^{-8}. Such impurity concentrations are not very practicable in metals. Therefore, electron-electron collisions play essentially no role in determining transport phenomena in most simple metals. By the time one has reached a sufficiently low temperature that they assume an importance equal to that of electron-phonon collisions, impurity scattering has become the dominant scattering mechanism. It is, in any case, the latter mechanism which is dominant for metals at temperatures below the liquid helium range; we therefore proceed to examine it in some detail.

STRUCTURE OF THE COLLISION INTEGRAL

In order to account for real collisions, one modifies the basic Landau transport equation by adding on the right-hand side a collision integral, $I(n)$, which describes the rate of change of the distribution function $n_{\mathbf{p}}$ due to scattering. In Section 1.8, we have discussed in some detail the structure of the collision integral for the mutual scattering of quasiparticles. We shall see that impurity scattering, although mathematically simpler, retains many of the physical features found for quasiparticle scattering.

We write the collision integral in the following form:

$$I(n_{\mathbf{p}}) = -\frac{2\pi}{\hbar}\sum_{\mathbf{p}'} W_{\mathbf{pp}'}\delta(\tilde{\epsilon}_{\mathbf{p}} - \tilde{\epsilon}_{\mathbf{p}'})\{n_{\mathbf{p}}(1 - n_{\mathbf{p}'}) - n_{\mathbf{p}'}(1 - n_{\mathbf{p}})\}. \qquad (3.126)$$

In this expression, the factor $(2\pi/\hbar)W_{\mathbf{pp}'}\delta(\tilde{\epsilon}_{\mathbf{p}} - \tilde{\epsilon}_{\mathbf{p}'})$ is the *transition probability* for the scattering of a quasiparticle from state $\mathbf{p}$ to state $\mathbf{p}'$. As in Section 1.8, we note that the conservation of energy involves the true *local* energy $\tilde{\epsilon}_{\mathbf{p}}$, instead of the equilibrium energy ϵ_p: a real transition must conserve the *total* energy, including the interaction energy with other excited quasiparticles. The last factor of Eq. (3.126) allows for the occupation of the initial and final states (taking the exclusion principle into account). The first term in brackets corresponds to scattering out of the state $\mathbf{p}$, the second term to scattering into that state.

In Eq. (3.126), the factor $W_{\mathbf{pp}'}$ appears as the squared modulus of the quasiparticle *scattering amplitude.* If the impurities are distributed *at random,* $W_{\mathbf{pp}'}$ is independent of the position of the impurities in the crystal. For dilute impurities, $W_{\mathbf{pp}'}$ will certainly be proportional to the density of impurities (as long as the duration of a collision remains short as compared to the time between two consecutive collisions). In practice, impurities are always of atomic size, and so give rise to large deflection angles and to momentum transfers of the order of p_F. The theoretical description of such processes lies beyond the scope of the Landau theory. We must therefore consider $W_{\mathbf{pp}'}$ as a phenomenological parameter, to be determined from experiment.

In the collision integral (3.126), let us replace the distribution function $n_{\mathbf{p}}$ by its expansion

$$n_{\mathbf{p}} = n^{0}(\tilde{\epsilon}_{\mathbf{p}}) + \delta\bar{n}_{\mathbf{p}}. \tag{3.127}$$

Owing to the conservation of energy, the contribution of n^0 cancels exactly. We are thus left with the linear collision integral

$$I(\delta\bar{n}) = -\frac{2\pi}{\hbar}\sum_{\mathbf{p}'} W_{\mathbf{pp}'}\delta(\epsilon_p - \epsilon_{p'})[\delta\bar{n}_{\mathbf{p}} - \delta\bar{n}_{\mathbf{p}'}]. \tag{3.128}$$

This integral is explicitly first order in $\delta\bar{n}$; hence to lowest order we can neglect the difference between ϵ_p and $\tilde{\epsilon}_{\mathbf{p}}$. Let us emphasize that the collision integral involves the local departure from equilibrium, $\delta\bar{n}$, rather than δn: this is a direct consequence of the conservation of $\tilde{\epsilon}_{\mathbf{p}}$ rather than ϵ_p in Eq. (3.126).

The collision integral (3.128) is quite general, and applies to any transport phenomenon at low temperature. At these temperatures, the electrons of interest are those near the Fermi surface; we can neglect any dependence of $W_{\mathbf{pp}'}$ on p and p'. For an isotropic system, $W_{\mathbf{pp}'}$ then depends only on the angle between $\mathbf{p}$ and $\mathbf{p}'$, a situation which leads to a considerable simplification in the calculation of $I(\delta\bar{n})$. In what follows, we shall always be interested in values of $\delta\bar{n}_{\mathbf{p}}$ of the form

$$\delta\bar{n}_{\mathbf{p}} = f(\epsilon_p, \sigma)g(\theta, \varphi), \tag{3.129}$$

where the polar angles (θ, φ) are used to define the direction of $\mathbf{p}$. We shall consider only collisions which conserve the spin σ of the quasi-particle. In that case, the summation over $\mathbf{p}'$ in Eq. (3.128) runs only over the momentum coordinates. Since the collisions also conserve energy, we may write Eq. (3.128) as

$$I(\delta\bar{n}) = -\frac{2\pi}{\hbar} f(\epsilon_p, \sigma) \sum_{\mathbf{p}'} W_{\mathbf{pp}'}\delta(\epsilon_p - \epsilon_{p'})[g(\theta, \varphi) - g(\theta', \varphi')]. \tag{3.130}$$

We next write $W_{\mathbf{pp}'}$ as a series of Legendre polynomials:

$$W_{\mathbf{pp}'} = \sum_{\ell} W_\ell P_\ell(\cos \xi), \tag{3.131}$$

where ξ is the angle between the directions (θ, φ) and (θ', φ'). Let us also expand $g(\theta, \varphi)$ in normalized spherical harmonics $Y_{\ell m}(\theta, \varphi)$. Using the addition theorem for spherical harmonics, we may write Eq. (3.130) in the following way:

$$I(\delta\bar{n}) = -\sum_{\ell m} \frac{\delta\bar{n}_{\ell m}}{\tau_\ell} \tag{3.132}$$

where $\delta\bar{n}_{\ell m}$ is that part of $\delta\bar{n}$ which has the symmetry (ℓ, m), and where the constants τ_ℓ are defined by

$$\frac{1}{\tau_\ell} = \frac{2\pi}{\hbar} \frac{1}{(2\pi\hbar)^3} \int p'^2\, dp'\, \delta(\epsilon_p - \epsilon_{p'}) 4\pi \left[W_0 - \frac{W_\ell}{2\ell + 1}\right]. \tag{3.133}$$

In practice, ϵ_p will always remain very close to μ; integration over p' therefore yields

$$\frac{1}{\tau_\ell} = \frac{3\pi}{\hbar} \frac{N}{m^* v_F{}^2} \left[W_0 - \frac{W_\ell}{2\ell + 1}\right] \tag{3.134}$$

This result is frequently written in the less explicit way

$$\frac{1}{\tau_\ell} = \frac{2\pi}{\hbar} \sum_{\mathbf{p}'} W_{\mathbf{pp}'}\delta(\epsilon_p - \epsilon_{p'})[1 - P_\ell(\cos \xi)]. \tag{3.135}$$

(For $\ell = 1$, we recognize the usual expression of the collision time for conductivity.) Note that $1/\tau_0$ is zero: this result expresses the conservation of the number of particles in the collisions.

Let us introduce the collision integral (3.132) into the Landau–

Silin transport equation. Equation (3.55) is thus replaced by

$$\mathbf{q} \cdot \mathbf{v}_{\mathbf{p}} \delta \bar{n}_{\mathbf{p}} - \omega \delta n_{\mathbf{p}} + ie\mathcal{E} \cdot \mathbf{v}_{\mathbf{p}} \delta(\epsilon_p - \mu) = i \sum_{\ell m} \frac{\delta \bar{n}_{\ell m}}{\tau_\ell}. \tag{3.136}$$

We shall solve this equation in a few simple cases.

RESPONSE TO A UNIFORM FIELD ($\mathbf{q} = 0$)

We know that there is no screening and therefore $\mathcal{E} = \mathfrak{D}$. From the form of the transport equation, (3.136), it follows that $\delta n_{\mathbf{p}}$ and $\delta \bar{n}_{\mathbf{p}}$ both have a vector symmetry; when both are expanded in spherical harmonics, the only contribution comes from spin symmetric components with $\ell = 1$. The relation between δn_{1m} and $\delta \bar{n}_{1m}$ is given in Eq. (1.34); thus we may write

$$\delta \bar{n}_{\mathbf{p}} = \left(1 + \frac{F_1{}^s}{3}\right) \delta n_{\mathbf{p}} = \delta n_{\mathbf{p}} \frac{m^*}{m} \tag{3.137}$$

[on making use of Eq. (1.100)]. We substitute Eq. (3.137) into (3.136), and find

$$\delta \bar{n}_{\mathbf{p}} = \frac{ie\mathcal{E} \cdot \mathbf{v}_{\mathbf{p}} \delta(\epsilon_p - \mu)}{(m\omega/m^*) + (i/\tau_1)}. \tag{3.138}$$

The current $\mathbf{J}$ is given by Eq. (1.89),

$$\mathbf{J} = \sum_{\mathbf{p}} \delta \bar{n}_{\mathbf{p}} \mathbf{v}_{\mathbf{p}},$$

from which we find the conductivity tensor

$$\sigma_{\alpha\beta} = \frac{Ne^2\tau}{m^*(1 - i\omega\tau m/m^*)} \delta_{\alpha\beta} \tag{3.139}$$

(for convenience, we have abbreviated τ_1 by τ). At zero frequency, we find that $\sigma = Ne^2\tau/m^*$, which agrees with the result Langer (1962) obtained by field-theoretic methods.

When the frequency ω increases, the conductivity changes from $Ne^2\tau/m^*$ to $iNe^2/m\omega$. The change from m^* to m looks like a real, observable many-body effect. Actually, the collision time τ is phenomenological, and we could equally well define an *effective* collision time

$$\tau_{\text{eff}} = m\tau/m^*, \tag{3.140}$$

in terms of which the conductivity becomes

$$\sigma_{\alpha\beta} = \frac{iNe^2}{m}\frac{1}{\omega + i/\tau_{\text{eff}}}\,\delta_{\alpha\beta}. \tag{3.141}$$

Equation (3.141) is similar to the result obtained for a noninteracting system: we cannot derive the ratio m^*/m from a measurement of $\sigma(\omega)$ alone.

SLOWLY VARYING EXTERNAL FIELDS ($0 < qv_F\tau \ll 1$)

We now consider the case $0 < qv_F\tau \ll 1$. We are then in the normal regime, for which the wavelength of the applied field is much longer than the mean free path of quasiparticles. Let us first consider a longitudinal field ($\mathbf{\mathcal{E}}\|\mathbf{q}$); if we choose the polar axis along $\mathbf{q}$, $\delta\bar{n}_\mathbf{p}$ has the symmetry $m = 0$. In Eq. (3.136), we can neglect the "diffusion term" $\mathbf{q}\cdot\mathbf{v}_\mathbf{p}\delta\bar{n}_\mathbf{p}$ as compared to the collision integral, except when $\ell = 0$; in the latter case one must include the diffusion term, since the collision integral vanishes. Within corrections of order $qv_F\tau$, $\delta\bar{n}$ and δn will only contain components $\delta\bar{n}_\ell$ and δn_ℓ with a symmetry $\ell = 0$ or $\ell = 1$. Collecting the corresponding contributions to Eq. (3.136), we find

$$\begin{aligned} -\omega\,\delta n_0 + \tfrac{1}{3}qv_F\delta\bar{n}_1 &= 0,\\ -\omega\,\delta n_1 + qv_F\delta\bar{n}_0 - \frac{i}{\tau}\,\delta\bar{n}_1 + ie\,\mathcal{E}v_F\delta(\epsilon_p - \mu) &= 0. \end{aligned} \tag{3.142}$$

According to Eq. (1.34), we have

$$\begin{aligned} \delta\bar{n}_0 &= (1 + F_0{}^s)\delta n_0,\\ \delta\bar{n}_1 &= \left(1 + \frac{F_1{}^s}{3}\right)\delta n_1. \end{aligned} \tag{3.143}$$

Upon elimination of δn_0 and $\delta\bar{n}_0$, we find [on making use of Eq. (3.137)],

$$\delta\bar{n}_1 = \frac{ie\mathbf{\mathcal{E}}\cdot\mathbf{v}_\mathbf{p}\delta(\epsilon_p - \mu)}{(m\omega/m^*) + (i/\tau) - (q^2v_F{}^2/3\omega)(1 + F_0{}^s)}. \tag{3.144}$$

From $\delta\bar{n}_1$, we easily obtain the conductivity. Let us introduce the sound velocity s, given by Eq. (1.61), and the effective collision time τ_{eff} defined in Eq. (3.140). The longitudinal conductivity may then be written as

$$\sigma = \frac{iNe^2}{m}\frac{1}{\omega + (i/\tau_{\text{eff}}) - (s^2q^2/\omega)} \tag{3.145}$$

(within corrections of order $qv_F\tau$). On comparing Eqs. (3.145) and (3.141) we see the way in which space charge effects act to modify σ.

It is instructive to express the longitudinal response in terms of $\epsilon(\mathbf{q}, \omega)$; one finds

$$\epsilon(\mathbf{q}, \omega) = 1 - \frac{4\pi Ne^2/m}{\omega^2 - s^2q^2 + (i\omega/\tau_{\text{eff}})}. \tag{3.146}$$

When $\omega \to 0$, $\epsilon(\mathbf{q}, \omega)$ goes to the static value (3.60): the dielectric constant in this limit is not modified by collisions with impurities. This is hardly surprising, since that limit corresponds to a *space charge regime*, in which the quasiparticles screen the applied charge; this is an essentially static process, which should not be sensitive to collisions. (It is interesting to note that this property follows directly from the fact that $1/\tau_0 = 0$, that is, from particle conservation in collisions.) On the contrary, when

$$\omega\tau_{\text{eff}} \gg (qs\tau_{\text{eff}})^2 \tag{3.147}$$

the term s^2q^2 in the denominator of Eq. (3.146) becomes negligible, and we obtain anew the result (3.141): we have passed into the *conduction regime*. Physically, the condition (3.147) means that the phase "seen" by a given particle changes faster due to the frequency ω than it does as a consequence of the random walk of the particle. Let us remark that, again, only τ_{eff} enters the result: many-body effects do not show up explicitly.

Consider now a *transverse* external field. In this case, the only components that play a role correspond to $m = \pm 1$ (the "pure" cases corresponding to circularly polarized waves). There is no $\ell = 0$ component (i.e., no space charge). Within corrections of order $qv_F\tau$, $\delta n_{\mathbf{p}}$ contains only terms of symmetry $\ell = 1$, and one finds the previous result (3.141).

In the limit $\omega\tau \ll 1$, the longitudinal conductivity may be written as

$$\sigma = \frac{iNe^2}{m} \frac{\omega\tau_{\text{eff}}}{i\omega - s^2\tau_{\text{eff}}q^2}. \tag{3.148}$$

Since we have also assumed $qv_F\tau \ll 1$, the result (3.148) corresponds to the *hydrodynamic* regime, in which the collisions act to bring about everywhere a state of local equilibrium. In that limit, it should be possible to describe the system's behavior (and thus to calculate σ) by using only *macroscopic* arguments. For that purpose, we notice that the electron-impurity collisions do *not* conserve momentum (as shown by the fact that $1/\tau_1 \neq 0$). As a consequence, the low-frequency motion of the electrons is essentially a diffusion process, controlled by the collisions with impurities. A density gradient $\nabla\rho$ gives rise to a *diffusion current*

$$\mathbf{J}_d = -D\nabla\rho, \tag{3.149}$$

where D is the diffusion coefficient.

When the electric field $\mathcal{E}$ applied to the system is transverse, there is no density fluctuation, hence no diffusion current. The current density $\mathbf{J}$ then reduces to the *conduction current*

$$\mathbf{J}_c = \sigma_0\mathcal{E} = \frac{Ne^2\tau_{\text{eff}}}{m}\,\mathcal{E}, \tag{3.150}$$

where σ_0 is the static conductivity. If on the other hand the field $\mathcal{E}$ is longitudinal, the induced current $\mathbf{J}$ gives rise to density fluctuations determined by the continuity equation

$$\operatorname{div}\mathbf{J} + \frac{\partial\rho}{\partial t} = 0. \tag{3.151}$$

Since a diffusion current is set up, the total current is equal to

$$\mathbf{J} = \mathbf{J}_c + \mathbf{J}_d. \tag{3.152}$$

On taking the Fourier transform of Eq. (3.149) to (3.152), we easily verify that

$$\sigma(\mathbf{q}, \omega) = \frac{e\mathbf{J}(\mathbf{q}, \omega)}{\mathcal{E}(\mathbf{q}, \omega)} = \frac{Ne^2\tau_{\text{eff}}}{m}\,\frac{i\omega}{i\omega - Dq^2}. \tag{3.153}$$

Equation (3.153) is identical to the result (3.148) obtained from the Landau theory, the diffusion coefficient D being equal to

$$D = s^2\tau_{\text{eff}}. \tag{3.154}$$

D may be obtained by a direct calculation based on the Landau–Silin transport equation (by a method similar to that used in Section 1.8 to obtain the thermal conductivity: see Problem 3.9). One can thus verify the expression (3.154).

We remark in passing the close relationship between D and the static *mobility:*

$$\mu = \frac{\sigma}{Ne} = \frac{e\tau_{\text{eff}}}{m}. \tag{3.155}$$

On comparing Eqs. (3.154) and (3.155), we see that

$$\frac{\mu}{D} = \frac{e}{ms^2}, \tag{3.156}$$

which is an extension of the famous Einstein relation.

Equation (3.148) provides us with a specific example of a response function in the hydrodynamic regime, in a case in which diffusion is the controlling factor. The expression of σ is simple because one is in the "extreme" hydrodynamic limit. In the opposite limit ($qv_F\tau \gg 1$), the collisions play an essentially negligible role, and one finds again the results of Sections 3.4 and 3.6. In the intermediate range, characterized by $qv_F\tau \sim 1$, the solution of the transport equation, (3.136), is difficult. Let us, however, remark that the equation itself remains valid as long as both q and ω have a "macroscopic" scale. For a given situation it may prove possible to obtain an approximate solution by cutting off the ℓ-expansion at some suitable value; we shall not treat such an approximation here.

3.8. ELECTRONS IN METALS

Throughout this chapter we have been concerned with an idealized physical system, the *electron liquid* in which the particles move in a uniform background of positive charge, subject only to their mutual Coulomb interaction. One does not encounter such electron liquids in nature; the conduction electrons in a metal see a nonuniform positive ionic charge distribution, which is of course periodic when the metal is in solid form. We now discuss briefly the applicability of the Landau–Silin theory to electrons in metals.

We have seen in Section 1.3 that the basic assumptions of the Landau theory remain valid in the presence of a periodic potential. The elementary excitations are still quasiparticles, with energy $\epsilon_{\mathbf{p}}$ and interaction $f_{\mathbf{pp}'}$. However, as a consequence of the anisotropy of the crystalline lattice, $\epsilon_{\mathbf{p}}$ depends on the direction of $\mathbf{p}$, while $f_{\mathbf{pp}'}$ depends on *both* $\mathbf{p}$ and $\mathbf{p}'$. The Fermi surface S_F is defined by the relation

$$\epsilon_{\mathbf{p}} = \mu,$$

where μ is the chemical potential. In most metals S_F has a rather complicated geometrical shape.

In some cases, for instance in the alkali metals, the Fermi surface is nearly spherical. The system's behavior is then approximately *isotropic;* as we have discussed in Section 1.3 the formulation of the theory is essentially the same as for the electron liquid. We may define an effective mass m^* on the Fermi surface, expand $f_{\mathbf{pp}'}$ in a series of Legendre polynomials, etc. There is, however, an important difference: the current $\mathbf{j}_{\mathbf{p}}$ carried by a quasiparticle with (pseudo) momentum $\mathbf{p}$ is no longer equal to $\mathbf{p}/m$, as was the case for the translationally invariant system.

We may instead write $\mathbf{j}_p$ in the form

$$\mathbf{j}_p = \frac{\mathbf{p}}{m_c}, \tag{3.157}$$

where m_c is a *crystalline mass*, which differs from the bare electron mass m because of the periodic potential acting on the electrons. m_c represents an additional parameter of the Landau–Silin theory (for an anisotropic metal, there are many more).

The real quasiparticle effective mass, m^*, is defined in terms of the velocity, $\mathbf{v}_p$, of a quasiparticle on the Fermi surface,

$$\mathbf{v}_p = \frac{\mathbf{p}}{m^*}. \tag{3.158}$$

The quantity m^* determines the density of states at the Fermi surface, and thus the specific heat. The difference between m^* and m_c reflects that between $\mathbf{v}_p$ and $\mathbf{j}_p$, and is thus a consequence of quasiparticle interaction. In order to relate m^* to m_c, we note that the relationship (1.91) between $\mathbf{j}_p$ and $\mathbf{v}_p$ does not depend on translational invariance. For an isotropic metal, we can therefore write

$$\mathbf{j}_p = \mathbf{v}_p + \sum_{p'} f_{pp'}\delta(\epsilon_{p'} - \mu)\mathbf{v}_{p'} = \mathbf{v}_p(1 + F_1{}^s/3), \tag{3.159}$$

from which it follows that

$$m^* = m_c(1 + F_1{}^s/3). \tag{3.160}$$

Equation (3.160) is similar to (1.100), the bare mass m being simply replaced by the crystalline mass m_c.

According to Eq. (3.160), the effective mass, m^*, appears as the product of the crystalline mass, m_c, and a factor representing the influence of the electron interaction, $(1 + F_1{}^s/3)$. In principle, the difference between m_c and the bare electron mass, m, should arise as a consequence of the periodic potential of the ions, but *not* of the Coulomb interaction between electrons (the latter modification being given by the Landau–Silin theory). Actually, such a decomposition is not in general possible, since in most calculations (e.g., the pseudo-potential method) the periodic potential is taken to be that of the bare ions, screened by the electron response. (We have calculated the long wavelength part of this screened electron-ion potential in Section 3.4.) Thus the periodic potential (and m_c) is influenced by electron interaction to a certain extent; the reverse influence, of the periodic potential on the quasiparticle interaction, may also be present. We shall give in Section 5.6 a further discussion of the extent to which characteristically "solid state" and "many-particle" effects may be disentangled.

It is straightforward to show, using the methods of Section 1.3, that the electronic compressibility and spin susceptibility of an isotropic solid are given by expressions of the same form as for the electron liquid [compare Eqs. (1.58) and (1.67)]:

$$\kappa = \frac{\nu(0)}{\rho N(1 + F_0^s)}, \tag{3.161}$$

$$\chi_P = \frac{\beta^2 \nu(0)}{1 + F_0^a}. \tag{3.162}$$

Here, $\nu(0)$ is the density of states per unit energy for an electron on the *metallic* Fermi surface. κ and χ_P are easily expressed in terms of the corresponding quantities, κ° and χ_P°, for the noninteracting free electron gas. On using Eq. (3.160), we may write these relations in the forms

$$\frac{\kappa}{\kappa^\circ} = \frac{m^*}{m} \frac{1}{1 + F_0^s} = \frac{m_c}{m} \frac{1 + F_1^s/3}{1 + F_0^s}, \tag{3.163}$$

$$\frac{\chi_P}{\chi_P^\circ} = \frac{m^*}{m} \frac{1}{1 + F_0^a} = \frac{m_c}{m} \frac{1 + F_1^s/3}{1 + F_0^a}. \tag{3.164}$$

Equations (3.163) and (3.164) show clearly the combined influence of a periodicity correction, m_c/m, and of the quasiparticle interaction, $f_{pp'}$. We further note that while the sound velocity s is unphysical (the usual laws of acoustics do not apply to electrons in metals), we can define a "pseudo" sound velocity, s, in much the same way as Eq. (3.160) may be regarded as defining a "pseudo" bare mass. Thus we may write

$$\kappa = \frac{1}{\rho m s^2}. \tag{3.165}$$

In practice, s always enters any calculation in the combination ms^2 (in other words, physical quantities depend on κ).

The suitably modified Landau–Silin theory provides a *formal* description of the influence of Coulomb interaction on electron motion in metals and, as such, is pleasing to the theoretical physicist. One may ask whether it offers any solace to the experimentalist engaged in measuring various metallic phenomena. In this respect, the most important application of the Landau theory is the assistance it provides concerning the extent to which each phenomenon of interest is affected by the interaction between electrons. As we have seen, this interaction is felt on two different levels: it modifies the energy and velocity of the quasiparticles, and gives rise to an interaction between excited quasiparticles. Phenomena that are not affected by quasiparticle interaction may be described in terms of a one-electron picture, with the proviso that the electron energies must be replaced by "renormalized" energies (modified by the electron interaction). In contrast, phenomena that are affected by the screened interaction $f_{\mathbf{pp}'}$ between excited quasiparticles cannot be described exactly in terms of a one-electron scheme.

In recent years, a number of elegant methods have been developed for the study of electrons on the Fermi surface in metals. [For a review of the experimental determinations of the Fermi surface see, for example, Pippard (1962).] The interpretation of the experimental results always relies on a one-electron approximation. Consequently, only those effects which are not affected by the interaction $f_{\mathbf{pp}'}$ can be used to obtain *exact* information concerning the Fermi surface. To the extent that the wavelength and frequency are such that the Landau theory is applicable, one can select these "well-behaved" physical effects on the basis of the Landau theory. The latter thus provides a theoretical check on the validity of a given experimental determination of the Fermi surface.

One important example of a quantity that is not modified by the interaction $f_{\mathbf{pp}'}$ is the transverse conductivity, $\sigma_{\perp}(\mathbf{q}, \omega)$, in the anomalous region $(0 < \omega \ll qv_F)$; cf. Eq. (3.122). As we have remarked, the anomalous skin effect may thus be used to determine directly the shape of a real Fermi surface [Silin (1958a)]. Another phenomenon which is unaffected by quasiparticle interaction is the cyclotron resonance as measured under the so-called Azbel–Kaner conditions (i.e., with the dc magnetic field parallel to a sample surface). Such an experiment can be described theoretically starting from the transport equation (3.118). The resonance is found to occur at the same frequency as in the absence of quasiparticle interaction. In the simple case of an isotropic system, the resonance frequency involves the same mass m^* as that found from specific heat measurements:

$$\omega_c^* = \frac{eH}{m^*c}. \tag{3.166}$$

Among other "well-behaved" effects which do not depend on $f_{\mathbf{pp}'}$, let us cite the de Haas–van Alphen effect and the low-frequency Hall constant in a high magnetic field. All those effects constitute valuable probes of the nature of the Fermi surface in metals.

We consider next phenomena which depend on the quasiparticle interaction, $f_{\mathbf{pp}'}$.

We have mentioned the compressibility and spin susceptibility, given by Eqs. (3.161) and (3.162), respectively. It is also straightforward to extend the various transport calculations based on the Landau theory to the case of conduction electrons in an isotropic metal. For example, the result (3.62) for the long wavelength static dielectric constant is still valid, provided κ is given by Eq. (3.161) or (3.163). Again, a general expression for the conductivity $\sigma(0, \omega)$ may be derived, as long as ω is less than any of the *interband excitation frequencies*; it is left as a problem for the reader to show that Eq. (3.139) is valid, provided one replaces m by m_c. A special case of this last result is of considerable importance: when ω is sufficiently large that one is in the collisionless regime, $\sigma(0, \omega)$ is given by the following modified version of Eq. (3.75) [Silin (1958b)]:

$$\sigma(0, \omega) = \frac{iNe^2}{m_c\omega} = \frac{iNe^2}{m^*\omega}\left(1 + \frac{F_1{}^s}{3}\right). \qquad (3.167)$$

[Equation (3.167) follows directly from the result (3.74) together with the expression (3.157) for the current.] By measuring $\sigma(0, \omega)$, one can therefore determine directly the crystalline mass m_c.

At this point it would be nice to present a table of the various values of $F_0{}^s$, $F_0{}^a$, $F_1{}^s$, m_c, etc. as determined experimentally for a number of metals. Unfortunately, this is not possible for a number of reasons. First, there is only widespread data available on the specific heat; it is not possible to measure the electronic compressibility, κ, directly, while χ_P has thus far been determined directly only for lithium and sodium. Second, as we have already mentioned in Section 1.3, the value for m^* obtained in a specific heat experiment includes the effect of electron-phonon interactions, as well as electron-electron and crystalline field effects. Third, optical experiments which permit a direct determination of m_c according to Eq. (3.167) are not easily carried out. The available experimental information is often concerned with frequencies comparable to the plasma frequency, for which Eq. (3.167) is not applicable (in that frequency range, interband transitions have a large effect). Recently, Ginzburg, Motulevitch, and Pitaevskii (1965) have used infrared reflection measurements to estimate $F_1{}^s$ in a number of transition metals. Their results are promising, although the values of $F_1{}^s$ they obtain are unexpectedly large. Thus we shall see in Section 5.6 that for the corresponding electron liquid, values of the order of a few tenths are to be expected while their results yield $F_1{}^s > 1$. Such a large quasiparticle interaction might possibly be traced to solid state effects; there are regions in the Brillouin zone where the Fermi surface touches the zone boundary (in those regions, the density of states is high and the quasiparticle interaction accordingly enhanced); it may also reflect the enhancement of the density of states by the electron-phonon interaction.

To some extent, microscopic considerations are useful in attempting to sort out all of the above effects. We shall discuss this possibility in

Section 5.6, and present in that section a discussion of our present knowledge of the influence of electron interaction on the spin susceptibility and specific heat.

Finally, it is likely that characteristically many-body effects will be found in other transport properties. Future methods of investigation might involve a detailed study of the transition from normal to anomalous skin effect, or of electromagnetic propagation in a very high magnetic field.

PROBLEMS

3.1. Show that the longitudinal solution of Eq. (3.40) has frequency ω_p when $\mathbf{q} \to 0$ (for a translationally invariant system).

3.2. Show that the longitudinal solution of Eq. (3.40) is such that $\epsilon(\mathbf{q}, \omega_q) = 0$.

3.3. By using the simple model of Chapter 1 in which $f_{\mathbf{pp}'}$ is a constant, derive the frequency of longitudinal collective modes as a function of $\mathbf{q}$. Study the transition from plasma oscillations to zero sound when $F_0 > 0$. Show that the transition occurs at a value of $q \sim \omega_p/v_F \sim q_{FT}$.

3.4. Calculate explicitly the dielectric constant $\epsilon(\mathbf{q}, \omega)$ in the Landau–Silin theory, using the simple model $F_{\mathbf{pp}'} = F_0$. Plot Im ϵ and Re ϵ as a function of ω/qv_F. Show that in the strong coupling limit, ϵ takes the form (3.77).

3.5. Verify that the transport equation, (3.116), is invariant under a Galilean transformation: one thereby justifies the Lorentz force with $\mathbf{v_p}$ instead of $\mathbf{j_p}$.

3.6. Calculate the first reactive term of $\sigma_\perp$ in an expansion in powers of ω/qv_F.

3.7. By using a model $F_{\mathbf{pp}'} = F_0 + F_1 \cos \xi$, calculate the transverse conductivity for arbitrary qv_F/ω. Discuss the behavior of transverse collective modes as a function of cq/ω_p.

3.8. Calculate the transverse conductivity in a dc magnetic field (a) when $\omega = 0$; (b) when $\mathbf{q} = 0$; (c) in the general case with an interaction $F_0 + F_1 \cos \xi$.

Discuss the behavior of σ (both its continuum aspects and the appearance of discrete poles) as a function of q.

Discuss the behavior of the transverse collective modes: (a) when $\omega_p/c \ll \omega_c/v_F$; (b) when $\omega_p/c \gg \omega_c/v_F$.

3.9. Calculate the diffusion coefficient D for an electron system. Compare your results with those of the usual kinetic theory, and interpret any differences.

3.10. Calculate the thermal conductivity $\mathbf{K}$; apply the result to a discussion of the Wiedemann–Franz ratio.

3.11. Study a neutral system with weak impurity collisions ($1/\tau_i$ small) and strong electron-electron collisions ($1/\tau_e$ large if $\ell \geq 2$). Study the transition from diffusion to first sound to zero sound. Calculate the diffusion damping of first sound.

REFERENCES

Debye, P. P. and Hückel, E. (1923), *Physik Z.* **24**, 185.
Friedel, J. (1954), *Advan. in Phys.* **3**, 446.
Ginzburg, V. L., Motulevitch, G. P. and Pitaevskii, L. P. (1965), *Doklady Acad. Nauk.* **163**, 1352.
Greene, M. P. and Kohn, W. (1965), *Phys. Rev.* **137**, A513.
Harrison, W. (1966), *Pseudopotentials in the Theory of Metals*, W. A. Benjamin, New York.
Heine, V., unpublished. See Heine, V. and Abarenkov, I. (1964), *Phil. Mag.* **9**, 451.
Langer, J. (1962), *Phys. Rev.* **127**, 5.
Langmuir, I. (1928), *Proc. Nat. Acad. Sci. U.S.* **14**, 627.
Mott, N. F. and Jones, H. (1936), *Theory of Metals and Alloys*, Dover, New York.
Pines, D. (1950), Ph.D. Thesis, Princeton Univ., unpublished.
Pines, D. and Bohm, D. (1952), *Phys. Rev.* **85**, 338.
Pines, D. (1963), *Elementary Excitations in Solids*, W. A. Benjamin, New York.
Pippard, A. B. (1962), *Dynamics of Conduction Electrons*, Gordon and Breach, New York.
Prange, R. E. and Kadanoff, L. P. (1964), *Phys. Rev.* **134**, A566.
Silin, V. P. (1957), *Sov. Phys. JETP* **6**, 387; *ibid.* **6**, 985.
Silin, V. P. (1958a), *Sov. Phys. JETP* **6**, 955.
Silin, V. P. (1958b), *Sov. Phys. JETP* **7**, 486.
Simkin, D. (1963), Ph.D. Thesis, Univ. of Illinois, unpublished.
Tonks, L. and Langmuir, I. (1929), *Phys. Rev.* **33**, 195.

CHAPTER 4

RESPONSE AND CORRELATION IN HOMOGENEOUS ELECTRON SYSTEMS

In the previous chapter we have used Poisson equations to define the frequency- and wave vector-dependent dielectric response function, $\epsilon(\mathbf{q}, \omega)$. Not surprisingly, there is an intimate connection between this measure of the longitudinal response and the density-density response function introduced in Chapter 2. We now proceed to develop and exploit this connection.

The aim of the present chapter is twofold: first, to establish the formal properties of dielectric response functions; second, to show how such functions both provide a compact characterization of, and offer considerable physical insight into, a number of phenomena of interest in electron systems. Specifically, we shall consider, in elementary fashion, the coupling of electrons with each other, their response to a fast external charged particle, to a phonon field, and to an external photon field.

The dielectric response of an electron liquid is characterized by a *tensor*, which relates the electric displacement $\mathfrak{D}$ to the electric field $\mathcal{E}$. For isotropic systems, there exist only two independent components of that tensor, the longitudinal dielectric function, $\epsilon(\mathbf{q}, \omega)$, and its transverse counterpart, $\epsilon_{\perp}(\mathbf{q}, \omega)$. Longitudinal and transverse phenomena are thereby decoupled, and may be treated separately. We shall assume this situation to apply throughout the present chapter. We develop the formal properties of $\epsilon(\mathbf{q}, \omega)$ in Section 4.1, those of $\epsilon_{\perp}(\mathbf{q}, \omega)$ in Section 4.6.

We have seen in Chapter 3 that $1/\epsilon(\mathbf{q}, \omega)$ offers a direct measure of the dielectric response of the electrons to an *external* longitudinal field which varies in space and time. In Section 4.1 we make the obvious connection between this measure of the system's longitudinal response and the density-density response function introduced in Chapter 2. In this way we can apply directly to the electron system many results of that chapter. For example, we may write down at once the spectral representation for $1/\epsilon(\mathbf{q}, \omega)$, as well as a sum rule which derives directly from the longitudinal f-sum rule.

In a sense, an electron system is "richer" than a system of neutral particles, in that there exist two density response functions of interest. In addition to $\chi(\mathbf{q}, \omega)$ [or $1/\epsilon(\mathbf{q}, \omega)$], which measures the response to an *external* longitudinal field, one may define a second, *screened* response function related to $\epsilon(\mathbf{q}, \omega)$, which measures the electron response to the sum of the external field and the polarization it produces. So to speak, $1/\epsilon(\mathbf{q}, \omega)$ measures the response to $\mathfrak{D}(\mathbf{q}, \omega)$, while $\epsilon(\mathbf{q}, \omega)$ measures the response to the effective, screened field, $\mathcal{E}(\mathbf{q}, \omega)$. Both response functions satisfy sum rules: we shall see that in the long wavelength limit there exist, in fact, four separate sum rules for $\epsilon(\mathbf{q}, \omega)$.

The nature of the density fluctuation excitation spectrum [and of $\epsilon(\mathbf{q}, \omega)$] is discussed in Section 4.2. It is shown with the aid of the sum rules that in the long wavelength limit screening and plasma oscillations are *exact* properties of an electron liquid at $T = 0$. In Section 4.3 this result is generalized to finite temperatures; the influence of the periodic lattice of ions in a solid is discussed. The energy loss by a fast electron (and plasmon excitation in solids) is described in Section 4.4; the self-consistent treatment of a coupled electron-phonon system is considered in Section 4.5. An elementary discussion of the effective electron interaction in simple metals is given in Section 4.6; polarization processes which involve phonons as well as electrons are taken into account.

The general question of the response of the electron liquid to an external electromagnetic field is considered in Section 4.7; the fundamental response function of interest is shown to be the transverse current-current correlation function, calculated with the neglect of magnetic interactions between the electrons. The appropriate sum rules are derived; the relationship between gauge invariance, charge conservation, and the f-sum rule is discussed in detail. An explicit quantum-mechanical treatment of the coupled electron-photon system is sketched; Raman scattering of light is likewise considered at the conclusion of the chapter.

4.1. DIELECTRIC RESPONSE FUNCTIONS

RESPONSE TO AN EXTERNAL FIELD: $1/\epsilon(\mathbf{q}, \omega)$

We consider anew the response of the electron system to an external charge of density $z\rho_e(\mathbf{r}, t)$. The interaction between the electron system and the test charge may be written in the form [compare Eq. (2.42)]

$$H_{\text{int}} = \lim_{\eta \to 0} \sum_{\mathbf{q}} V_q \int_{-\infty}^{\infty} \frac{d\omega}{2\pi} \rho_{\mathbf{q}}^{+} \rho_e(\mathbf{q}, \omega) e^{-i\omega t} e^{\eta t}, \tag{4.1}$$

where

$$V_q = \frac{4\pi z e}{q^2}. \tag{4.2}$$

The response of the electron system may be described in terms of the quantity $1/\epsilon(\mathbf{q}, \omega)$. If we combine Eqs. (3.19) and (3.23), we obtain

$$\frac{1}{\epsilon(\mathbf{q}, \omega)} = 1 + \frac{e\langle\rho(\mathbf{q}, \omega)\rangle}{z\rho_e(\mathbf{q}, \omega)}. \tag{4.3}$$

On making use of Eq. (4.2), we can write (4.3) in the form

$$\frac{1}{\epsilon(\mathbf{q}, \omega)} \equiv 1 + \frac{4\pi e^2}{q^2} \frac{\langle\rho(\mathbf{q}, \omega)\rangle}{V_q \rho_e(\mathbf{q}, \omega)}. \tag{4.4}$$

The relation to the density-density response function is then clear. The latter was defined as

$$\chi(\mathbf{q}, \omega) = \frac{\langle\rho(\mathbf{q}, \omega)\rangle}{V_q \rho_e(\mathbf{q}, \omega)}. \tag{4.5}$$

We have therefore

$$\frac{1}{\epsilon(\mathbf{q}, \omega)} = 1 + \frac{4\pi e^2}{q^2} \chi(\mathbf{q}, \omega). \tag{4.6}$$

With the aid of Eq. (4.6) and the results of Chapter 2, we may establish directly a number of useful expressions and relationships involving $1/\epsilon(\mathbf{q}, \omega)$. Making use of Eq. (2.57), we express $1/\epsilon(\mathbf{q}, \omega)$ in terms of the exact eigenstates of the electron system:

$$\frac{1}{\epsilon(\mathbf{q}, \omega)} = 1 + \frac{4\pi e^2}{q^2} \sum_n \left| (\rho_{\mathbf{q}}^{+})_{no} \right|^2 \left\{ \frac{1}{\omega - \omega_{no} + i\eta} - \frac{1}{\omega + \omega_{no} + i\eta} \right\}. \tag{4.7}$$

We see that $1/\epsilon(\mathbf{q}, \omega)$ is analytic in the upper half of the complex ω-plane, as befits a causal response function. Thus one can write down Kramers–Kronig relations which relate the real and imaginary parts of

$1/\epsilon(\mathbf{q}, \omega)$. One may derive in a similar way the spectral representation [cf. Eq. (2.58)]:

$$\frac{1}{\epsilon(\mathbf{q}, \omega)} = 1 + \frac{4\pi e^2}{q^2}\int_0^\infty d\omega'\, S(\mathbf{q}, \omega')\left\{\frac{1}{\omega - \omega' + i\eta} - \frac{1}{\omega + \omega' + i\eta}\right\}. \tag{4.8}$$

The dynamic form factor $S(\mathbf{q}, \omega)$ thus serves as a spectral density for $1/\epsilon(\mathbf{q}, \omega)$. Let us write ϵ in the form

$$\epsilon(\mathbf{q}, \omega) = \epsilon_1(\mathbf{q}, \omega) + i\epsilon_2(\mathbf{q}, \omega) \tag{4.9}$$

(where ϵ_1 and ϵ_2 are both real). According to Eq. (4.8), we have

$$\frac{\epsilon_2(\mathbf{q}, \omega)}{|\epsilon(\mathbf{q}, \omega)|^2} = \frac{4\pi^2 e^2}{q^2}\{S(\mathbf{q}, \omega) - S(\mathbf{q}, -\omega)\}. \tag{4.10}$$

There exists a longitudinal sum rule for $\epsilon(\mathbf{q}, \omega)$ which follows directly from the f-sum rule, (2.28); on making use of Eqs. (2.28) and (4.10) we find

$$\int_0^\infty d\omega\, \omega \frac{\epsilon_2(\mathbf{q}, \omega)}{|\epsilon(\mathbf{q}, \omega)|^2} = \frac{\pi}{2}\omega_p^{\,2} = \frac{2\pi^2 N e^2}{m}. \tag{4.11}$$

At first sight, one might think it equally easy to write down the analog of the compressibility sum rule, (2.85). Actually (2.85) does not apply for the dynamic form factor of an electron system. The reason is that, as discussed in Chapter 3, the compressibility of the latter system is defined as the long wavelength limit of the electron response to a *screened* external field, that is, to $\mathcal{E}$ rather than to $\mathfrak{D}$.

RESPONSE TO A SCREENED FIELD: $\epsilon(\mathbf{q}, \omega)$

We define the response of an electron liquid to a screened external field by analogy with Eq. (4.5):

$$\chi_{sc}(\mathbf{q}, \omega) = \frac{\langle \rho(\mathbf{q}, \omega)\rangle}{V_q\{\rho_e(\mathbf{q}, \omega)/\epsilon(\mathbf{q}, \omega)\}} = \epsilon(\mathbf{q}, \omega)\chi(\mathbf{q}, \omega). \tag{4.12}$$

$\chi_{sc}(\mathbf{q}, \omega)$ measures the density fluctuation induced by a *screened* external charge, $\rho_e(\mathbf{q}, \omega)/\epsilon(\mathbf{q}, \omega)$. We may combine Eq. (4.12) with (4.6) to write

$$\epsilon(\mathbf{q}, \omega) = 1 - \frac{4\pi e^2}{q^2}\chi_{sc}(\mathbf{q}, \omega). \tag{4.13}$$

We note the similarity between Eqs. (4.13) and (4.6).

The dielectric constant may be written as [see Eq. (3.32)]

$$\epsilon(\mathbf{q}, \omega) = 1 + \frac{4\pi i}{\omega} \sigma(\mathbf{q}, \omega). \tag{4.14}$$

We see that the longitudinal conductivity $\sigma(\mathbf{q}, \omega)$ is simply related to $\chi_{sc}(\mathbf{q}, \omega)$:

$$\sigma(\mathbf{q}, \omega) = \frac{i\omega e^2}{q^2} \chi_{sc}(\mathbf{q}, \omega). \tag{4.15}$$

The screened response function is therefore the quantity one calculates directly in the usual elementary outlook, in which one computes the current $e\mathbf{J}$ induced by the net electric field $\mathbf{\mathcal{E}}$. This same quantity may be obtained from the transport equation for the charged Fermi liquid, since there too one determines the response to $\mathbf{\mathcal{E}}$ rather than to $\mathbf{\mathfrak{D}}$.

If we compare Eq. (4.13) with (3.56), we see that $\chi_{sc}(\mathbf{q}, \omega)$ is equal to the quantity $\chi_n(\mathbf{q}, \omega)$ which we introduced in Chapter 3. It corresponds to the density-density response function for a *fictitious* neutral system with the same law of interaction as the screened Coulomb interaction obtaining between the electrons.

There is an important difference between $\chi_{sc}(\mathbf{q}, \omega)$ and a "genuine" density-density response function, as there is between $\epsilon(\mathbf{q}, \omega)$ and $1/\epsilon(\mathbf{q}, \omega)$. $\chi_{sc}(\mathbf{q}, \omega)$ [as well as $\sigma(\mathbf{q}, \omega)$ and $\epsilon(\mathbf{q}, \omega)$] is a *construct:* it does not measure the system response to an *external* field, but to a *screened* field which is internal to the system. As a result, causality arguments are of no avail in the problem we consider next, that of determining the analytic behavior of $\chi_{sc}(\mathbf{q}, \omega)$ or $\epsilon(\mathbf{q}, \omega)$.

ANALYTIC BEHAVIOR OF $\epsilon(\mathbf{q}, \omega)$

The dielectric constant, $\epsilon(\mathbf{q}, \omega)$, will be analytic in the upper half of the complex ω-plane provided $\epsilon^{-1}(\mathbf{q}, \omega)$ has no zeros there. Let us consider first ω to be real. In that case, $\text{Im}\, \epsilon^{-1}(\mathbf{q}, \omega) = 0$ implies $S(\mathbf{q}, \omega) = 0$: the poles of $\epsilon(\mathbf{q}, \omega)$ are only found outside the range of frequencies covered by the dynamic form factor. Let us consider a *normal* charged Fermi liquid. According to the Landau theory, $\epsilon(\mathbf{q}, \omega)$ may in principle have a pole lying beyond the continuum of single-pair excitations (see Section 3.4). However, we have also seen that if $\mathbf{q}$ is finite, $S(\mathbf{q}, \omega)$ contains a contribution from multipair excitations, which extends from $\omega = 0$ to infinity. When these are taken into account, the possible pole in $\epsilon(\mathbf{q}, \omega)$ is damped, and thereby moved below the real axis. It follows that for a normal Fermi liquid $\epsilon(\mathbf{q}, \omega)$ is analytic on the real axis. Matters are different in superconductors, for which there is a gap in the quasiparticle excitation spectrum. In that case, $\epsilon(\mathbf{q}, \omega)$ possesses an undamped pole; one may, however, decide to locate this pole at an infinitesimal distance *below* the real axis: $\epsilon(\mathbf{q}, \omega)$ thus remains analytic *on* the axis.

Let us now search for zeros of $\epsilon^{-1}(\mathbf{q}, \omega)$ among complex values of ω:

$$\omega = \omega_1 + i\omega_2 \qquad (\omega_2 > 0). \tag{4.16}$$

For such values of ω, we may write Eq. (4.8) in the following form:

$$\frac{1}{\epsilon(\mathbf{q}, \omega)} = 1 + \frac{8\pi e^2}{q^2} \int_0^\infty d\omega'\, S(\mathbf{q}, \omega') \frac{\omega'}{\omega^2 - (\omega')^2}. \tag{4.17}$$

If we take the imaginary parts of both sides of Eq. (4.17), we find

$$\operatorname{Im} \frac{1}{\epsilon(\mathbf{q}, \omega)} = -\frac{16\pi e^2}{q^2} \omega_1 \omega_2 \int_0^\infty d\omega' \frac{S(\mathbf{q}, \omega')\omega'}{\{(\omega_1 + i\omega_2)^2 - (\omega')^2\}\{(\omega_1 - i\omega_2)^2 - (\omega')^2\}} \tag{4.18}$$

Since $S(\mathbf{q}, \omega')$ is positive definite, the integral is positive. Thus $\operatorname{Im} \epsilon^{-1}$ can be zero only if $\omega_1\omega_2 = 0$. As ω_2 is nonzero by assumption, zeros of $\epsilon^{-1}(\mathbf{q}, \omega)$ can be found only on the imaginary axis of the ω-plane.

Let $\omega = i\alpha$, where α is real. On making use of Eq. (4.17), we see that $\epsilon(\mathbf{q}, \omega)$ will have a singularity at $\omega = i\alpha$ provided

$$1 = \frac{8\pi e^2}{q^2} \int_0^\infty d\omega' \frac{S(\mathbf{q}, \omega')\omega'}{(\omega')^2 + \alpha^2}. \tag{4.19}$$

Can this condition be satisfied for a nonzero value of α? Since the integral on the right-hand side of Eq. (4.19) is a monotonically decreasing function of α, approaching zero as $\alpha \to \infty$, there will be a solution to Eq. (4.19) if and only if

$$1 < \frac{8\pi e^2}{q^2} \int_0^\infty d\omega' \frac{S(\mathbf{q}, \omega')}{\omega'}. \tag{4.20}$$

On the other hand, we see directly from Eq. (4.17) that

$$\epsilon(\mathbf{q}, 0) = \left[1 - 8\pi e^2/q^2 \int_0^\infty d\omega' \frac{S(\mathbf{q}, \omega')}{\omega'}\right]^{-1}. \tag{4.21}$$

The condition (4.20) can therefore be expressed as

$$\epsilon(\mathbf{q}, 0) < 0. \tag{4.22}$$

The analyticity of $\epsilon(\mathbf{q}, \omega)$ in the upper half of the complex ω-plane thus depends on the sign of $\epsilon(\mathbf{q}, 0)$. What we must do, then, is to seek a guiding principle that will enable us to decide whether or not $\epsilon(\mathbf{q}, 0) > 0$.

Such a principle will be of physical rather than mathematical origin, for we have exhausted the applicable mathematical arguments in arriving at Eq. (4.22).

Let us first consider the simple model of an electron liquid immersed in a uniform background of positive charge. Thus far, we have assumed that the system is stable against small density fluctuations of the positive background. In fact, this need not be so! Let us suppose that this background density is modulated by an amount $\delta\rho(\mathbf{q})$. Such a fluctuation may be considered as a *static* test charge acting on the electrons. As a result of the perturbation, the total energy of the system (electrons plus positive background) is modified by an amount

$$\delta E = \frac{2\pi e^2}{q^2} \frac{|\delta\rho(\mathbf{q})|^2}{\epsilon(\mathbf{q}, 0)}$$

(the demonstration of this result is left as an exercise to the reader; cf. Problem 4.1). We see that $\epsilon(\mathbf{q}, 0) < 0$ implies $\delta E < 0$: density fluctuations of the positive background are then energetically favorable. If $\epsilon(\mathbf{q}, 0)$ is negative for some value of $\mathbf{q}$, the positive background will be unstable against the development of *spontaneous* density fluctuations of this wave vector.

Such an instability is particularly simple in the long wavelength imit. In that case $\epsilon(\mathbf{q}, 0) < 0$ corresponds to a negative compressibility κ of the "electron + background" system [see Eq. (3.62)]. Since $1/\kappa$ is essentially the second derivative of the ground state energy with respect to the density [see Eq. (1.47)], it is clear that $\kappa > 0$ implies a maximum in E, corresponding to an unstable state.

In conclusion, the condition $\epsilon(\mathbf{q}, 0) > 0$ follows from the requirement that the *positive background be stable under the influence of the electron liquid*. Under such circumstances, $\epsilon(\mathbf{q}, \omega)$ is analytic in the upper half of the complex ω-plane. Let us emphasize that this result depends on our assumption of an infinitely *heavy* background of positive charge. (We have implicitly assumed that there was no kinetic energy associated with background fluctuations.) One may therefore question the applicability of the above results to the real case of electrons in metals (in which the ions have a finite mass), and even more to electrons and holes in semiconductors (in which the two types of carriers have comparable masses). We believe that $\epsilon(\mathbf{q}, 0)$ must remain positive even in these more complicated cases. Otherwise, the system would *overcompensate* the field of an external test charge (the net field in the system would be of opposite sign to that created by the probe alone). That a system will, so to speak, overshoot in response to a static disturbance seems a highly unlikely state of affairs. In the language of electrical network theory, the system would be considered as "active," while we expect it to display "passive" behavior with respect to static disturbances. It thus appears reasonable to impose the physical requirement that $\epsilon(\mathbf{q}, 0)$ be positive definite for any system.

Since $\epsilon(\mathbf{q}, \omega)$ and $\epsilon^{-1}(\mathbf{q}, \omega)$ have the same analytic properties, we may at once write down Kramers–Kronig relations which relate the real and imaginary parts of $\epsilon(\mathbf{q}, \omega)$. We can, moreover, derive a conductivity sum rule. From the definition of $\epsilon(\mathbf{q}, \omega)$, Eq. (4.7), and the longitudinal f-sum rule, Eq. (2.29), it follows that at large frequencies

$$\epsilon(\mathbf{q}, \omega) = 1 - \frac{\omega_p^2}{\omega^2} \qquad (\omega \to \infty). \tag{4.23}$$

Given this asymptotic behavior and the analytic properties of $\epsilon(\mathbf{q}, \omega)$, one can show that (cf. Eq. (4.11))

$$\int_0^\infty d\omega\, \omega\epsilon_2(\mathbf{q}, \omega) = \frac{\pi}{2}\omega_p^2 = \frac{2\pi^2 N e^2}{m}. \tag{4.24}$$

This sum rule is often expressed as a sum rule for the dissipative part of the longitudinal conductivity, $\sigma_1(\mathbf{q}, \omega)$. One has, using Eq. (4.14),

$$\int_0^\infty d\omega\, \sigma_1(\mathbf{q}, \omega) = \frac{\omega_p^2}{8} = \frac{\pi N e^2}{2m}. \tag{4.25}$$

LONG WAVELENGTH SUM RULES FOR $\epsilon^{-1}(\mathbf{q}, \omega)$ AND $\epsilon(\mathbf{q}, \omega)$

There exist two additional long wavelength sum rules, which follow from the analytic behavior of ϵ^{-1} and ϵ, and from the definition of the compressibility of an electron system. We note first that in the limit of long wavelengths it is χ_{sc} which reduces to the compressibility, according to

$$\lim_{\mathbf{q}\to 0} \chi_{sc}(\mathbf{q}, 0) = -\frac{N}{ms^2}. \tag{4.26}$$

Indeed Eq. (4.26) is simply another way of writing the long wavelength limit of the static dielectric constant

$$\epsilon(\mathbf{q}, 0) = 1 + \frac{\omega_p^2}{s^2 q^2} \qquad (\mathbf{q} \to 0). \tag{4.27}$$

If we now apply the Kramers–Kronig relation [cf. Eq. (2.78a)],

$$\epsilon_1(\mathbf{q}, \omega) = 1 + \frac{2}{\pi}\int_0^\infty d\omega'\, \{\epsilon_2(\mathbf{q}, \omega')\}\, P\left(\frac{\omega'}{(\omega')^2 - \omega^2}\right) \tag{4.28}$$

(where P means that principal parts are to be taken in the integrand), we find

$$\lim_{q\to 0} \int_0^\infty d\omega \, \frac{\epsilon_2(\mathbf{q}, \omega)}{\omega} = \frac{\pi}{2} \frac{\omega_p^2}{s^2 q^2}. \tag{4.29}$$

The analogous sum rule for Im $\epsilon^{-1}(\mathbf{q}, \omega)$ follows from the long wavelength response of the electrons to a static external field. To obtain it, we make use of the Kramers–Kronig relation for ϵ^{-1}:

$$\frac{1}{\epsilon(\mathbf{q}, 0)} = 1 - \frac{2}{\pi} \int_0^\infty \frac{d\omega'}{\omega'} \frac{\epsilon_2(\mathbf{q}, \omega')}{|\epsilon(\mathbf{q}, \omega')|^2}. \tag{4.30}$$

As we have seen [cf. Eq. (4.27)], in the long wavelength limit the electron gas exhibits "perfect" screening:

$$\lim_{q\to 0} \frac{1}{\epsilon(\mathbf{q}, 0)} = 0,$$

from which it follows that

$$\lim_{q\to 0} \int_0^\infty \frac{d\omega'}{\omega'} \frac{\epsilon_2(\mathbf{q}, \omega')}{|\epsilon(\mathbf{q}, \omega')|^2} = +\frac{\pi}{2}. \tag{4.31}$$

This second long wavelength sum rule is independent of the macroscopic system parameters.

These sum rules are useful in checking the consistency of a given calculation of $\epsilon(\mathbf{q}, \omega)$. They also permit one to make exact statements about the behavior of the dynamic form factor in the long wavelength limit.

4.2. DENSITY-FLUCTUATION EXCITATIONS

SUM RULES AND GENERAL CHARACTER OF $S(\mathbf{q}, \omega)$

The dynamic form factor for the electron gas will, in general, contain contributions from single "quasiparticle-quasihole" pair excitations, from multipair excitations (involving two or more quasiparticle-quasihole pairs), and from the longitudinal collective modes, the plasmons. For the time being, we consider only translationally invariant systems; the case of electrons in real solids is briefly discussed in Section 4.3. As was the case for the neutral fermion system, for moderate and large values of $\mathbf{q}$ the dynamic form factor is spread more or less uniformly over all possible excitation frequencies, up to energies comparable to ϵ_F. The contributions from the different modes of excitation cannot, of course, be disentangled. On the other hand, in the long wavelength limit it does become possible to disentangle the different excitations, and to estimate the relative importance of their contributions to $S(\mathbf{q}, \omega)$. The four sum rules which govern the behavior of $S(\mathbf{q}, \omega)$ in the limit of $\mathbf{q} \to 0$ are of great assistance in this respect.

With the aid of Eq. (4.10), we can rewrite the sum rules of the preceding section in the following form:

$$\int_0^\infty d\omega\, \omega S(\mathbf{q}, \omega) = \frac{Nq^2}{2m}, \tag{4.32}$$

$$\lim_{\mathbf{q}\to 0} \int_0^\infty d\omega\, \frac{S(\mathbf{q}, \omega)}{\omega} = \frac{q^2}{8\pi e^2}, \tag{4.33}$$

$$\int_0^\infty d\omega\, \omega S(\mathbf{q}, \omega) |\epsilon(\mathbf{q}, \omega)|^2 = \frac{Nq^2}{2m}, \tag{4.34}$$

$$\lim_{\mathbf{q}\to 0} \int_0^\infty d\omega\, \frac{S(\mathbf{q}, \omega)}{\omega} |\epsilon(\mathbf{q}, \omega)|^2 = \frac{N}{2ms^2}. \tag{4.35}$$

The first of these is the familiar f-sum rule. The second is the long wavelength "perfect screening" sum rule, (4.31). Equation (4.34) is the "conductivity" sum rule, (4.25), while Eq. (4.35) is the compressibility sum rule, (4.29). The fact that we have four sum rules for the electron dynamic form factor, as contrasted with the two sum rules found for its neutral counterpart, is a direct consequence of the existence of two distinct response functions of interest for the electron gas.*

We consider plasmons first. Plasmons make an important contribution to the first two sum rules, (4.32) and (4.33). To see how this comes about, we write Eq. (4.10) in the following way:

$$S(\mathbf{q}, \omega) = \frac{q^2}{4\pi^2 e^2} \frac{\epsilon_2(\mathbf{q}, \omega)}{|\epsilon(\mathbf{q}, \omega)|^2} \qquad (\omega > 0). \tag{4.36}$$

As we have seen, the complex plasmon frequency, ω_q, obeys the dispersion relation

$$\epsilon(\mathbf{q}, \omega_q) = 0. \tag{4.37}$$

Let us suppose, for the moment, that the plasmons are not damped.

* The alert reader may remark that in the neutral system we could have introduced a screening function, analogous to the dielectric function, in which $4\pi e^2/q^2$ is simply replaced by V_q, the Fourier transform of the particle interaction. We could then discuss four sum rules in fashion similar to the discussion given here. Such an approach is not, however, very useful. The point is that the singular nature of the Coulomb interaction *requires* introduction of a screening or dielectric function, and in so doing, leads to an essential simplification of the resulting physical and mathematical treatments. For a system with short-range interactions, polarization processes need not be placed on a special footing—nor is it, in general, convenient to single them out for such special treatment.

Then ω_q is real; Eq. (4.37) implies that

$$\epsilon_1(\mathbf{q}, \omega_q) = \epsilon_2(\mathbf{q}, \omega_q) = 0. \tag{4.38}$$

The plasmon contribution to $S(\mathbf{q}, \omega)$ is thus highly singular. We can deal with this limiting case with the aid of our causal boundary conditions, which require that $\epsilon^{-1}(\mathbf{q}, \omega)$ be analytic in the upper half of the complex ω-plane. Thus in the vicinity of $\omega = \omega_q$, we may write

$$\begin{aligned} S(\mathbf{q}, \omega) &= -\frac{q^2}{4\pi^2 e^2}\,\mathrm{Im}\,\frac{1}{\epsilon(\mathbf{q}, \omega)} \\ &= -\frac{q^2}{4\pi^2 e^2}\,\mathrm{Im}\left\{\lim_{\eta\to 0}\frac{1}{\epsilon_1(\mathbf{q}, \omega_q) + (\partial\epsilon_1/\partial\omega)_{\omega=\omega_q}(\omega - \omega_q + i\eta)}\right\} \\ &= \frac{q^2}{4\pi e^2}\,\frac{1}{(\partial\epsilon_1/\partial\omega)_{\omega=\omega_q}}\,\delta(\omega - \omega_q). \end{aligned} \tag{4.39}$$

We see that undamped plasmons give rise to a δ-function singularity in $S(\mathbf{q}, \omega)$, just as did zero sound modes for the neutral Fermi liquid.

As we shall see, this "no-damping" case applies only to translationally invariant systems in the limit $\mathbf{q} \to 0$, where the plasmons possess an energy ω_p. At finite values of $\mathbf{q}$, plasmons are damped; $\epsilon_2(\mathbf{q}, \omega_q)$ no longer vanishes, and the plasmon energy, as obtained from Eq. (4.37), is given by

$$\omega_q = \omega_1 - i\omega_2 \qquad (\omega_2 > 0).$$

For small values of $\mathbf{q}$, that damping will not be large. Because of the large plasmon energy, ω_p, there can be no overlap between the plasmon spectrum and that of single quasiparticle-quasihole pairs, which has an upper bound qv_F; damping arises only in consequence of the overlap between plasmons and the multipair excitation spectrum. Such a damping may be shown to be of order q^2; it increases fairly slowly with increasing wave number. Single-pair damping becomes possible when

$$q = q_c \cong \omega_p/v_F. \tag{4.40}$$

For $q > q_c$, this Landau damping mechanism is quite effective; the inverse lifetime of a plasmon becomes comparable to the real part of its energy, and the plasmon ceases to be a well-defined excitation.

To the extent that the plasmons are well defined, they will not contribute to the sum rules (4.34) and (4.35), since $|\epsilon(\mathbf{q}, \omega)|^2$ must be very small in order for the plasmon contribution to $S(\mathbf{q}, \omega)$ to be appreciable. These sum rules may be regarded, then, as sum rules on the second kind of contribution to $S(\mathbf{q}, \omega)$, that coming from the quasiparticle excitations. They are reminiscent of the two sum rules found earlier for a neutral system. The analogy may be made explicit if we split the screened response function, χ_{sc} [defined in Eqs. (4.12) and (4.13)], into its real and imaginary parts:

$$\chi_{sc} = \chi'_{sc} + i\chi''_{sc}.$$

We may write, on making use of Eq. (4.13), the sum rules (4.34) and (4.35) as

$$-\frac{1}{\pi}\int_{-\infty}^{\infty} d\omega\, \chi''_{sc}(\mathbf{q}, \omega)\omega = \frac{Nq^2}{m}, \tag{4.41}$$

$$\lim_{\mathbf{q}\to 0}\left[-\frac{1}{\pi}\int_{-\infty}^{\infty} d\omega \frac{\chi''_{sc}(\mathbf{q}, \omega)}{\omega}\right] = \frac{N}{ms^2}, \tag{4.42}$$

in which form they are directly analogous to Eqs. (2.173) and (2.175).*

What is the physical meaning of χ''_{sc}? As we have seen, it is a "construct," which measures the response to the total screened field, $\mathcal{E} = \mathfrak{D} + \mathcal{E}_p$. It must be calculated with some care, in order to avoid including those effects of the Coulomb interaction which have already been incorporated in $\mathcal{E}_p$. (This problem of double counting has already been discussed in Section 3.6.) In the limit of long wavelengths and low frequencies, $\chi''_{sc}(\mathbf{q}, \omega)$ may be obtained directly with the aid of the Landau–Silin transport equation, (3.38), in which the screened field $\mathcal{E}$ is introduced explicitly. For microscopic values of $\mathbf{q}$ and ω, double counting is avoided if one uses the appropriate field-theoretic techniques.†

The character of χ''_{sc} follows from its close relationship to $\chi''_n(\mathbf{q}, \omega)$, the density-density response function for an equivalent neutral system. There will be contributions from single-pair and multipair excitations; provided F_0^s is sufficiently large and positive, there will likewise be a pole in $\chi''_{sc}(\mathbf{q}, \omega)$ which is of essentially "collective" origin. Let us emphasize that for an isotropic system, such a pole never shows up in the density fluctuation excitation spectrum, that is, in $S(\mathbf{q}, \omega)$ or in $\chi''(\mathbf{q}, \omega)$. To see this, we take the imaginary part of both sides of Eq. (4.12), and find

$$\chi''(\mathbf{q}, \omega) = \frac{\chi''_{sc}(\mathbf{q}, \omega)}{|\epsilon(\mathbf{q}, \omega)|^2}. \tag{4.43}$$

According to Eq. (4.13), a pole in χ''_{sc} is likewise a pole in $\epsilon(\mathbf{q}, \omega)$; it follows that any such pole corresponds to a *zero* in $\chi''(\mathbf{q}, \omega)$ or $S(\mathbf{q}, \omega)$.

* We have written Eqs. (4.41) and (4.42) in a form which is valid both at $T = 0$ and at finite temperatures; the reader may easily verify that our derivation of Eqs. (4.34) and (4.35) goes through in the same way at finite temperatures, provided s is taken to be the *isothermal* sound velocity.

† For the benefit of the initiated, we note that one simply sums over all *proper* polarization diagrams, using a self-consistent dynamically screened electron-electron interaction.

The physical relationship between $\chi''(\mathbf{q}, \omega)$ and $\chi''_{sc}(\mathbf{q}, \omega)$ is clearly apparent in Eq. (4.43). Let us assume first that $\epsilon(\mathbf{q}, \omega)$ has no collective pole on the real axis; $\chi''_{sc}(\mathbf{q}, \omega)$ then extends over the same range of frequencies as $\chi''(\mathbf{q}, \omega)$. On using Eqs. (2.63) and (2.12), we may write $\chi''_{sc}(\mathbf{q}, \omega)$ in the form

$$\chi''_{sc}(\mathbf{q}, \omega) = -\pi \sum_n |(\rho_{\mathbf{q}}^{(1)})_{no}|^2 \delta(\omega - \omega_{no}) \qquad (\omega > 0), \tag{4.44}$$

where we have set

$$[\rho_{\mathbf{q}}^{(1)}]_{no} = (\rho_{\mathbf{q}})_{no} \epsilon(\mathbf{q}, \omega_{no}). \tag{4.45}$$

In this form, the only difference between the unscreened and screened response functions is the replacement of $\rho_{\mathbf{q}}$ by $\rho_{\mathbf{q}}^{(1)}$ in Eq. (4.44) (as a consequence, the plasmon pole found in χ'' does not appear in χ''_{sc}).

It is instructive to look at Eq. (4.45) the other way around. Let us assume that we know how to calculate $\chi''_{sc}(\mathbf{q}, \omega)$ in terms of the appropriately screened particle interaction. We thereby arrive directly at an expression of the form (4.44), in which the matrix elements and excitation frequencies are those appropriate to the corresponding neutral system described in Chapter 2. In order to pass from $\chi''_{sc}(\mathbf{q}, \omega)$ to the real response function $\chi''(\mathbf{q}, \omega)$, we must divide the matrix element $(\rho_{\mathbf{q}}^{(1)})_{no}$ by the dielectric constant $\epsilon(\mathbf{q}, \omega_{no})$. Such a procedure is equivalent to *screening* the response of the "equivalent neutral system."

We note that the response function for the screened field, χ''_{sc}, involves the bare, unscreened density fluctuation $\rho_{\mathbf{q}}^{(1)}$, while the response to the unscreened field, χ'', involves the screened density fluctuation $\rho_{\mathbf{q}}$. Put another way, screening may be taken into account either in the definition of the field or in the matrix element of the density fluctuation, but not in both.

Physically, we may picture the bare quasiparticles (of our equivalent neutral system) as surrounded by the appropriate dynamic screening cloud of other quasiparticles. Mathematically, one can compute a density fluctuation spectrum ignoring all polarization processes, and then put these in by means of Eq. (4.45). That this could be done in the case of the random phase approximation was recognized some time ago by the authors [Nozières and Pines (1958)]; the present considerations show that superposing appropriately "dressed" quasiparticles provides a correct account of the density fluctuation spectrum under all circumstances.*

* The above considerations are not confined to $T = 0$; they apply equally well to finite temperature quantum and classical plasmas. Indeed the method of superposition of dressed single-particle excitations has proved of considerable assistance in calculating fluctuations in classical plasmas [Thompson and Hubbard (1960), Ichimaru (1962, 1965)].

The simple relation (4.45) raises some difficulties when $\epsilon(\mathbf{q}, \omega)$ possesses a collective pole. The latter contributes to χ''_{sc}, but not to χ'' (in much the same way as the plasmon state contributes to χ'' and not to χ''_{sc}). The expansion (4.44) then contains an additional "collective term," with frequency $\omega_{no}^{(1)}$, which does not exist in the expression of χ''. Such a term cannot be assigned to a real excitation of the system, since the collective pole of ϵ does not correspond to a well-defined excited state. However, one must be careful not to miss it in the calculation of χ''_{sc}.

We may summarize the above discussion in the following way. There exist single-pair and multipair excitations, plus poles of collective origin in $\chi''_{sc}(\mathbf{q}, \omega)$. The relative importance of these various contributions is exactly the same as for the analogous neutral system considered in Chapter 2. For long wavelengths, the various contributions can be disentangled; we shall see that the multipair excitations play no role. Moreover, for low frequencies ($\omega \ll qv_F$), $\chi''_{sc}(\mathbf{q}, \omega)$ is proportional to ω in the vicinity of $\omega = 0$, as long as $q < 2p_F$; for $q > 2p_F$, $\chi''_{sc}(\mathbf{q}, \omega)$ is at least of order ω^2 when $\omega \to 0$.

On passing to $\chi''(\mathbf{q}, \omega)$, the contributions from single-pair and multipair excitations are found at the same excitation frequencies; the respective contributions are, however, *screened* by a factor $|\epsilon(\mathbf{q}, \omega_{no})|^{-2}$. Such screening is very important at long wavelengths and low frequencies; it eliminates altogether any contribution to $\chi''(\mathbf{q}, \omega)$ from the collective poles in $\chi''_{sc}(\mathbf{q}, \omega)$. On passing from $\chi''_{sc}(\mathbf{q}, \omega)$ to $\chi''(\mathbf{q}, \omega)$, one picks up the plasmons, which correspond to the zeros of $\epsilon(\mathbf{q}, \omega)$, whose behavior we have discussed at the beginning of this subsection. Because of the screening of single-pair excitations, and because the multipair excitations become important only for values of $q \cong p_F$, plasmons of reasonably well-defined energy dominate the density-fluctuation excitation spectrum of electron liquids at metallic densities for values of $q \lesssim \omega_p/v_F$. In the opposite limit of large q ($q \gg \omega_p/v_F$), one finds a continuous energy spectrum made up of single-pair and multipair excitations.

LONG WAVELENGTH EXCITATIONS

As was the case for the neutral Fermi liquid, we can make use of the various long wavelength sum rules to identify the physical excitations of importance in the long wavelength limit. We shall see that in this limit, the four sum rules, (4.32)–(4.35), actually separate into pairs of two; the first two [Eqs. (4.32) and (4.33)] determine the density-fluctuation excitation spectrum, which consists entirely in plasmons. The second two govern the screened density-fluctuation excitation spectrum, according to Eqs. (4.41) and (4.42).

We first consider the sum rules (4.32) and (4.33), which involve the dynamic form factor $S(\mathbf{q}, \omega)$. In the long wavelength limit, where plasmons are not damped, we may write $S(\mathbf{q}, \omega)$ as a sum of two terms:

$$S(\mathbf{q}, \omega) = S_{pl}(\mathbf{q}, \omega) + S_{\text{inc}}(\mathbf{q}, \omega), \tag{4.46}$$

where S_{pl} is the spectral density for the plasmon mode. The other term, S_{inc}, represents the essentially incoherent contribution from quasiparticle excitations.

We begin with the quasiparticle part. A first contribution comes from the excitation of a single quasiparticle-quasihole pair. In translationally invariant systems, the corresponding excitation energy, ω_{no}, is of order qv_F. (In real solids, ω_{no} may be finite, corresponding to an interband transition; this more general case is discussed briefly in Section 4.3.) The long wavelength matrix element, $(\rho_{\mathbf{q}}^{(1)})_{no}$, is of order 1 for the states that contribute; the number of such states is of order q/p_F. One sees readily that such excitations contribute to the sum rules, (4.34) and (4.35). On the other hand, as a result of the dielectric screening of this very low-frequency transition, we have from Eq. (4.45)

$$\lim_{\mathbf{q}\to 0} (\rho_{\mathbf{q}})_{no} \cong \frac{(\rho_{\mathbf{q}}^{(1)})_{no}}{\epsilon(\mathbf{q}, 0)} \cong \frac{q^2 s^2}{\omega_p^2}. \tag{4.47}$$

It follows that single-pair excitations contribute a term of order q^6 to the f-sum rule, and of order q^4 to the screening sum rule. The single-pair contribution to these sum rules is thus negligible in the long wavelength limit.

The contribution from multipair states (containing at least two quasiparticles and quasiholes) to the sum rules may be estimated in similar fashion, by analogy with the neutral system of Chapter 2. In the long wavelength limit, and for a translationally invariant system, one has

$$\lim_{\mathbf{q}\to 0} \langle\omega_{no}\rangle \sim \bar{\omega},$$
$$(\rho_{\mathbf{q}}^{(1)})_{no} \sim q^2,$$

where $\bar{\omega}$ is finite (of order μ). It follows that $\epsilon(\mathbf{q}, \bar{\omega}) \sim 1$, so that the density fluctuation matrix element, $(\rho_{\mathbf{q}})_{no}$, is likewise of order q^2 in this limit. We have no Pauli principle restrictions on the available phase space in either case. We conclude, therefore, that multipair excitations contribute terms of order q^4 to all four sum rules in the long wavelength limit, and hence may be altogether neglected.

We return to the f-sum rule, (4.32), and the perfect screening sum rule, (4.33); the above arguments tell us that S_{pl} must exhaust both sum rules in the long wavelength limit. Let us look for a contribution of the form

$$S_{pl}(\mathbf{q}, \omega) = A\delta(\omega - \omega_q). \tag{4.48}$$

Substitution of Eq. (4.48) into Eqs. (4.32) and (4.33) determines both A and ω_q; one finds

$$\lim_{\mathbf{q}\to 0} S_{pl}(\mathbf{q}, \omega) = \frac{Nq^2}{2m\omega_p}\,\delta(\omega - \omega_p). \tag{4.49}$$

Looked at from the vantage point of the f-sum rule, plasmons at energy ω_p are seen to be the dominant long wavelength excitation mode of a homogeneous electron system, no matter what its density.

If we now make use of the relationship (4.10) between $S(\mathbf{q}, \omega)$ and the imaginary part of $\epsilon^{-1}(\mathbf{q}, \omega)$, we find

$$\lim_{\mathbf{q}\to 0} \epsilon(\mathbf{q}, \omega) = 1 - \omega_p^2/\omega^2. \tag{4.50}$$

The result, (4.50), is our promised justification of the calculation based on the Landau theory which was presented in Chapter 3. A little thought shows that it is valid for values of $\mathbf{q}$ such that

$$qv_F \ll \omega_p.$$

We note that the matrix element for plasmon excitation is

$$\lim_{\mathbf{q}\to 0} \{(\rho_{\mathbf{q}}{}^{+})_{no}\}_{pl} = \sqrt{N/2m\omega_p}\,q. \tag{4.51}$$

On applying longitudinal particle conservation, we find that

$$\lim_{\mathbf{q}\to 0} \{(\mathbf{J}_{\mathbf{q}}{}^{+})_{no}\}_{pl} = \sqrt{N\omega_p/2m}. \tag{4.52}$$

The latter result contradicts our earlier argument that in this limit $\mathbf{J}_{\mathbf{q}}$ must be of order $\mathbf{q}$ (an argument based on the fact that for a translationally invariant system the total current $\mathbf{J}$ is a good quantum number). The contradiction is resolved if we realize that translation of the electrons alone does produce a physical effect, since the positive charge is fixed; one has, as a result, a charge separation. In the long wavelength limit, such a translation gives rise to a plasma oscillation, whence the finite value of $(\mathbf{J}_{\mathbf{q}})_{no}$.

We return to our consideration of the screened density-fluctuation spectrum or, what is equivalent, $\epsilon(\mathbf{q}, \omega)$. We have seen that multipair excitations do not contribute to $\chi''_{sc}(\mathbf{q}, \omega)$ in the long wavelength limit. There remain contributions to $\chi''_{sc}(\mathbf{q}, \omega)$ from single-pair excitations, and from collective poles. In general both contributions are present, and cannot be disentangled; one therefore can make no precise statements about the behavior of $\epsilon(\mathbf{q}, \omega)$. In the strong coupling limit, $F_0^s \gg 1$, the collective pole may be expected to dominate the behavior of χ''_{sc}. Under these circumstances, and in the limit of long wavelengths, $\epsilon(\mathbf{q}, \omega)$ will be given by [cf. Eq. (3.77)]

$$\lim_{\mathbf{q}\to 0} \epsilon(\mathbf{q}, \omega) = 1 - \frac{\omega_p^2}{(\omega + i\eta)^2 - s_0^2 q^2}, \qquad F_0^s \gg 1. \tag{4.53}$$

Equation (4.53) reduces to our earlier result, (4.50), as $\mathbf{q} \to 0$, and to the static result (4.27) as $\omega \to 0$.

We note that our earlier result, (4.50), implies that when $\mathbf{q} \equiv 0$, the real part of the conductivity, σ_1, must possess a δ-function at $\omega = 0$. On making use of the Kramers–Kronig relations which relate the real and imaginary parts of $\epsilon(\mathbf{q}, \omega)$, one obtains

$$\lim_{\mathbf{q}\to 0} \sigma_1(\mathbf{q}, \omega) = \frac{Ne^2}{m} \pi\delta(\omega). \tag{4.54}$$

Such a pole is not surprising. We are considering the "collisionless" regime, in which the electron liquid should display perfect conductivity under an applied uniform field; the pole in σ is the mathematical characterization of that perfect conductivity.

Once one introduces collisions, matters are, of course, different. In the limit of very low frequencies and wave vectors, one enters the hydrodynamic regime; χ''_{sc} can no longer be described in terms of elementary excitations. Assume first that the important collisions are those between quasiparticles. The hydrodynamic limit of $\epsilon(\mathbf{q}, \omega)$ is then given by Eq. (4.53), s_0 being replaced by the first sound velocity s [compare with Eq. (2.182)]. In the case where impurity scattering is dominant, $\epsilon(\mathbf{q}, \omega)$ is instead given by

$$\lim_{\mathbf{q}\to 0} \epsilon(\mathbf{q}, \omega) = 1 - \frac{\omega_p^2}{\omega^2 - s^2 q^2 + i\omega/\tau_{\text{eff}}} \tag{4.55}$$

[see Eq. (3.146)].

In this section, we are only concerned with the structure of $\chi''_{sc}(\mathbf{q}, \omega)$ in the collisionless regime. The results we have obtained for the matrix elements and excitation frequencies of the different kinds of excitations are summarized in Table 4.1. We have also included in this table the estimates one obtains for the static form factor, $S(\mathbf{q})$, in the long wavelength limit. We recall that

$$S(\mathbf{q}) = \sum_n |(\rho_{\mathbf{q}}^+)_{n0}|^2/N. \tag{4.56}$$

TABLE 4.1. *Matrix Elements, Excitation Frequencies, and Sum Rule Contributions of Density Fluctuation Excitations in the Long Wavelength Limit*

	Plasmon	Single pair	Multipair
$(\rho_{\mathbf{q}}^{(1)})_{no}$	—	1	q^2
$(\rho_{\mathbf{q}}^{+})_{no}$	$q(N/2m\omega_p)^{1/2}$	q^2	q^2
ω_{no}	ω_p	qv_F	$\bar{\omega}$
Pauli principle restrictions	None	q/p_F	None
$\Sigma_n\omega_{no}\|(\rho_{\mathbf{q}}^{+})_{no}\|^2$	$Nq^2/2m$	q^6	q^4
$\Sigma_n\|(\rho_{\mathbf{q}}^{+})_{no}\|^2/\omega_{no}$	$q^2/8\pi e^2$	q^4	q^4
$\Sigma_n\|(\rho_{\mathbf{q}}^{+})_{no}\|^2\omega_{no}\|\epsilon(\mathbf{q}, \omega_{no})\|^2$	—	$Nq^2/2m$	q^4
$\Sigma_n\|(\rho_{\mathbf{q}}^{+})_{no}\|^2\|\epsilon(\mathbf{q}, \omega_{no})\|^2/\omega_{no}$	—	$N/2ms^2$	q^4
$\Sigma_n\|(\rho_{\mathbf{q}}^{+})_{no}\|^2/N$	$q^2/2m\omega_p$	q^5	q^4

In the limit of small **q**, the static form factor is determined entirely by the plasmons, and is given by

$$\lim_{\mathbf{q}\to 0} S(\mathbf{q}) = \frac{q^2}{2m\omega_p}. \tag{4.57}$$

4.3. SCREENING AND PLASMA OSCILLATION

We have discussed in Chapter 3 the physical basis for the twin phenomena characteristic of homogeneous electron systems: screening and plasma oscillation. The methods of the present chapter enable us to prove that screening at a characteristic length

$$\lambda_c = s/\omega_p \tag{4.58}$$

and plasma oscillation at a frequency

$$\omega = \omega_p \tag{4.59}$$

are *exact* properties of *any* homogeneous electron plasma in the long wavelength limit. Thus no matter what the density or temperature might be, we can write the static dielectric constant as in Eq. (4.27),

$$\lim_{\mathbf{q}\to 0} \epsilon(\mathbf{q}, 0) = 1 + \frac{\omega_p^2}{s^2 q^2},$$

where s is the isothermal sound velocity, related to the isothermal compressibility, κ_{iso}, by

$$ms^2 = \frac{1}{\rho\kappa_{iso}} = N\left(\frac{\partial\mu}{\partial N}\right)_{T,\Omega} \tag{4.60}$$

Correspondingly, the dissipative part of the density-density response function possesses the limiting value

$$\lim_{\mathbf{q}\to 0} \chi''(\mathbf{q}, \omega) = -\frac{\pi}{2}\frac{Nq^2}{\omega_p}\{\delta(\omega - \omega_p) - \delta(\omega + \omega_p)\} \tag{4.61}$$

or, what is equivalent,

$$\lim_{\mathbf{q}\to 0} \epsilon(\mathbf{q}, \omega) = 1 - \frac{\omega_p^{\ 2}}{\omega^2}.$$

FINITE TEMPERATURES

Thus far we have established the above limiting behavior only in the case of an electron liquid at $T = 0$. [Equation (4.61) is easily derived on making use of Eqs. (2.167) and (4.49).] The validity of these relations at finite temperatures is readily established. The result, (4.27), is almost trivial, depending as it does only on the proper definition of the compressibility for a system of electrons immersed in a uniform background of positive charge. The proof of Eq. (4.61) at finite temperatures proceeds from the finite temperature version of the f-sum rule,

$$\int_{-\infty}^{\infty} d\omega\, \chi''(\mathbf{q}, \omega)\omega = \frac{Nq^2}{m}, \tag{4.62}$$

and the "perfect screening" sum rule,

$$\int_{-\infty}^{\infty} d\omega\, \frac{\chi''(\mathbf{q}, \omega)}{\omega} = \frac{q^2}{4\pi e^2}, \tag{4.63}$$

By making use of the definition of $\chi''(\mathbf{q}, \omega)$ at finite temperatures, Eq. (2.165), and by applying matrix element arguments directly analogous to those used at $T = 0$, one easily establishes the fact that only plasma oscillations contribute to Eqs. (4.62) and (4.63). The result, (4.61), follows at once.

The above results apply to classical electron plasmas, as well as to quantum plasmas. For example, in the case of a low-density, weakly interacting, classical plasma, s^2 may be replaced by its value for a non-interacting particle system,

$$s^2 = \kappa T/m. \tag{4.64}$$

One finds then the usual Debye–Hückel screening,

$$\lim_{\mathbf{q}\to 0} \epsilon(\mathbf{q}, 0) = 1 + \frac{4\pi N e^2}{q^2 \kappa T}. \tag{4.65}$$

Equation (4.27) permits one to obtain corrections to that screening for arbitrary densities and temperatures.

ELECTRONS IN SOLIDS

It is interesting to consider to what extent the above results apply to electrons in solids, where the periodic array of positive ions destroys the translational invariance. In this case, quasiparticle excitations are labeled by their momentum $\mathbf{p}$ (in the first Brillouin zone) and their band index n (see Section 1.3); single-pair excitations are obtained by transferring a quasiparticle from the state $(\mathbf{p}, n)$ to the state $(\mathbf{p} + \mathbf{q}, n')$. *Intraband* transitions ($n = n'$) occur in all solids except insulators; they are very similar to the single-pair excitations found for uniform systems. On the other hand, *interband* transitions ($n \neq n'$) are a new feature of real solids; they have a finite energy when $\mathbf{q} \to 0$, and in a sense characterize the lack of translational invariance. The corresponding matrix element is of order q in the limit of $\mathbf{q} \to 0$ (as compared with the value $\sim q^2$ found for a finite frequency excitation in the uniform system).

We consider first the applicability of our four sum rules, Eqs. (4.32) to (4.35). Since particle conservation takes the same form in a solid as in the electron liquid, the f-sum rule, (4.32), continues to be valid; in the long wavelength limit, there will be contributions to it from both plasmon and interband excitations. The conductivity sum rule, (4.24) or (4.34), is equally valid, and involves both intraband and interband excitations. On the other hand, the remaining two sum rules depend on the existence of *free* charges (electrons in the conduction band or holes in the valence band).

In metals, semimetals, or semiconductors, where free charges are found, there continues to exist *perfect* screening in the long wavelength limit. The screening length is, however, different, and will, in general, depend on direction. For a sufficiently isotropic metal (such as Na), one can still define an isothermal electron compressibility, which is related to $(\partial N/\partial \mu)$ as in Eq. (4.60); one can, moreover, use Eq. (4.60) to *define* an effective s^2 (see Section 3.8). It is this value of s^2 which appears in the solid state compressibility sum rule, (4.35). We note that while interband transitions contribute to the perfect screening sum rule, (4.33), they make no contribution to the compressibility sum rule.

For an insulator, on the other hand, there is no "perfect" screening of a static external charge; the sum rules, (4.33) and (4.35), do not apply. An external charge will be reduced in magnitude, by an amount depending on the static dielectric constant

$$\lim_{\mathbf{q}\to 0} \epsilon(\mathbf{q}, 0) = \epsilon_0 \qquad \text{(insulator).}$$

ϵ_0 is determined by the interband oscillator strength and excitation frequencies; a convenient expression may be obtained with the aid of the Kramers–Kronig relations.

We consider next the frequency dependence of ϵ, confining our attention to the long wavelength limit. Detailed calculations may be carried out in limiting cases; here, we only present a brief summary of the results, and of their implications for the existence of plasmons in solids. We distinguish between three different frequency regimes.

(*i*) $qv_F \ll \omega \ll \bar{\omega}_{\min}$ (*a minimum energy for interband excitation*) The contribution to the dielectric function from interband transitions is frequency independent, equal to a constant ϵ_i. The full dielectric function may be written in the form

$$\lim_{\mathbf{q}\to 0} \epsilon(\mathbf{q}, \omega) = \epsilon_i - \frac{4\pi N_c e^2}{m_c \omega^2}, \tag{4.66a}$$

where N_c is the total density of free *charge carriers* (electrons or holes), and m_c their "crystalline" effective mass. We have seen in Section 3.8 that m_c differs from the bare mass m because of the lack of translational invariance. The difference between m and m_c may indeed be expressed in terms of the interband oscillator strengths for the conduction electrons (or holes)—see Problem 4.2.

In many cases (such as semimetals or semiconductors), there exist several types of carriers, with respective densities $N_{c\alpha}$, and crystalline masses $m_{c\alpha}$. The dielectric function is then obtained by adding the polarizabilities of each group of carriers, Eq. (4.66a) being replaced by

$$\lim_{\mathbf{q}\to 0} \epsilon(\mathbf{q}, \omega) = \epsilon_i - \sum_\alpha \frac{4\pi N_{c\alpha} e^2}{m_{c\alpha} \omega^2}. \tag{4.66b}$$

In the more general case of strongly anisotropic energy bands, there is no simple expression for $\epsilon(\mathbf{q}, \omega)$.

According to Eq. (4.66b), plasma oscillations may occur at a frequency

$$\omega = \left[\sum_\alpha \frac{4\pi N_{c\alpha} e^2}{\epsilon_i m_{c\alpha}}\right]^{1/2}$$

[to the extent that this frequency lies in the region where Eq. (4.66b) is valid]. Such "intraband" plasma oscillations are usually found in semiconductors and semimetals, where $N_{c\alpha}$ is relatively small: the plasma frequency then lies in the far-infrared region.

In this range of frequencies, the behavior of $\epsilon(\mathbf{q}, \omega)$ is dominated by intraband transitions; it may be studied experimentally by measuring optical properties of the solid in the infrared and far infrared regions. Such experiments have been carried out on metals. They were briefly described in Section 3.8.

(*ii*) $\omega \sim \bar{\omega}$(*a typical interband excitation energy*) The form of ϵ is complicated, and depends on the details of the interband excitation spectrum. If zeros of ϵ exist, they occur at a frequency different from ω_p. Moreover, the plasmons are damped, since the plasmon peak is immersed in the continuum of interband excitations.

(*iii*) $\omega \gg \bar{\omega}_{\max}$ (*the maximum frequency for which interband transitions of valence electrons possess an appreciable oscillator strength*) In this limit, the dielectric function takes the form

$$\lim_{\mathbf{q}\to 0} \epsilon(\mathbf{q}, \omega) = 1 - \frac{4\pi N_v e^2}{m\omega^2}, \tag{4.67}$$

where m is the bare electron mass and N_v the total number of electrons in the valence band (not necessarily equal to the number of carriers, N_c). The behavior of ϵ is unchanged from the uniform system, essentially because the electrons move so fast that they do not feel the influence of the periodic potential: they behave as if they were free. Under such circumstances, one expects to see "free electron-like" plasma oscillations of the valence electrons.

We discuss the experimental evidence for the observation of plasmons in the following section; we refer the interested reader to the literature [Pines (1956), Pines (1963), Chapter 4] for a detailed discussion of plasma oscillation, screening, and the optical properties of solids.

4.4. ENERGY LOSS BY A FAST ELECTRON

DIELECTRIC DESCRIPTION OF ENERGY LOSS

The concept of a dielectric response function may be applied to the scattering of fast electrons by an electron system. As we have mentioned already, a fast electron beam constitutes a suitable weakly coupled probe when the Born approximation may be used to describe the scattering act. If the incoming electron (which we assume to be nonrelativistic) has a momentum $\mathbf{P}_e$, the condition for validity of the Born approximation is

$$P_e \gg p_F$$

(since the momentum transfers of interest are such that $q < p_F$). Under these circumstances, we may neglect the recoil of the fast electron; it therefore may be characterized by a density fluctuation

$$\rho_e(\mathbf{r}, t) = \delta(\mathbf{r} - \mathbf{R}_e - \mathbf{V}_e t)$$

whose Fourier transform is

$$\rho_e(\mathbf{q}, \omega) = e^{i\mathbf{q}\cdot\mathbf{R}_e} 2\pi\delta(\omega - \mathbf{q}\cdot\mathbf{V}_e).$$

In other words, the fast electron behaves like a test charge of frequency $\omega = \mathbf{q}\cdot\mathbf{V}_e$, strength $2\pi e^{i\mathbf{q}\cdot\mathbf{R}_e}$.

The probability per unit time that the electron transfer momentum $\mathbf{q}$ and energy ω to the electron gas is, according to Eq. (2.11),

$$\mathcal{P}(\mathbf{q}, \omega) = 2\pi \left(\frac{4\pi e^2}{q^2}\right)^2 S(\mathbf{q}, \omega). \tag{4.68}$$

On making use of Eq. (4.36), we can write

$$\mathcal{P}(\mathbf{q}, \omega) = \frac{8\pi e^2}{q^2} \left\{ - \operatorname{Im} \frac{1}{\epsilon(\mathbf{q}, \omega)} \right\} = \frac{8\pi e^2}{q^2} \frac{\epsilon_2(\mathbf{q}, \omega)}{|\epsilon(\mathbf{q}, \omega)|^2}. \tag{4.69}$$

The result (4.69) has been used for some time as the departure for a macroscopic treatment of the stopping power of an electron gas.* It could, in fact, have been obtained directly in a macroscopic calculation based on Ohm's law. One would expect such a calculation to yield correct results in the limit of

$$q \ll p_F, \qquad \omega \ll \epsilon_F.$$

The present calculation serves to extend these results into the microscopic regime ($q \sim p_F$, $\omega \sim \epsilon_F$).

A quantity of more experimental interest is $\mathcal{P}(\Upsilon, \omega)\, d\Upsilon$, the probability per unit time that the electron will be scattered into a solid angle $d\Upsilon = 2\pi \sin\theta\, d\theta$ while suffering an energy loss ω (θ is the scattering angle shown in Fig. 4.1). We can easily obtain $\mathcal{P}(\Upsilon, \omega)$ from $\mathcal{P}(\mathbf{q}, \omega)$, since the momentum transfer $\mathbf{q}$ is small compared to $\mathbf{P}_e$ and nearly perpendicular to it. Let us introduce cylindrical coordinates, $q_\parallel$, $q_\perp$

* Such an approach was first developed by Fermi (1940). Applications to the electron gas may be found in the papers of Kramers (1947), Bohr (1948), Hubbard (1955), Fröhlich and Pelzer (1955), and Ritchie (1957).

(with $\mathbf{P}_e$ as the cylindrical axis); we may write

$$q_\perp = P_e\theta,$$
$$\omega = q_\parallel V_e,$$
$$d^3\mathbf{q} = P_e^2\, d\Upsilon\, dq_\parallel. \tag{4.70}$$

According to Eq. (2.16), we have

$$\mathcal{P}(\Upsilon, \omega) = \frac{P_e^2}{(2\pi)^3}\int dq_\parallel \mathcal{P}(\mathbf{q}, \omega)\delta(\omega - q_\parallel V_e).$$

On substituting Eq. (4.69) into (4.71), we obtain

$$\mathcal{P}(\Upsilon, \omega) = \frac{1}{\pi^2 a_0 P_e}\frac{1}{\theta^2 + (\omega/mV_e^2)^2}\left\{-\operatorname{Im}\frac{1}{\epsilon(q_1, \omega)}\right\} \tag{4.71}$$
$$= \frac{4e^2}{a_0 P_e^3}\frac{1}{\{\theta^2 + (\omega/mV_e^2)^2\}^2}S(q_1, \omega), \tag{4.72}$$

where

$$q_1 = [q_\perp^2 + (\omega/V_e)^2]^{1/2} = P_e[\theta^2 + (\omega/mV_e^2)^2]^{1/2} \tag{4.73}$$

and where a_0 is the Bohr radius, $\hbar^2/me^2$. One sees clearly from Eq. (4.73) that when the energy transfer is small compared to mV_e^2, the

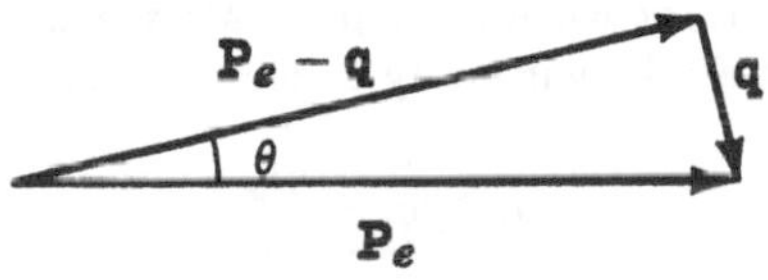

FIGURE 4.1. *Geometry for description of electron scattering.*

scattering takes place in very nearly the forward direction, as one should expect.

Equation (4.72) gives the relation between $\mathcal{P}(\Upsilon, \omega)$ and $S(\mathbf{q}, \omega)$. On the basis of our discussion in Section 4.2, we may divide the region of scattering angles (and momentum transfers) into essentially three different parts:

(i) A collective, or plasmon region, in which $S(\mathbf{q}, \omega)$ is completely dominated by the plasmons.

(ii) A transition region, in which the character of $S(\mathbf{q}, \omega)$ changes gradually from being purely plasmon to one in which single-pair and multipair excitations play an increasingly important role.

(iii) An incoherent excitation region, in which plasmons no longer exist as a well-defined excitation mode; the density fluctuation spectrum consists in single-pair and multipair excitations, more or less uniformly distributed in energy.

The pure plasmon regime is best defined as that region of wave vectors (and scattering angles) for which the plasmons effectively exhaust the f-sum rule. Under these circumstances one has, from Eq. (4.32),

$$S(\mathbf{q}_1, \omega) = S_{pl}(\mathbf{q}_1, \omega) = \frac{Nq_1^2}{2m\omega_1}\,\delta(\omega - \omega_1), \tag{4.74}$$

while

$$\mathcal{P}(\Upsilon, \omega) = \mathcal{P}_{pl}(\Upsilon, \omega) = \frac{\omega_p^2}{2\pi a_0 P_e}\left[\frac{1}{\theta^2 + (\omega_1/mV_e^2)^2}\right]\frac{1}{\omega_1}\,\delta(\omega - \omega_1). \tag{4.75}$$

The plasmon frequency ω_1 may be written as

$$\omega_1 \cong \omega_p + \alpha q_1^2/m, \tag{4.76}$$

where α is of order unity. In the "plasmon" regime, the q_1^2 term of Eq. (4.76) is a small correction (it arises only when the plasmon does *not* exhaust the sum rule).

The boundary of such a plasmon regime can only be defined in qualitative fashion. We have seen that the contribution of multipair excitations to the sum rule became important when $q_1 \sim p_F$. Moreover, the plasmons will certainly disappear when single-pair decay becomes possible, i.e., when

$$q_1 \geq q_c \sim \frac{\omega_p}{v_F}. \tag{4.77}$$

In practice, $q_c < p_F$. Thus, we shall certainly be in the plasmon regime if $q_1 \ll q_c$, say if $q_1 \leq q_c/4$. This corresponds to scattering angles θ such that

$$\theta \leq \theta_1 = \frac{1}{4}\frac{\omega_p}{v_F P_e}. \tag{4.78}$$

We may expect the transition region to span wave vectors ranging from $q_c/4$ to, say, $3q_c$. As q_1 increases, the plasmon energy becomes increasingly q-dependent, while plasmons become more and more damped; the probability of plasmon excitation begins to fall off, since single-pair and multipair excitations play an increasingly important role. By the time one has reached q_c, the plasmon is no longer a well-defined mode of excitation. However, between q_c and, say, $3q_c$, there may continue to be important correlation effects on the single-pair and multipair excitation spectrum, arising from the Coulomb interactions. (One does not, after all, have a *sharp* transition from essentially collective behavior and a broad plasmon peak, to completely incoherent single-particle-like behavior.)

In the transition region and beyond, $\theta > \theta_1$, the scattering probability decreases very rapidly with scattering angle. In that region, the energy transfers are of order μ, the Fermi energy. Since

$$\theta_1 \gg \mu/mV_e^2, \tag{4.79}$$

it follows that

$$q_1 \approx P_e\theta. \tag{4.80}$$

We therefore find

$$\mathcal{P}(\Upsilon, \omega) = \frac{4e^2}{a_0 P_e} \frac{1}{\theta^4} S(P_e\theta, \omega) \qquad (\theta > \theta_1). \tag{4.81}$$

The scattering probability falls off as θ^{-4}. In practice, one will see mostly plasmons, there being some slight chance of studying experimentally the correlated single-pair and multipair excitation spectrum between q_c and $3q_c$.*

We may easily calculate the total probability per unit time for scattering into a given solid angle:

$$\mathcal{P}(\Upsilon) = \int_0^\infty d\omega\, \mathcal{P}(\Upsilon, \omega). \tag{4.82}$$

(The usual complications, arising from a dependence of q on ω, are here avoided.) On carrying out the integration over ω, we find,

$$\mathcal{P}(\Upsilon) = \frac{\omega_p}{2\pi a_0 P_e} \frac{1}{\theta^2 + (\omega_p/mV_e^2)^2}, \qquad \theta \leq \theta_1, \tag{4.83a}$$

$$= \frac{4Ne^2}{P_e^3 a_0 \theta^4} S(P_e\theta), \qquad \theta \geq \theta_1. \tag{4.83b}$$

Electron scattering experiments thus offer a direct measure of the static form factor.

The mean free path for plasmon production may be estimated from Eq. (4.83a), by assuming that this expression applies up to the maximum wave vector q_c for which plasmons exist as a well-defined excitation. On integrating Eq. (4.83a) over solid angles, we obtain

$$\mathcal{P}_{pl} = 2\pi \int_0^{\omega_p/P_e v_F} d\theta\, \theta \mathcal{P}_{pl}(\Upsilon) = \frac{\omega_p}{a_0 P_e} \ln \frac{V_e}{v_F}, \tag{4.84}$$

The mean free path for plasmon production is therefore (to logarithmic accuracy),

$$\lambda_{pl} = \frac{V_e}{\mathcal{P}_{pl}} = 2a_0 \left(\frac{E_e}{\omega_p}\right) \frac{1}{\ln V_e/v_F}. \tag{4.85}$$

We can also calculate in straightforward fashion the energy transfer per unit time from the fast electron to the system. The energy transfer

* For a detailed study of this regime, carried out in the random phase approximation, see Glick and Ferrell (1960).

per unit time, accompanied by scattering into a solid angle, $d\Upsilon$, is

$$\frac{dE(\Upsilon)}{dt} = \int_0^\infty d\omega\, \omega\, \mathcal{P}(\Upsilon, \omega). \tag{4.86}$$

On making use of Eqs. (4.75) and (4.81), and of the f-sum rule, (4.32), we find

$$\frac{dE(\Upsilon)}{dt} = \frac{\omega_p^2}{2\pi a_o P_e} \frac{1}{\theta^2 + (\omega_p/mV_e^2)^2}, \qquad \theta \leq \theta_1, \tag{4.87}$$

$$\frac{dE(\Upsilon)}{dt} = \frac{\omega_p^2}{2\pi a_o P_e} \frac{1}{\theta^2}, \qquad \theta \geq \theta_1.$$

The total energy transfer per unit time is then

$$\left(\frac{dE}{dt}\right) = \frac{\omega_p^2}{a_o P_e} \left\{ \int_0^{\theta_1} d\theta \frac{\theta}{\theta^2 + (\omega_p/mV_e^2)^2} + \int_{\theta_1}^{\theta_{\max}} \frac{d\theta}{\theta} \right\} \tag{4.88}$$

$$\cong \frac{\omega_p^2}{a_o P_e} \ln \frac{q_{\max} V_e}{\omega_p},$$

where $q_{\max} = P_e \theta_{\max}$ is the maximum momentum transfer of interest (of the order of p_F). The result (4.88) has been derived in essentially all papers on the subject.

PLASMON EXCITATION IN SOLIDS*

The preceding discussion is only applicable to an electron liquid. Nonetheless, it provides a semiquantitative description of electron energy losses in real solids, including the essential feature of plasmon excitation. As we mentioned in Section 4.3, in a real solid the incoming particle may excite interband transitions, which compete with plasmon excitation even when $\theta \ll \theta_1$. Such interband excitations are of essentially two kinds:

(i) From the valence to higher bands; typical energies of transitions (with an appreciable oscillator strength) for metals and semiconductors are of the order of 1–5 eV.

(ii) From bands lying below the valence band to unoccupied higher band states. Such transitions (of energy 30 to 50 eV) are not much affected when the free atoms are brought together to make a solid, since changes in atomic energy levels upon forming a solid are small compared to 30 eV.

* This subsection follows closely the treatment given in Pines (1963).

The energy of these transitions should be compared to the "unperturbed" plasma frequency of valence electrons

$$\omega_{pv} = \left(\frac{4\pi N_v e^2}{m}\right)^{1/2} \tag{4.89}$$

(where N_v is the density of valence electrons).

In many solids, ω_{pv} falls between the two regions of interband processes. In this case, the high-energy transitions do not affect the plasma oscillation (in physical terms, one may neglect the polarizability of the ion cores at the frequency ω_{pv}). On the other hand, the low-energy transitions are screened by the Coulomb interaction; because of the f-sum rule, they do not modify the high-frequency dielectric constant. One then expects to observe a plasma oscillation at a frequency nearly equal to ω_{pv}. (At this frequency, the electrons move so rapidly that the lattice structure is blurred out; the ions appear as a uniform background of positive charge.)

Such an approximation is satisfactory for alkali metals, alkaline earths, valence semiconductors (Ge, Si), etc. On the other hand, it breaks down for transition and noble metals (in which the interband excitations from the d-band possess energies of order ω_{pv}), and also in insulators with a substantial gap between the valence and conduction bands.

We turn now to a consideration of the experimental results. In the so-called "characteristic energy losses experiments,"* one measures the energy transfer of a beam of kilovolt electrons to a solid. One may distinguish between

(i) reflection experiments, in which one observes the energy spectrum of fast electrons after they are reflected by a solid surface;
(ii) transmission experiments, in which one measures the spectrum of electrons which emerge from a thin solid film (~1000 Å) thick.

Reflection experiments possess the great advantage that one can work with freshly cleaned solid surfaces; as Swan and his collaborators [Powell and Swan (1960), Powell (1960), Robins and Swan (1960)] have shown, this permits the unambiguous identification of loss lines as being characteristic of the solid itself, rather than due to surface contamination. Transmission experiments, on the other hand, provide

* For a recent review of such experiments, see Klemperer and Shepherd (1963). A detailed comparison of theory and experiment was first given by Pines (1956); for a recent discussion (in more detail than that presented here), see Pines (1963).

in principle much more information, since one can measure the angular distribution of the inelastically scattered electrons; one thus measures $S(\mathbf{q}, \omega)$, and traces out the energy vs. momentum curve for the excitations of the solid. In a reflection experiment, the incident fast electron may suffer a Bragg reflection (with no energy loss) either before or after it loses energy; a knowledge of the angle at which it emerges provides no direct information on the momentum transfer associated with the energy transfer.

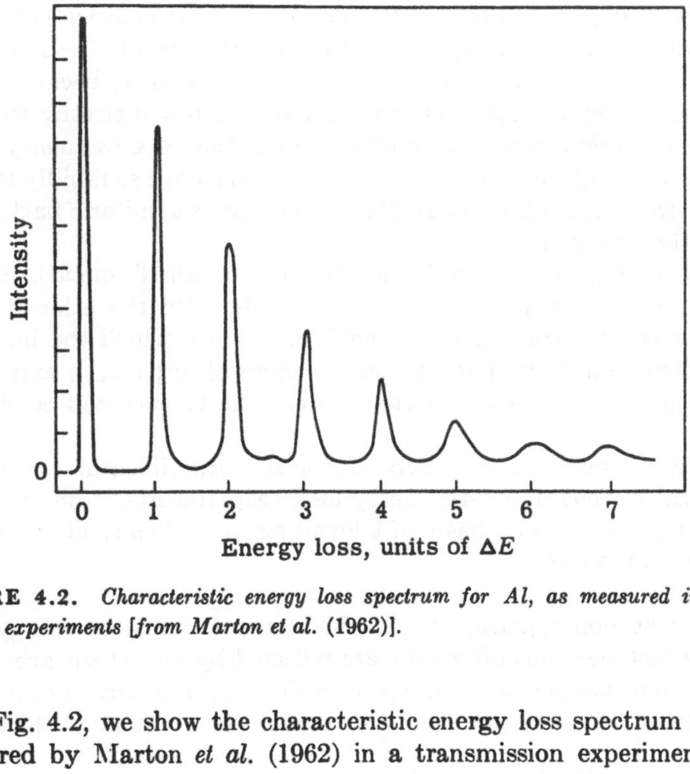

FIGURE 4.2. *Characteristic energy loss spectrum for Al, as measured in transmission experiments [from Marton et al.* (1962)].

In Fig. 4.2, we show the characteristic energy loss spectrum for Al, measured by Marton *et al.* (1962) in a transmission experiment. In Fig. 4.3, we reproduce results of a typical recent reflection experiment, that of Robins (1961), for gallium, indium, and thallium. The presence of two sharp peaks in the loss spectrum is a general feature of reflection experiments: one is associated with the volume plasma oscillation we have been considering; the second is due to "surface" plasmons. These surface plasma oscillations were first discussed by Ritchie (1957): their existence was confirmed by Powell and Swan (1960). According to Stern and Ferrell (1960), they involve a wave of charge *bound* at the interface between the solid under study and the neighboring dielectric medium. If the latter has a static dielectric constant, ϵ_B, the energy of a surface plasmon is roughly

$$\omega_s \cong \frac{\omega_{pv}}{(1+\epsilon_B)^{1/2}}. \tag{4.90}$$

Observation of surface plasmons together with the volume plasmons makes possible the unambiguous identification of a loss line as plasmon excitation for the solid in question.

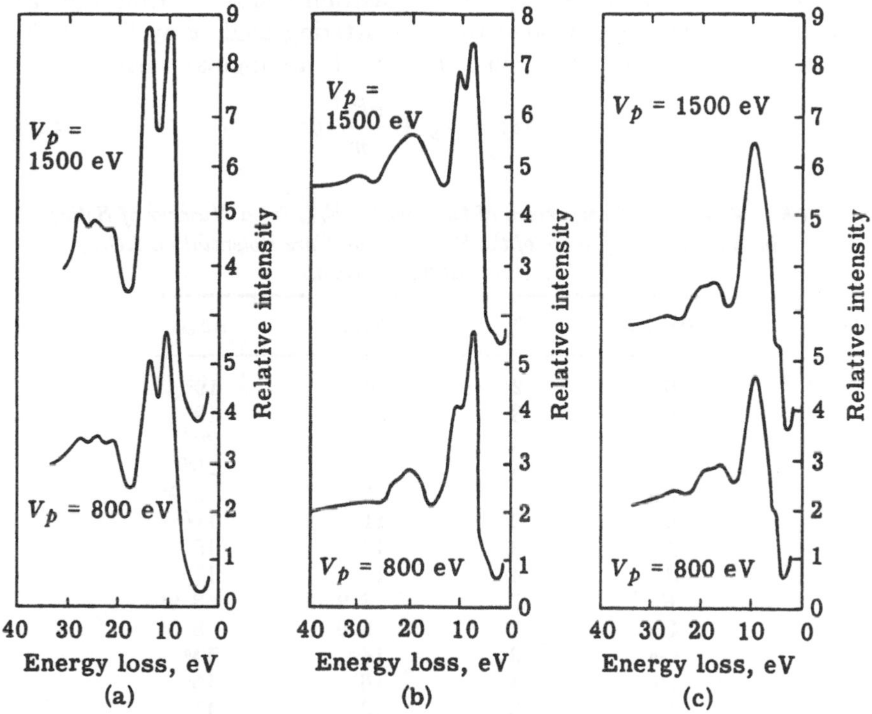

FIGURE 4.3. *Typical reflection measurements of the characteristic energy loss spectrum [from Robins* (1961)]: *(a) gallium, (b) indium, (c) thallium.*

In Tables 4.2 and 4.3 we give a comparison, for a number of solids, between $\hbar\omega_{pv}$ and that measured loss, ΔE_{obs}, which has been identified as a volume plasmon loss. The agreement between the two values is seen to be remarkably good. Indeed, as was remarked some time ago [Pines, (1956)], what puzzles exist have to do with why the agreement is so good, rather than with explaining existing disagreements. By contrast, for the solids in which interband excitations are expected to play an important role, one finds broad loss lines, at energies which do not correspond in any simple way to the calculated values of ω_{pv}.

Watanabe (1956) was the first to measure the angular distribution of inelastically scattered electrons emerging from a thin solid film. A schematic representation of the results of his experiment for aluminum is given in Fig. 4.4. To interpret his experimental results, we note that the plasmon energy depends on momentum; to a good degree of approximation, we have

$$\omega_1(q) = \omega_p + \alpha q^2/m. \tag{4.91}$$

For a fast incident electron, the momentum transfer $\mathbf{q}$ from the particle to the plasmon is related to the scattering angle θ by Eq. (4.80); if the energy loss is to plasmons, one expects an angular variation,

$$\Delta E(\theta) = \omega_p + \frac{\alpha P_e^{\,2}}{m}\,\theta^2. \tag{4.92}$$

TABLE 4.2. *A Comparison of $\hbar\omega_{pv}$ with ΔE_{obs} for a Number of Solids in Which the Influence of the Periodic Ion-Core Potential on ΔE_{obs} Is Expected To Be Small*

Element	Z[a]	$\hbar\omega_{pv}$	ΔE_{obs}
Be	2	19	19[b]
B	3	24	19[b]
C_1[c]	4	19	20[d]
C_2[c]	4	25	25[d]
Na	1	5	5.9[e]
Mg	2	11	11[f]
Al	3	16	15[e]
Si	4	17	17[e]
K	1	3.9	3.9[e]
Ca	2	8.0	8.8[e]
Ga	3	14	14[g]
Ge	4	16	16[g]
In	3	13	11[g]
Sn	4	14	14[g]
Sb	5	15	16[g]
Ba	2	6.5	6.5[e]
Tl	3	12	9.6[e]
Pb	4	14	14[g]
Bi	5	14	15[g]

[a] Z denotes the number of valence electrons per atom which are assumed to take part in plasma oscillation.

[b] L. Marton and L. B. Leder, *Phys. Rev.* **94**, 203 (1954).

[c] C_1 and C_2 refer to evaporated and amorphous carbon, respectively.

[d] L. B. Leder and J. C. Suddeth, private communication.

[e] J. L. Robins and P. E. Best, *Proc. Phys. Soc.* (*London*) **79**, 110 (1962).

[f] C. J. Powell, *Proc. Phys. Soc.* (*London*) **76**, 593 (1960).

[g] J. L. Robins, *Proc. Phys. Soc.* (*London*) **79**, 119 (1962).

TABLE 4.3. *A Comparison of $\hbar\omega_{pv}$ with ΔE_{obs} for a Number of Compounds*[a]

Compound	Z^a	$\hbar\omega_{pv}$	ΔE_{obs}
ZnS	4	17	17[b]
PbS	5	16	15[b]
SbS_3	5.6	18	19[b]
MoS_2	6	23	21[c]
PbTe	5	14	15[b]
PbSe	5	15	15[b]
Mica	4.7	24	25[d]
BeO	4	29	29[c]
MgO	4	25	25[c]
Li_2CO_3	4	24	24[c]
$Ca(OH)_2$	3.2	21	22[c]
MoO_3	6	24	25[c]
SiO_2	5.3	25	25[c]
Al_2O_3	4.8	27	25[c]
TeO_2	6	23	23[c]
SnO_2	4	26	18[b]
KBr	4	13	20[c]
KCl	4	14	13[b]
NaCl	4	16	16[b]

[a] Z denotes the average number of valence electrons per atom we assume to take part in plasma oscillation.

[b] L. Marton and L. B. Leder, *Phys. Rev.* **94**, 203 (1954).

[c] H. Watanabe, *J. Phys. Soc. Japan* **9**, 1035 (1954).

[d] G. Möllenstedt, *Optik* **5**, 499 (1949).

In Table 4.4 we give a summary of Watanabe's experimental results for α for several metals, together with a theoretical calculation of this quantity for an equivalent free electron gas.* The order-of-magnitude agreement is good; the close agreement for Be can scarcely be regarded as significant, in view of the approximations made in calculating α.

TABLE 4.4

Element	α_{exp}	α_{free}
Be	0.42 ± 0.04	0.41
Mg	0.62 ± 0.04	0.37
Al	0.50 ± 0.05	0.40

* The theoretical calculations neglect periodic effects, but take into account exchange corrections to the plasmon dispersion relation as calculated in the RPA. See Pines (1960).

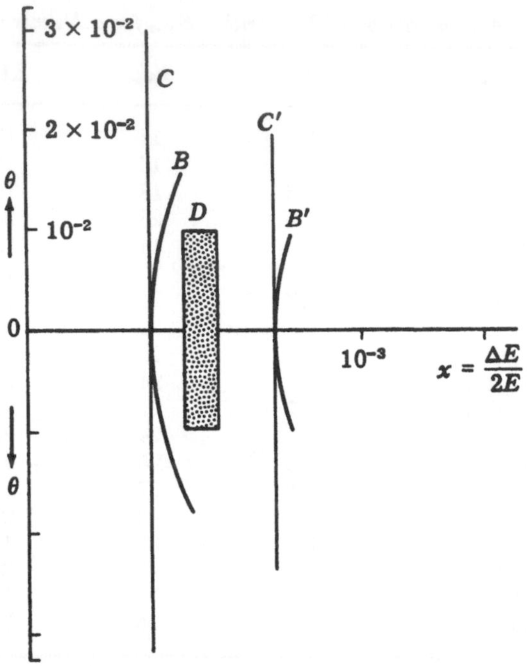

FIGURE 4.4. *Angular distribution of inelastically scattered electrons [after Watanabe* (1956)]. *The ordinate is the scattering angle in radians; the abscissa is the energy loss divided by twice the incident electron energy.*

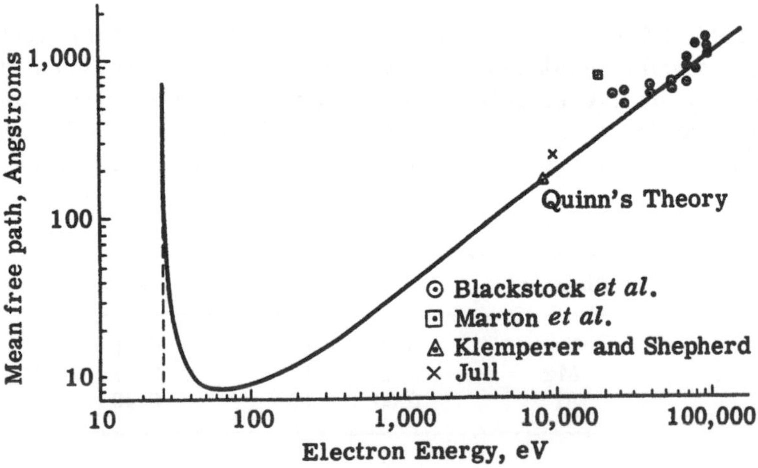

FIGURE 4.5. *Comparison between theory and experiment for the mean free path for plasmon production [from Klemperer and Shepherd* (1963)].

Such an experiment also provides the maximum wave vector beyond which the plasmon does not exist as a well-defined elementary excitation. Watanabe found a cut-off of 15–18 milliradians for 25-keV electrons incident on aluminum. Ferrell (1957) calculated a cut-off of 16 milliradians by considering only single-pair damping. Such agreement suggests that the damping due to multipair or interband transitions is negligible (the latter should not depend very much on wave vector).

In conclusion, we note that measurements for the mean free path for plasmon production are in good agreement with the estimate, (4.85), and with the more detailed calculations of Quinn (1962). A comparison between theory and experiment for Al is given in Fig. 4.5.

4.5. RESPONSE TO A PHONON FIELD

ELECTRON-PHONON INTERACTION

We now consider the consequences of the coupling between a quantum plasma in its ground state and an external longitudinal phonon field. The problem is closely related to those studied earlier in this chapter; it serves as a model for the coupling between phonons and conduction electrons in a metal. Being a model it lacks many of the "solid state" effects associated with the periodic nature of the ion array; nonetheless it reveals many of the interesting features of the coupled systems.

The coupling of the electrons to a longitudinal phonon field is described by the following Hamiltonian*:

$$H = H_{\text{ph}} + H_{\text{elec}} + H_{\text{int}}. \tag{4.93}$$

Here H_{ph} is the Hamiltonian for the phonons in the absence of their coupling to the electrons and H_{elec} is that for the electrons in the quantum plasma, while H_{int} describes the coupling between the electrons and the phonons. The phonon Hamiltonian is that for a collection of harmonic oscillators.

$$H_{\text{ph}} = \frac{1}{2} \sum_{\mathbf{q}} [P_{\mathbf{q}}{}^{+} P_{\mathbf{q}} + \Omega_{\mathbf{q}}{}^{2} Q_{\mathbf{q}}{}^{+} Q_{\mathbf{q}}]. \tag{4.94}$$

$P_{\mathbf{q}}$ and $Q_{\mathbf{q}}$ are the momentum and coordinate of the **q**th normal mode, of wave vector **q** and frequency $\Omega_{\mathbf{q}}$; they satisfy the Hermiticity condition

$$P_{\mathbf{q}}{}^{+} = P_{-\mathbf{q}}, \qquad Q_{\mathbf{q}}{}^{+} = Q_{-\mathbf{q}}.$$

The phonon-electron coupling is, to a first approximation, linear in the

* For a derivation of Eq. (4.93) see, for example, Pines (1963), Chapter 5.

phonon coordinates, and takes the form

$$H_{\text{int}} = \sum_{\mathbf{q}} v_{\mathbf{q}}{}^{i} \rho_{\mathbf{q}}{}^{+} Q_{\mathbf{q}}. \tag{4.95}$$

The phonons, being longitudinal, are coupled directly to the electronic density fluctuations. The matrix elements for that coupling satisfy the relation

$$(v_{\mathbf{q}}{}^{i})^{+} = v_{-\mathbf{q}}{}^{i}. \tag{4.96}$$

Let us first consider, in qualitative fashion, some of the consequences of the electron-phonon interaction. At finite temperatures, that which is perhaps most familiar is the scattering of electrons by phonons, an important cause of resistance in metals. In such a process an electron is scattered from some state $\mathbf{p}$ to a state $\mathbf{p} + \mathbf{q}$, with absorption of a phonon of momentum $\mathbf{q}$ or emission of a phonon of momentum $-\mathbf{q}$. One sees this clearly in the representation of second quantization (see the Appendix), in which H_{int} takes the form

$$H_{\text{int}} = \sum_{\mathbf{q}} \frac{v_{\mathbf{q}}{}^{i}}{\sqrt{2\Omega_{\mathbf{q}}}} c_{\mathbf{p}+\mathbf{q}}^{+} c_{\mathbf{p}} (A_{\mathbf{q}} + A_{-\mathbf{q}}^{+}), \tag{4.97}$$

where $A_{\mathbf{q}}$ and $A_{\mathbf{q}}{}^{+}$ are the destruction and creation operators for the phonon field. A second, closely related, phenomenon is phonon attenuation, the absorption of a sound wave by the electron gas, which is likewise an obvious consequence of Eq. (4.97). In addition to these two real processes, there exist three kinds of virtual processes that can affect the motion of the phonons and the electrons. First, the phonon field will act to polarize the electron gas; this polarization in turn gives rise to a shift in the phonon frequencies, a shift which may be thought of as resulting from an altered effective interaction between the ions. Second, in the presence of the phonon field the quasiparticle properties of the electron gas will be changed. A quasiparticle now consists of an electron plus a phonon cloud, in addition to the polarization cloud of other electrons. Finally, there is the possibility of a new mechanism of interaction between the electrons. A given electron acts to polarize the ions (by inducing a phonon displacement) and the polarization in turn affects a second electron; this interaction may be viewed as arising from an exchange of virtual phonons between the two electrons. It is of great importance in the theory of superconductivity.

In the present section we shall confine our attention primarily to phonon attenuation, the phonon energy shift, and to the effective electron interaction. We shall be interested in electron-phonon scattering only to show how the electrons act to screen the bare electron-phonon interaction. We shall not treat the change in quasiparticle properties; at the present time any such treatment requires the introduction of field theoretic techniques which are beyond the purview of this book.

SELF-CONSISTENT TREATMENT OF COUPLED ELECTRON-PHONON SYSTEMS

The treatment of the coupled electron-phonon systems is made much simpler by the fact that a typical phonon frequency is very small compared to a typical electron frequency. As is well known (and as we shall demonstrate explicitly), the ratio of the frequencies is of the order of $\sqrt{m/M}$, where M is the ionic mass. Because phonon frequencies are so low, in many problems the coupling of the electrons to the phonons may be treated as a polarization process, in which the phonons are regarded as an external field which is only weakly coupled to the electron gas. One neglects thereby any corrections to the electronic polarizability arising from virtual phonon processes: such corrections may be shown to be of order $(m/M)^{1/2}$ relative to the terms one has kept. This important result is, in a certain sense, a reflection of the well-known adiabatic theorem of Born and Oppenheimer (1928); it has, however, only recently been proved explicitly by Migdal (1958). As was the case with the Fermi liquid theory, the justification of the theory we shall present requires considerable use of field-theoretic techniques. We therefore refer the interested reader to Migdal's important paper for that justification.

The electrons and phonons are coupled via H_{int}; as a result the phonon field acts to induce a certain density fluctuation in the electron system. That induced density fluctuation in turn affects the phonon field, since it serves to alter the phonon frequencies and, as well, provides a mechanism for the direct absorption of the sound wave field. The calculation of the induced density fluctuations and the phonon dispersion relation should therefore be carried out in self-consistent fashion. Thus we should assume the existence of a sound wave of some frequency $\omega_{\mathbf{q}}$, say, and then calculate the induced density fluctuation which leads to a wave of that frequency.

The calculation may be simply carried out by following the equation of motion of the phonon oscillator amplitude, $Q_{\mathbf{q}}$. One finds easily from Eq. (4.93) the quantum-mechanical operator equation,

$$\ddot{Q}_{\mathbf{q}} + \Omega_{\mathbf{q}}^2 Q_{\mathbf{q}} = -v_{\mathbf{q}}{}^i \rho_{\mathbf{q}}. \tag{4.98}$$

Let us now take the expectation value of both sides of this operator equation with respect to the electron coordinates. In the absence of a phonon field, one has

$$\langle \rho_{\mathbf{q}} \rangle = 0$$

for our translationally invariant system. In the presence of a field with frequency $\omega_{\mathbf{q}}$, the electron response may be specified with the aid of the response function, $\chi(\mathbf{q}, \omega_{\mathbf{q}})$, introduced in Chapter 2 (to the extent that the electrons respond linearly). From Eqs. (2.49) and (4.95), we see directly that the density fluctuation induced by a longitudinal field of average amplitude $\langle Q_{\mathbf{q}} \rangle$ is simply

$$\langle \rho_{\mathbf{q}} \rangle = v_{\mathbf{q}}{}^{i} \chi(\mathbf{q}, \omega_{\mathbf{q}}) \langle Q_{\mathbf{q}} \rangle. \tag{4.99}$$

On substituting Eq. (4.99) into (4.96), one finds the phonon frequency is given by

$$\omega_{\mathbf{q}}{}^{2} = \Omega_{\mathbf{q}}{}^{2} + |v_{\mathbf{q}}{}^{i}|^{2} \chi(\mathbf{q}, \omega_{\mathbf{q}}). \tag{4.100}$$

The phonon dispersion relation, (4.100), is deceptively simple, in that it seems unlikely that anything so simple could also be correct. It was obtained by assuming that the electrons respond linearly to the phonon field, so that $\chi(\mathbf{q}, \omega)$ is the electronic polarizability calculated in *the absence of phonons.* This assumption is correct provided the phonon frequencies are such that we are in the collisionless regime. Equation (4.100) is then accurate to order $(m/M)^{1/2}$ (essentially because phonon frequencies are so very small compared to typical electron frequencies). In the opposite, low-frequency limit, in which electron-phonon collisions can restore local thermodynamic equilibrium, it is clear that one cannot ignore phonons in calculating the response function $\chi(\mathbf{q}, \omega)$. Equation (4.100) is then modified; we shall not discuss that modification here.

PHONON ENERGY SHIFT AND DAMPING

We make use of the definition of the dielectric constant, Eq. (4.6), to cast the phonon dispersion relation in slightly different form:

$$\omega_{\mathbf{q}}{}^{2} = \Omega_{\mathbf{q}}{}^{2} + \frac{|v_{\mathbf{q}}{}^{i}|^{2} q^{2}}{4\pi e^{2}} \left\{ \frac{1}{\epsilon(\mathbf{q}, \omega_{\mathbf{q}})} - 1 \right\}. \tag{4.101}$$

Since $1/\epsilon(\mathbf{q}, \omega_{\mathbf{q}})$ is in general complex, the corresponding phonon frequency will likewise be complex. Let us introduce the real and imaginary parts of the phonon frequency, according to

$$\omega_{\mathbf{q}} = \omega_1 - i\omega_2 \tag{4.102}$$

and, further, assume that

$$\omega_2 \ll \omega_1.$$

Equation (4.101) transforms to the following two equations:

$$\omega_1^2 = \Omega_{\mathbf{q}}^2 - \frac{q^2}{4\pi e^2} |v_{\mathbf{q}}^i|^2 \left\{ 1 - \frac{1}{\epsilon(\mathbf{q}, 0)} \right\}, \tag{4.103}$$

$$\frac{\omega_2}{\omega_1} = \frac{q^2}{8\pi e^2} \frac{|v_{\mathbf{q}}^i|^2}{\omega_1^2} \left\{ - \operatorname{Im} \frac{1}{\epsilon(\mathbf{q}, \omega_1)} \right\}. \tag{4.104}$$

In Eq. (4.103) we have assumed that $\omega_1 \ll qv_F$, and have therefore introduced the static dielectric constant, $\epsilon(\mathbf{q}, 0)$. Equation (4.103) describes the shift in the phonon frequencies as a result of phonon-electron interaction, while Eq. (4.104) describes the damping of a sound wave brought about by that interaction.

A good order-of-magnitude estimate of phonon frequencies and phonon attenuation in metals may be obtained with the aid of the so-called "jellium" model. In this model one neglects all effects of periodicity on the phonon-electron coupling and on the phonon frequencies, and further one takes the ion-ion and ion-electron interactions to be purely Coulombic in nature. The "bare" phonon frequency then reduces to the appropriate ionic plasma frequency,

$$\Omega_{\mathbf{q}} = \Omega_p = \left(\frac{4\pi N Z^2 e^2}{M} \right)^{1/2}, \tag{4.105}$$

where Z is the ion valency; the phonon-electron coupling is given by*

$$v_{\mathbf{q}}^i = - \frac{4\pi Z e^2 i\mathbf{q}}{q^2} \left(\frac{N}{M} \right)^{1/2}. \tag{4.106}$$

In this case, Eq. (4.103) reduces to

$$\omega_1^2 = \frac{\Omega_p^2}{\epsilon(\mathbf{q}, 0)}. \tag{4.107}$$

In the long wavelength limit, $\epsilon(\mathbf{q}, 0)$ is given by Eq. (4.27); we thus find

$$\omega_1^2 = s_{\text{ph}}^2 q^2 \qquad (q \ll p_F), \tag{4.108}$$

where the phonon sound velocity is given by

$$s_{\text{ph}} = \sqrt{\frac{m}{M}}\, Z\, s. \tag{4.109}$$

On comparing Eq. (4.105) with (4.108), we see that the electron response changes the frequency of a long wavelength longitudinal

* For a derivation of Eqs. (4.105) and (4.106) see, for example, Pines (1963), Chapter 5.

phonon from a constant value to the familiar form linear in $\mathbf{q}$. One may think of this change as resulting from the screening of the ion-ion interaction [since the latter is reduced in magnitude by a factor of $\epsilon^{-1}(\mathbf{q}, 0)$]. In the Fermi–Thomas, or weak coupling approximation, s is given by its free-electron value, $v_F/\sqrt{3}$, and the phonon sound velocity becomes

$$s^{(1)} = \sqrt{\frac{m}{M}\frac{Z}{3}}\, v_F, \tag{4.110}$$

a result which was first obtained by Bohm and Staver (1952). The result, (4.110), is in good order-of-magnitude agreement with experiments for metals; it shows clearly that typical sound wave frequencies, $s_{ph}q$, are of order $(m/M)^{1/2}$ compared to the corresponding electron frequencies, qv_F.

In the jellium model, the expression for phonon damping, (4.104), becomes

$$\frac{\omega_2}{\omega_1} = \frac{\epsilon_2(\mathbf{q}, \omega_1)}{2\epsilon(\mathbf{q}, 0)}, \tag{4.111}$$

where $\epsilon_2(\mathbf{q}, \omega_1)$ is the imaginary part of the dielectric constant. In the long wavelength limit, this expression simplifies with the aid of Eqs. (3.72) and (3.60) to

$$\frac{\omega_2}{\omega_1} = \frac{\pi}{12}\sqrt{\frac{m}{M}}\, Z\, \frac{p_F}{ms}, \tag{4.112}$$

a result which is valid in the long wavelength "collisionless" regime.

We see that phonon damping as a consequence of the electron response is indeed small. It corresponds to "Landau damping" by those electrons that move with the phase velocity of the sound wave. The damping is small because of the great mismatch in phonon and electron frequencies [$\sim(m/M)^{1/2}$]. Together with the exclusion principle, this mismatch limits the number of electrons which contribute to the damping process. Again, Eq. (4.112) may be expected to furnish a good qualitative account of the collisionless damping of phonons by conduction electrons in a metal.

In more realistic treatments of electron-phonon interaction in metals, certain features of the jellium model remain. Thus while both $v_{\mathbf{q}}{}^i$ and $\Omega_{\mathbf{q}}$ are changed as a result of the influence of the periodic array of ions, such periodicity corrections vanish in the long wavelength limit. To lowest order in $\mathbf{q}$, one still has in Eq. (4.103) the cancellation between $\Omega_{\mathbf{q}}{}^2$ and $(q^2/4\pi e^2)|v_{\mathbf{q}}{}^i|^2$ which is characteristic of the jellium model. The frequency of the long wavelength longitudinal sound wave thus remains proportional to q.

SCREENING OF THE ELECTRON-PHONON INTERACTION

The matrix element, $v_{\mathbf{q}}{}^{i}$, that appears in Eq. (4.95) describes the "bare" electron-phonon interaction, that which would be observed if there were no Coulomb interaction between electrons. When the latter is taken into account, the "effective" coupling between electrons and phonons is changed, since a given electron sees not only the external phonon field, but also the electron polarization that it produces. This polarization field acts to screen the bare electron-ion coupling. As a result, we might expect the bare matrix element, $v_{\mathbf{q}}{}^{i}$, to be replaced by the screened matrix element,

$$v_{\mathbf{q}}^{\text{eff}} = v_{\mathbf{q}}{}^{i}/\epsilon(\mathbf{q}, \omega_{\mathbf{q}}) \cong v_{\mathbf{q}}{}^{i}/\epsilon(\mathbf{q}, 0) \tag{4.113}$$

(static screening being sufficiently accurate for the low-frequency phonons). A result of the form (4.113) was first obtained by Bardeen (1937) in his classic calculation of the conductivity of the alkali metals.

Actually, this result is not quite correct, as it allows only for an average dielectric screening of the bare electron-ion interaction. In addition, there will be a correction coming from exchange scattering of the electron against the screening cloud around the ion. We have used the Landau theory to calculate the combined influence of these two corrections (in the long wavelength limit) on the scattering of an electron by a static impurity (see Section 3.4); we saw that the two corrections tend to cancel one another. The considerations of that section can easily be generalized to cover the present case; we find [compare Eq. (3.65a)]:

$$\lim_{\mathbf{q}\to 0} v_{\mathbf{q}}^{\text{eff}} \cong \frac{v_{\mathbf{q}}{}^{i}}{\epsilon(\mathbf{q}, 0)} \frac{1}{1 + F_0{}^{s}}, \tag{4.114}$$

which, upon making use of Eq. (4.27), may be put in the form

$$\lim_{\mathbf{q}\to 0} v_{\mathbf{q}}^{\text{eff}} = \lim_{\mathbf{q}\to 0} \frac{\pi q^2}{4 p_F m^* e^2} v_{\mathbf{q}}{}^{i}. \tag{4.115}$$

As a result of screening, the interaction of electrons with long wavelength phonons is considerably reduced.

The effective electron-phonon interaction, (4.115), takes a very simple form in the case of jellium. The bare coupling, $v_{\mathbf{q}}{}^{i}$, is then given by

Eq. (4.106). On substituting this expression into Eq. (4.115), we find

$$\lim_{\mathbf{q}\to 0} v_{\mathbf{q}}^{\text{eff}} = -\frac{iq}{3}\frac{p_F^2}{m^*}\frac{1}{(NM)^{1/2}} = -iq\left[\frac{N}{M}\right]^{1/2}\frac{Z}{\nu(0)}, \tag{4.116}$$

where N is, as usual, the number of atoms per unit volume, and $\nu(0)$ is the density of states per unit energy at the Fermi surface. In the long wavelength limit, $v_{\mathbf{q}}^{\text{eff}}$ is proportional to q, rather than being inversely proportional to it, as was the case for $v_{\mathbf{q}}^{i}$.

On comparing Eq. (4.116) with (3.65b), we see that

$$v_{\mathbf{q}}^{\text{eff}} = -iq\mathfrak{V}_{\mathbf{q}}^{\text{eff}}\left(\frac{N}{M}\right)^{1/2}, \tag{4.117}$$

where $\mathfrak{V}_{\mathbf{q}}^{\text{eff}}$ is the effective matrix element for the scattering of an electron by a static point ion. The result, (4.117), is not specific to jellium: it remains valid for an arbitrary law of interaction between ions and electrons. In this way one can make connection with the pseudo-potential approach to the calculation of resistance in metals [see for example, Sham and Ziman (1964), Harrison (1966)].

In conclusion, it should be emphasized that the corrections which appear in Eq. (4.116) are automatically included in our calculation of the real and imaginary parts of the phonon frequency, Eqs. (4.103) and (4.104). Consider, for example, the result, (4.104), we have obtained for the decay of a phonon into a quasiparticle-quasihole pair. The *intensity* of the phonon field varies as $\exp(-2\omega_2 t)$: the probability for decay of a single phonon is therefore equal to $2\omega_2$. In terms of the effective electron-phonon interaction, we would obtain this transition probability by using the golden rule of second-order perturbation theory,

$$2\omega_2 = 2\pi\sum_{\mathbf{p}}\frac{1}{2\omega_{\mathbf{q}}}|v_{\mathbf{q}}^{\text{eff}}|^2 n_{\mathbf{p}}[1 - n_{\mathbf{p}+\mathbf{q}}]\delta(\omega_{\mathbf{q}} - \epsilon_{\mathbf{p}+\mathbf{q}} + \epsilon_{\mathbf{p}}) \tag{4.104a}$$

(the factor $1/2\omega_{\mathbf{q}}$ corresponding to the matrix element of the phonon operator, $Q_{\mathbf{q}}$: see Eq. (4.97) and the Appendix). On using the expression, (3.72), for the quasistatic dielectric constant, it is straightforward to show that Eq. (4.104) is identical to (4.104a), provided the effective electron-phonon interaction is given by Eq. (4.115). (This result does not depend on the choice of $v_{\mathbf{q}}^{i}$ or $\Omega_{\mathbf{q}}$, and is thus completely general.) The detailed comparison between Eqs. (4.104) and (4.104a) is left as an exercise to the reader.

4.6. SCREENING

EFFECTIVE ELECTRON–ELECTRON INTERACTION

As we have frequently remarked, the effective interaction between a pair of electrons in a quantum plasma will be screened as a result of the polarization cloud which surrounds any given electron. We now consider the nature of that screening in some detail. We shall treat first the case of the pure quantum plasma, and then go on to discuss the quantum plasma coupled to a longitudinal phonon field.

Let us adopt a momentum space approach, and consider the scattering of two electrons of momentum and spin $\mathbf{p} + \mathbf{q}$, σ and $\mathbf{p}'$, σ' to new states $\mathbf{p}$, σ and $\mathbf{p}' + \mathbf{q}$, σ'. In the scattering act, momentum and spin are conserved. In order to avoid complications with exchange effects,

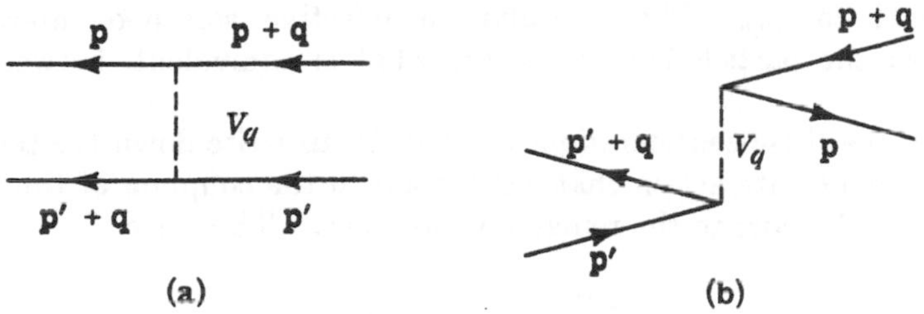

FIGURE 4.6. *Coulomb scattering of electrons and of particle-hole pairs.*

we assume that the two electrons have opposite spins; in that case in the absence of any polarization effects the scattering matrix element is

$$V_q = \frac{4\pi e^2}{q^2}.$$

In the representation of second quantization, the interaction would be described by the following term in the Coulomb interaction:

$$V_q c_{\mathbf{p}}^+ c_{\mathbf{p}'+\mathbf{q}}^+ c_{\mathbf{p}'} c_{\mathbf{p}+\mathbf{q}} \tag{4.118}$$

(where we have now suppressed the spin indices). A diagrammatic representation of this interaction is given in Fig. 4.6(a). We wish to consider the modification to Eq. (4.117) that arises because the electrons in question are not in free space, but rather embedded in a system of interacting electrons.

In order to treat this modification by the methods of the present chapter, it is convenient first of all to write Eq. (4.118) in a slightly different form. Let us define the electron-hole pair operator

$$\rho_{\mathbf{qp}} = c_{\mathbf{p}}^{+} c_{\mathbf{p+q}} \tag{4.119}$$

which represents a particular component of the density fluctuation, $\rho_{\mathbf{q}}$. The interaction (4.118) may then be expressed as an interaction between particle-hole pairs,

$$V_q \rho_{\mathbf{qp'}}^{+} \rho_{\mathbf{qp}}, \tag{4.120}$$

as shown on Fig. 4.6(b).

Let us further consider the interaction between the pair operator, $\rho_{\mathbf{qp}}$, and the remaining electrons in the system. To the extent that the response of the other electrons to $\rho_{\mathbf{qp}}$ is *linear*, $\rho_{\mathbf{qp}}$ may be regarded as an "internal" test charge which acts to polarize the electron gas. The pair operator $\rho_{\mathbf{qp'}}^{+}$ interacts then not only with $\rho_{\mathbf{qp}}$ directly via Eq. (4.120), but indirectly via the polarization field induced by, and proportional to, $\rho_{\mathbf{qp}}$. There results an effective screened interaction between the particle-hole pairs, or, what is equivalent, between the electrons.

To make these considerations explicit, let us write down the terms in the Coulomb interaction that correspond to the coupling of the above particle-hole pairs to the remaining electrons. These are

$$\frac{4\pi e^2}{q^2} \{\rho_{\mathbf{q}}^{+} \rho_{\mathbf{qp}} + \rho_{\mathbf{qp'}}^{+} \rho_{\mathbf{q}}\}, \tag{4.121}$$

where $\rho_{\mathbf{q}}$ is understood not to contain any contributions from the pairs in question. Through the first term in Eq. (4.121), $\rho_{\mathbf{qp}}$ acts to polarize the electron gas, while through the second, $\rho_{\mathbf{qp'}}^{+}$ feels the effect of that polarization. To calculate the induced polarization we need to know the frequency at which $\rho_{\mathbf{qp}}$ acts. Here we must distinguish between real scattering processes, for which energy is conserved, and virtual interactions, for which energy need not be conserved. In a real scattering process, which we consider first, there will be an energy transfer from one particle to the other which is given by

$$\omega_{\mathbf{qp}} = \epsilon_{\mathbf{p+q}} - \epsilon_{\mathbf{p}}. \tag{4.122}$$

Under those circumstances the pair operator, $\rho_{\mathbf{qp}}$, resembles the test charge of Section 4.4 and must be regarded as possessing a frequency $\omega_{\mathbf{qp}}$. Within the linear response approximation, the density fluctuation that it induces is given by

$$\langle \rho_{\mathbf{q}} \rangle = \frac{4\pi e^2}{q^2} \chi(\mathbf{q}, \omega_{\mathbf{qp}}) \rho_{\mathbf{qp}}. \tag{4.123}$$

The action of this polarization field on $\rho^+_{\mathbf{q}\mathbf{p}'}$ is, according to (4.121), given by

$$H_{\text{pol}} = \left(\frac{4\pi e^2}{q^2}\right)^2 \chi(\mathbf{q}, \omega_{\mathbf{q}\mathbf{p}})\rho^+_{\mathbf{q}\mathbf{p}'}\rho_{\mathbf{q}\mathbf{p}}. \tag{4.124}$$

On combining Eq. (4.124) with the original interaction, (4.120), we arrive at the effective interaction

$$H_{\text{eff}} = \frac{4\pi e^2}{q^2}\frac{1}{\epsilon(\mathbf{q}, \omega_{\mathbf{q}\mathbf{p}})}\rho^+_{\mathbf{q}\mathbf{p}'}\rho_{\mathbf{q}\mathbf{p}} = \frac{4\pi e^2}{q^2\epsilon(\mathbf{q}, \omega_{\mathbf{q}\mathbf{p}})} c_{\mathbf{p}}{}^+ c^+_{\mathbf{p}'+\mathbf{q}} c_{\mathbf{p}'} c_{\mathbf{p}+\mathbf{q}} \tag{4.125}$$

which describes the dielectric screening of a real scattering process.

The result, Eq. (4.125), is scarcely surprising; it was nearly obvious once we decided to treat the modification in particle interactions as a polarization process. It is clear that in arriving at Eq. (4.125) we have *not* taken into account *all* possible modifications in the interaction between a pair of electrons which result from their coupling to the other electrons; all that we have included is that set of modifications which corresponds to a dielectric screening of the interaction. Nonlinear effects in the response to $\rho_{\mathbf{q}\mathbf{p}}$ are completely neglected, as well as any exchange interaction between the scattering particles and the rest of the liquid. Equation (4.125) should thus be considered as an *approximate* result, yet a very enlightening one.

In arriving at Eq. (4.125), we have singled out one particle-hole pair, $\rho_{\mathbf{q}\mathbf{p}}$, as the "polarizer," and the other, $\rho^+_{\mathbf{q}\mathbf{p}'}$, as the "analyzer"; it should not, of course, matter which pair one chooses to focus attention on first, nor does it. One arrives at the same result by taking $\rho_{\mathbf{q}\mathbf{p}'}$ as the polarizer and $\rho_{\mathbf{q}\mathbf{p}}$ as the analyzer, since in a real process the frequency $\omega_{\mathbf{q}\mathbf{p}'}$ must be equal to the frequency $\omega_{\mathbf{q}\mathbf{p}}$ by energy conservation. (Remark that, in dealing with their mutual interaction, both pairs cannot simultaneously be regarded as giving rise to polarization clouds, for in that way one would treat the same polarization processes twice.)

The treatment of virtual scattering processes proceeds along the same lines, except for the fact that it is not clear what frequency ω should be attributed to a given particle-hole pair. The choice of ω depends on the problem under study. We may regard a given electron as a test charge, which is surrounded with a wave vector and frequency-dependent polarization cloud. The qth Fourier component of the effective electron interaction is then

$$V^{\text{eff}}_{\mathbf{q}} = \frac{4\pi e^2}{q^2\epsilon(\mathbf{q}, \omega)}. \tag{4.126}$$

For virtual processes, ω is independent of $\mathbf{q}$, while for real scattering processes, it is related to $\mathbf{q}$ by an expression like Eq. (4.122).

Wherever $\omega \ll qv_F$ (for an interaction which involves a momentum transfer $\mathbf{q}$), the screening is essentially static. It is then unnecessary to know ω exactly: in Eq. (4.126), we may replace $\epsilon(\mathbf{q}, \omega)$ by $\epsilon(\mathbf{q}, 0)$. Where ω is comparable to qv_F, the screening becomes "dynamic." It is then instructive to consider the effective interaction between electrons from a space-time point of view. Consider an electron at the origin at time $t = 0$. We may associate with it a screened space and time-dependent potential which is given by the Fourier transform of Eq. (4.126):

$$V_{\text{eff}}(\mathbf{r}, t) = \sum_{\mathbf{q}} \int \frac{d\omega}{2\pi} \left(\frac{4\pi e^2}{q^2}\right) \frac{1}{\epsilon(\mathbf{q}, \omega)} \exp i(\mathbf{q} \cdot \mathbf{r} - \omega t). \quad (4.127)$$

$V_{\text{eff}}(\mathbf{r}, t)$ is the potential experienced by a second electron at a distance $\mathbf{r}$ and at a subsequent time t. (Since we have imposed causal boundary conditions in our definition of $\epsilon(\mathbf{q}, \omega)$, $V_{\text{eff}}(\mathbf{r}, t) = 0$ for $t < 0$.) In the absence of dielectric screening, we have of course,

$$V_{\text{eff}}(\mathbf{r}, t) = \frac{e^2}{r} \delta(t). \quad (4.128)$$

The potential is long range in space and instantaneous in time. With dielectric screening matters change. Suppose first that $\epsilon(\mathbf{q}, \omega)$ is given for all $\mathbf{q}$ and ω by its long wavelength static value, (4.27); we then find

$$V_{\text{eff}}(\mathbf{r}, t) = \delta(t) \frac{e^2}{r} \exp(-\omega_p r/s). \quad (4.129)$$

The interaction is still instantaneous, but is now "localized" in space. If now we allow ϵ to vary with ω, $V_{\text{eff}}(\mathbf{r}, t)$ becomes a *retarded* interaction: it extends over a finite interval of time, of the order of the inverse of the frequency at which $\epsilon(\mathbf{q}, \omega)$ begins to display any appreciable variation. Such frequencies are of the order of ϵ_F for the quantum plasma: the retarded interaction then lasts for a very short time. This is no longer true for the coupling of the quantum plasma to a longitudinal phonon field, as we shall shortly demonstrate.

Thus far, we have confined our attention to the screening of the interaction between electrons of antiparallel spins. In fact, similar considerations can be applied to the interaction between electrons of parallel spins. The one complication which appears in the latter case is that in addition to *direct* scattering, with a momentum transfer $\mathbf{q}$, matrix element V_q, there may also occur *exchange* scattering involving a momentum transfer $(\mathbf{p}' - \mathbf{p})$, with matrix element $-V_{|\mathbf{p}'-\mathbf{p}|}$. The two scattering processes must be separately screened, with the frequency and wave vector dielectric constant appropriate to each. We note that dielectric screening of the direct scattering process is far more important, since for small **q**, $V_q = 4\pi e^2/q^2$ is strongly divergent, while $V_{\mathbf{p}'-\mathbf{p}}$ is independent of **q** and quite well behaved. The dielectric screening of the direct interaction drastically changes its character, while that of the exchange scattering modifies the latter considerably less.

DYNAMIC SCREENING WITH PHONONS

We now extend our study of the effective interaction between electrons to the case of a quantum plasma coupled to a longitudinal phonon field. The problem is specified by the Hamiltonian (4.93). Again, we select two electron-hole pairs, $\rho_{\mathbf{qp}}$ and $\rho^+_{\mathbf{qp}'}$, and we try to calculate their effective interaction. As we have remarked, the phonons provide a new mechanism for interaction between the electrons, and will be shown to modify profoundly that interaction for low excitation frequencies.

In order to obtain a first notion of that effect, we consider only the coupling of the selected pairs to the phonons, and neglect both the coupling of the phonons to the other electrons and the direct Coulomb interaction between the electrons. The relevant interaction terms, taken from Eq. (4.95), are

$$H' = v_{-\mathbf{q}}^{i}\rho_{\mathbf{qp}}Q_{\mathbf{q}}^{+} + v_{\mathbf{q}}^{i}\rho^{+}_{\mathbf{qp}'}Q_{\mathbf{q}}. \tag{4.130}$$

Let the pair $\rho_{\mathbf{qp}}$ possess the frequency ω. In that case, because of the first term in the interaction, the equation of motion of the amplitude, $Q_{\mathbf{q}}$, will read

$$\ddot{Q}_{\mathbf{q}} + \Omega_{\mathbf{q}}^2 Q_{\mathbf{q}} = -v_{-\mathbf{q}}^{i}\rho_{\mathbf{qp}} = (\Omega_{\mathbf{q}}^2 - \omega^2)Q_{\mathbf{q}}. \tag{4.131}$$

The pair thus acts to induce a fluctuation in the phonon field which is

$$\langle Q_{\mathbf{q}}\rangle = \frac{v_{-\mathbf{q}}^{i}\rho_{\mathbf{qp}}}{\omega^2 - \Omega_{\mathbf{q}}^2 + i\eta} \tag{4.132}$$

(on choosing the appropriate retarded boundary conditions). This "forced" phonon oscillation acts on $\rho^+_{\mathbf{qp}'}$ via the second term in Eq. (4.130); the result is a new interaction between the electrons, given by

$$H'_{\text{eff}} = \frac{|v_{\mathbf{q}}^{i}|^2}{\omega^2 - \Omega_{\mathbf{q}}^2 + i\eta}\,\rho^+_{\mathbf{qp}'}\rho_{\mathbf{qp}}. \tag{4.133}$$

The phonon-induced interaction, (4.133), is attractive for frequencies $\omega < \Omega_{\mathbf{q}}$, and repulsive for higher frequencies. It corresponds to an exchange of virtual phonons between the electrons, pictured as in Fig.

4.7. There we see an electron of momentum $(\mathbf{p} + \mathbf{q})$ virtually emitting a phonon of momentum $\mathbf{q}$ (at a frequency ω), the latter then being absorbed by an electron of momentum $\mathbf{p}'$. At very high frequencies $(\omega \gg \Omega_\mathbf{q})$, the ions will not be able to follow the electron motion at all, and the phonon-induced interaction becomes negligible.

We now consider the modification in Eq. (4.133) owing to the interaction between the phonons and the other electrons, and of the electrons with each other. To do this we single out from the full Hamiltonian, (4.93), the interactions that involve the pairs, $\rho_{\mathbf{qp}}$ and $\rho_{\mathbf{qp}'}^+$; these are given by (4.121) and (4.120) respectively. We next calculate the average density fluctuation $\langle\rho_\mathbf{q}\rangle$, and phonon field amplitude, $\langle Q_\mathbf{q}\rangle$, induced by one pair $\rho_{\mathbf{qp}}$ at frequency ω. We substitute the resulting

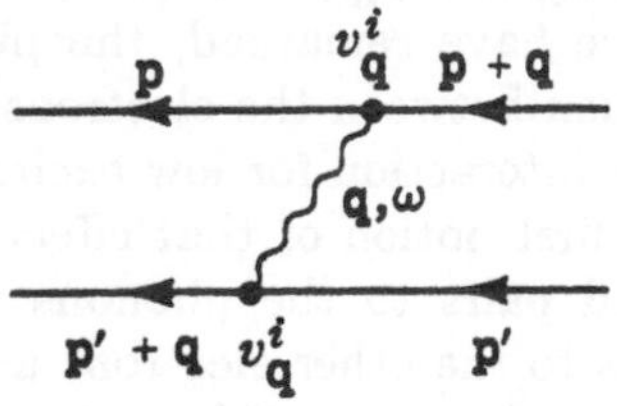

FIGURE 4.7. *Phonon-induced interaction between electrons. The small dot corresponds to the matrix element $v_\mathbf{q}^i$, while the wavy line denotes the phonon which is virtually exchanged between the electrons.*

values into the terms of the Hamiltonian which couple to $\rho_{\mathbf{qp}'}^+$, and obtain thereby the net effective interaction. The calculation is somewhat more involved than those we have carried out in this section, because the induced electron and phonon density fluctuations, $\langle\rho_\mathbf{q}\rangle$ and $\langle Q_\mathbf{q}\rangle$, are themselves coupled via Eq. (4.95).

The equation of motion of the sound wave amplitude reads

$$\ddot{Q}_\mathbf{q} + \Omega_\mathbf{q}^2 Q_\mathbf{q} = -v_{-\mathbf{q}}^i(\langle\rho_\mathbf{q}\rangle + \rho_{\mathbf{qp}}). \tag{4.134}$$

We have separated out, for explicit consideration, the contribution to $\langle\rho_\mathbf{q}\rangle$ from $\rho_{\mathbf{qp}}$. We solve this equation to obtain a relation between the induced phonon amplitude, $\langle Q_\mathbf{q}\rangle$, the induced density fluctuation, $\langle\rho_\mathbf{q}\rangle$, and $\rho_{\mathbf{qp}}$:

$$\langle Q_\mathbf{q}\rangle = \frac{v_{-\mathbf{q}}^i(\langle\rho_\mathbf{q}\rangle + \rho_{\mathbf{qp}})}{\omega^2 - \Omega_\mathbf{q}^2}. \tag{4.135}$$

A second relation is obtained by computing the linear electron response to $\rho_{\mathbf{qp}}$; we have

$$\langle\rho_{\mathbf{q}}\rangle = \chi(\mathbf{q}, \omega)\left\{v_{\mathbf{q}}^{i}\langle Q_{\mathbf{q}}\rangle + \frac{4\pi e^2}{q^2}\rho_{\mathbf{qp}}\right\}. \tag{4.136}$$

These two equations serve to determine the response of the coupled electron-phonon system to $\rho_{\mathbf{qp}}$.

On substituting Eq. (4.136) into (4.135), we find

$$\langle Q_{\mathbf{q}}\rangle = \frac{v_{-\mathbf{q}}^{i}}{\epsilon(\mathbf{q}, \omega)}\frac{1}{\omega^2 - \Omega_{\mathbf{q}}^2 - |v_{\mathbf{q}}^{i}|^2\chi(\mathbf{q}, \omega)}\rho_{\mathbf{qp}}. \tag{4.137}$$

This expression may be further simplified, since for the low excitation frequencies ω for which the phonon response is of importance, $\chi(\mathbf{q}, \omega)$ may be replaced by $\chi(\mathbf{q}, 0)$. On making use of the phonon dispersion relation (4.100), and further neglecting the very small phonon damping, we see that Eq. (4.137) may be written in the form

$$\langle Q_{\mathbf{q}}\rangle \cong \frac{v_{-\mathbf{q}}^{i}}{\epsilon(\mathbf{q}, 0)}\frac{1}{\omega^2 - \omega_{\mathbf{q}}^2}\rho_{\mathbf{qp}} \qquad (\omega \ll qv_F). \tag{4.138}$$

The effect on $\langle Q_{\mathbf{q}}\rangle$ of taking into account the full electron-electron and electron-phonon interactions may be seen by comparing Eq. (4.138) with (4.132); the phonon frequency is shifted from $\Omega_{\mathbf{q}}$ to $\omega_{\mathbf{q}}$, while the effective matrix element that determines the phonon response is screened by the static dielectric constant, as in Eq. (4.113).

To obtain the net effective electron interaction, we combine Eq. (4.136) with the interaction terms in (4.121) and (4.130) which involve $\rho_{\mathbf{qp}'}^{+}$. On adding the "bare" interaction term, (4.120), we obtain the following expression for the sum of terms coupled to $\rho_{\mathbf{qp}'}^{+}$:

$$\frac{4\pi e^2}{q^2\epsilon(\mathbf{q}, \omega)}\rho_{\mathbf{qp}'}^{+}\rho_{\mathbf{qp}} + \frac{v_{\mathbf{q}}^{i}\rho_{\mathbf{qp}'}^{+}}{\epsilon(\mathbf{q}, \omega)}\langle Q_{\mathbf{q}}\rangle. \tag{4.139}$$

On making use of Eq. (4.138) we find the effective electron interaction is given by

$$H_{\text{eff}}(\mathbf{q}, \omega) = \left\{\frac{4\pi e^2}{q^2\epsilon(\mathbf{q}, \omega)} + \frac{|v_{\mathbf{q}}^{i}|^2}{|\epsilon(\mathbf{q}, 0)|^2}\frac{1}{\omega^2 - \omega_{\mathbf{q}}^2}\right\}\rho_{\mathbf{qp}'}^{+}\rho_{\mathbf{qp}}. \tag{4.140}$$

The first term (in brackets) in Eq. (4.140) is the screened Coulomb interaction between the electrons, while the second is the modified phonon-induced interaction. As might have been anticipated, the latter interaction resembles Eq. (4.133); it differs in that the phonon-electron matrix element is screened, and the true phonon frequency appears. The net effective screened interaction may be pictured as in

Fig. 4.8. The result, (4.140), is of considerable importance in the theory of superconductivity. As we shall see in Chapter 7, the low-frequency attractive phonon-induced interaction makes possible the transition to the superconducting state.

To study the phonon-induced interaction in more detail, we return to the jellium model for the electron-phonon coupling. On substituting Eqs. (4.106) and (4.107) into (4.140), we find*

$$\frac{4\pi e^2}{q^2\epsilon(\mathbf{q}, 0)} \frac{\omega_{\mathbf{q}}^2}{\omega^2 - \omega_{\mathbf{q}}^2} \rho_{\mathbf{q}\mathbf{p}'}^{+} \rho_{\mathbf{q}\mathbf{p}}. \tag{4.141}$$

At low frequencies ($\omega \lesssim \omega_{\mathbf{q}}$), the interaction is comparable in size to the direct screened Coulomb interaction between the electrons; at high

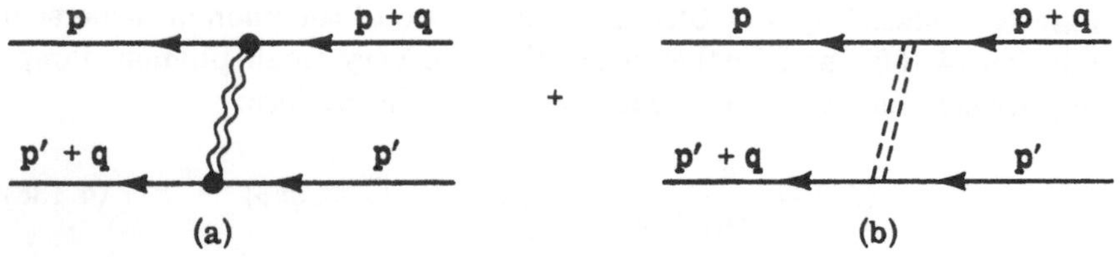

FIGURE 4.8. *The net effective screened interaction between electrons consists of the phonon-induced interaction (a) plus the dielectrically screened Coulomb interaction (b). The large dot denotes a screened electron-phonon matrix element, while the double wavy line denotes the appropriately renormalized phonon which is virtually exchanged between the electrons [compare Fig. 4.8(a) with Fig. 4.7, which shows the virtual exchange of a "bare" phonon, governed by the unscreened matrix elements, $v_{\mathbf{q}}{}^{i}$].*

frequencies ($\omega \gg \omega_{\mathbf{q}}$), it is negligible. The net effective interaction is given by

$$V_{\text{eff}}(\mathbf{q}, \omega) = \frac{4\pi e^2}{q^2\epsilon(\mathbf{q}, 0)} \frac{\omega^2}{\omega^2 - \omega_{\mathbf{q}}^2} \rho_{\mathbf{q}\mathbf{p}'}^{+} \rho_{\mathbf{q}\mathbf{p}} \qquad (\omega \lesssim \omega_{\mathbf{q}}). \tag{4.142}$$

It vanishes at zero excitation frequency, where the phonon-induced interaction cancels the screened Coulomb interaction in the jellium model; it is attractive for frequencies $\omega \leq \omega_{\mathbf{q}}$, repulsive for larger frequencies.

The net interaction, (4.142), varies slowly in **q** space, but quite rapidly in ω space. Thus, $V_{\text{eff}}(\mathbf{r}, t)$ will be localized in space, but will be noticeably retarded in time. The effective field produced by an electron at the origin may be expected to affect a nearby electron for times as long as

* This result was first obtained by Bardeen and Pines (1955) for the special case of real scattering processes.

$1/\omega_{\mathbf{q}}$. This behavior is in sharp contrast with that found for the quantum plasma alone.*

It is also possible to define a dielectric constant for the coupled electron-phonon system, by considering the coupling of a test charge to the ions as well as to the electrons. The definition of that dielectric constant, ϵ_{tot}, and its calculation are given as a problem at the end of this chapter. Here we remark that in the special case of jellium, the effective interaction between the electrons, (4.140), takes the form

$$V_{eff} = \frac{4\pi e^2}{q^2 \epsilon_{tot}(\mathbf{q}, \omega)} \rho^+_{\mathbf{q}\mathbf{p}'} \rho_{\mathbf{q}\mathbf{p}}. \tag{4.143}$$

More generally, as may be seen from Problem 4.3, this will not be the case.

4.7. RESPONSE TO AN ELECTROMAGNETIC FIELD

We now apply the methods of the present chapter to the coupling between the electron liquid and a transverse electromagnetic field. We consider first the microscopic generalization of the semiclassical approach of Section 3.6 to this problem, in which the electron response is specified in terms of a conductivity or dielectric tensor. We give a microscopic derivation of the appropriate transverse response functions, and discuss their limiting behavior in several cases of physical interest.

In this first part, no explicit account is taken of the quantum nature of the electromagnetic field; the latter is treated as a "classical" source. This approximation is certainly valid for the fields encountered in practice; it is nevertheless of interest to consider the coupled electron-photon system from a quantum point of view. We present a brief account of such a treatment, which serves as well to justify certain approximations made in the response function calculation. We use the explicit quantum approach to derive the appropriate expressions for the Raman scattering of light by an electron liquid.

The treatment in the present section tends to be somewhat formal and mathematical; it is certainly not necessary for those many problems for which the Landau–Silin theory suffices. It does, however, serve to illuminate the differences and similarities in the treatments of longitudinal and transverse electron response; moreover, it lays the groundwork for a particularly simple treatment of the Meissner effect in superconducting systems.

* For a discussion of the phonon-induced interaction using a more realistic model than that of jellium, see Pines (1963), Chapter 5.

TRANSVERSE ELECTRON RESPONSE

The transverse electromagnetic properties of a charged system are governed by the usual Maxwell equations:

$$\text{curl}\, \mathcal{E}(\mathbf{r}, t) = -\frac{1}{c}\frac{\partial \mathcal{H}(\mathbf{r}, t)}{\partial t}, \tag{4.144}$$

$$\text{curl}\, \mathcal{H}(\mathbf{r}, t) = +\frac{1}{c}\frac{\partial \mathcal{E}(\mathbf{r}, t)}{\partial t} + \frac{4\pi}{c}\{\langle \mathcal{J}(\mathbf{r}, t)\rangle + \mathcal{J}_{\text{ext}}(\mathbf{r}, t)\}. \tag{4.145}$$

In Eq. (4.145), $\langle \mathcal{J}\rangle$ is the electrical current induced in the system by an external current source, whose strength is $\mathcal{J}_{\text{ext}}$. $\mathcal{E}$ and $\mathcal{H}$ are the self-consistent effective, or local fields, which are determined in part by the currents they induce.

It is convenient to specify the fields in terms of vector potentials. The local potential $\mathcal{A}(\mathbf{r}, t)$ is defined in the usual way:

$$\mathcal{H} = \text{curl}\, \mathcal{A}, \qquad \mathcal{E} = -\frac{1}{c}\frac{\partial \mathcal{A}}{\partial t}. \tag{4.146}$$

It is the sum of the external vector potential, $\mathcal{A}_{\text{ext}}$, and the polarization field produced by the latter:

$$\mathcal{A} = \mathcal{A}_{\text{ext}} + \mathcal{A}_{\text{pol}}. \tag{4.147}$$

On taking the Fourier transforms in space and time of Maxwell's equations, one finds the following relations between the currents and vector potentials:

$$(\omega^2 - c^2q^2)\mathcal{A}_{\text{ext}}(\mathbf{q}, \omega) = -4\pi c \mathcal{J}_{\text{ext}}(\mathbf{q}, \omega), \tag{4.148}$$

$$(\omega^2 - c^2q^2)\mathcal{A}(\mathbf{q}, \omega) = -4\pi c\{\mathcal{J}_{\text{ext}}(\mathbf{q}, \omega) + \langle \mathcal{J}(\mathbf{q}, \omega)\rangle\}. \tag{4.149}$$

We make the usual approximation of weak external electromagnetic fields, in which case the induced current is proportional to the local field $\mathcal{E}$; the conductivity tensor is defined by

$$\langle \mathcal{J}_\mu(\mathbf{q}, \omega)\rangle = \sigma_{\mu\nu}(\mathbf{q}, \omega)\mathcal{E}_\nu(\mathbf{q}, \omega). \tag{4.150}$$

Equation (4.150) represents the extension to a microscopic level of the usual macroscopic definition of $\sigma_{\mu\nu}$. It is often convenient to introduce the displacement field, $\mathcal{D}$, defined by

$$\frac{\partial \mathcal{D}}{\partial t} = \frac{\partial \mathcal{E}}{\partial t} + 4\pi\langle \mathcal{J}\rangle \tag{4.151}$$

and the dielectric tensor, $\epsilon_{\mu\nu}$,

$$\mathcal{D}_\mu = \epsilon_{\mu\nu}\mathcal{E}_\nu. \tag{4.152}$$

The relation between $\epsilon_{\mu\nu}$ and $\sigma_{\mu\nu}$ is, of course,

$$\epsilon_{\mu\nu}(\mathbf{q}, \omega) = 1 + \frac{4\pi i}{\omega} \sigma_{\mu\nu}(\mathbf{q}, \omega). \tag{4.153}$$

For an isotropic electron liquid, the conductivity and dielectric tensors are diagonal. In each, the two transverse components are equal, respectively, to $\sigma_\perp$ and $\epsilon_\perp$. It follows from Eq. (4.153) that

$$\epsilon_\perp(\mathbf{q}, \omega) = 1 + \frac{4\pi i}{\omega} \sigma_\perp(\mathbf{q}, \omega). \tag{4.154}$$

In most cases, $\sigma_\perp$ and $\epsilon_\perp$ differ from the longitudinal components $\sigma_\parallel$ and $\epsilon_\parallel$.

The natural frequencies of propagation of electromagnetic waves correspond to a self-consistent solution of Maxwell's equations in the absence of any external sources. On making use of Eq. (4.152), one finds readily (for the isotropic system)

$$\omega^2 = \frac{c^2 q^2}{\epsilon_\perp(\mathbf{q}, \omega)}. \tag{4.155}$$

More generally, on combining Eqs. (4.148), (4.149), and (4.150), one finds

$$\mathcal{A}(\mathbf{q}, \omega) = \frac{\mathcal{A}_{\text{ext}}(\mathbf{q}, \omega)}{1 + 4\pi i \omega \sigma_\perp(\mathbf{q}, \omega)/(\omega^2 - c^2 q^2)}. \tag{4.156}$$

Equation (4.156) relates the effective field and its external counterpart, and thus specifies the transverse screening action of the electron system.

CALCULATION OF THE TRANSVERSE RESPONSE TENSOR

Since we deal with a linear response problem, it is obviously possible to calculate $\sigma_{\mu\nu}$ or $\epsilon_{\mu\nu}$ in terms of exact microscopically defined correlation functions. As for the longitudinal fields considered earlier in this chapter, one has a choice in the definition (and calculation) of response functions. The alternatives are those presented in Section 3.6:

(1) We may compute the response to the *external* field (e.g., $\mathcal{A}_{\text{ext}}$) taking into account the full interaction between the electrons (Coulomb interactions plus their transverse magnetic interactions).

(2) We may treat the polarization field (e.g., $\mathcal{A}_{\text{pol}}$) as an added "external" field, and so compute the response to the effective local field ($\mathcal{A} = \mathcal{A}_{\text{ext}} + \mathcal{A}_{\text{pol}}$). In following this second course, we must, however,

be careful to avoid counting polarization processes twice; this we may do by screening the original magnetic interaction.

The second alternative is preferable here, essentially because the screened magnetic interaction is negligible. The latter, first calculated by Bohm and Pines (1951), may be viewed as arising from the exchange of photons between electrons. The corresponding matrix element for a transition in which two electrons with momenta $\mathbf{p}_1$ and $\mathbf{p}_2$ exchange a photon of momentum $\mathbf{q}$ takes the approximate form

$$-4\pi e^2 \sum_{\mu} \frac{(\mathbf{p}_1 \cdot \mathbf{n}_{q\mu})(\mathbf{p}_2 \cdot \mathbf{n}_{q\mu})}{\omega_p{}^2 + c^2q^2}, \tag{4.157}$$

where $\mathbf{n}_{q\mu}$ is the polarization vector of the photon transferred from one electron to the other. To order v^2/c^2, (4.157) may be neglected compared to the matrix element of the Coulomb interaction, $4\pi e^2/q^2$.

Put another way, to order v^2/c^2 the only consequences of the transverse magnetic interactions between electrons are the polarization processes (we give an explicit demonstration of this point later in the section). Once these are taken into account by regarding the polarization field as an added external field, the only remaining interaction of importance is the Coulomb interaction between the electrons. We shall only consider the second alternative in this section; exposition of the first is left to the reader, via a series of problems at the end of the chapter.

The coupling between the electrons and the effective field $\mathcal{A}$ is specified by the following interaction:

$$H_e = -\lim_{\eta \to 0} \frac{e}{c} \sum_{q\mu} \mathbf{J}_q \cdot \mathcal{A}_q e^{i(\mathbf{q}\cdot\mathbf{r} - \omega t)} e^{\eta t}, \tag{4.158}$$

where $\mathbf{J}_q$ is the usual particle current density fluctuation [see Eq. (2.22)],

$$\mathbf{J}_q = \frac{1}{2m} \sum_i \{\mathbf{p}_i e^{-i\mathbf{q}\cdot\mathbf{r}_i} + e^{-i\mathbf{q}\cdot\mathbf{r}_i}\mathbf{p}_i\},$$

and $\mathcal{A}_q$ is the Fourier transform in space and time of the self-consistent transverse vector potential, $\mathcal{A}(\mathbf{r}, t)$. The use of Eq. (4.158) as an effective interaction Hamiltonian will be justified later in this section.

The calculation of the linear system response to $\mathcal{A}$ is a straightforward application of the perturbation-theoretic techniques of Chapter 2. Let us calculate the conductivity tensor $\sigma_{\mu\nu}(\mathbf{q}, \omega)$, defined by Eq.

(4.150). We note first that the total electric current $\mathfrak{J}(\mathbf{r})$ is given by

$$\mathfrak{J}(\mathbf{r}) = \frac{e}{2} \sum_i \{\mathbf{v}_i \delta(\mathbf{r} - \mathbf{r}_i) + \delta(\mathbf{r} - \mathbf{r}_i)\mathbf{v}_i\}, \tag{4.159}$$

where $\mathbf{v}_i$ is the particle velocity in the presence of the field, equal to $[\mathbf{p}_i - (e/c)\mathfrak{A}(\mathbf{x}_i)]$. On taking the Fourier transform of Eq. (4.159) we find

$$\mathfrak{J}_\mathbf{q} = e\mathbf{J}_\mathbf{q} - \frac{e^2}{mc} \sum_\mathbf{k} \mathfrak{A}_\mathbf{k}(t)\rho_{\mathbf{q}-\mathbf{k}}, \tag{4.160}$$

where $\mathbf{J}_\mathbf{q}$ is the current operator (2.22) *in the absence of the field.* Since we are interested only in the linear response of the electron system, we can replace $\rho_{\mathbf{q}-\mathbf{k}}$ by its expectation value in the absence of the field: the only contribution arises from the term $\mathbf{q} = \mathbf{k}$. On taking the Fourier transform with respect to time, we find

$$\langle \mathfrak{J}(\mathbf{q}, \omega) \rangle = e\langle \mathbf{J}(\mathbf{q}, \omega) \rangle - \frac{Ne^2}{mc} \mathfrak{A}(\mathbf{q}, \omega). \tag{4.161}$$

The induced currents may then be written as

$$\langle J_\mu(\mathbf{q}, \omega) \rangle = -\frac{e}{c} \chi_{\mu\nu}(\mathbf{q}, \omega)\mathfrak{A}_\nu(\mathbf{q}, \omega), \tag{4.162a}$$

$$\langle \mathfrak{J}_\mu(\mathbf{q}, \omega) \rangle = -\frac{e^2}{c} \{\chi_{\mu\nu} + (N/m)\delta_{\mu\nu}\} \mathfrak{A}_\nu(\mathbf{q}, \omega), \tag{4.162b}$$

where $\chi_{\mu\nu}$ is the transverse current-current response tensor, given by

$$\chi_{\mu\nu}(\mathbf{q}, \omega) = \sum_n \left\{ \frac{(J_{\mathbf{q}\mu})_{on}(J^+_{\mathbf{q}\nu})_{no}}{\omega - \omega_{no} + i\eta} - \frac{(J^+_{\mathbf{q}\nu})_{on}(J_{\mathbf{q}\mu})_{no}}{\omega + \omega_{no} + i\eta} \right\}. \tag{4.163}$$

Let us emphasize that the matrix elements and excitation frequencies that appear in Eq. (4.163) are those appropriate to a system of electrons *with* Coulomb interactions, but *without* any transverse magnetic interactions. On making use of Eq. (4.146) we find our desired expression:

$$\sigma_{\mu\nu} = \frac{ie^2}{\omega} \left\{ \chi_{\mu\nu} + \frac{N}{m} \delta_{\mu\nu} \right\}. \tag{4.164}$$

The corresponding expression for the transverse dielectric tensor is, from Eq. (4.153),

$$\epsilon_{\mu\nu} = 1 - \frac{{\omega_p}^2}{\omega^2} \delta_{\mu\nu} - \frac{4\pi e^2}{\omega^2} \chi_{\mu\nu}. \tag{4.165}$$

For an isotropic system, there are no longitudinal current fluctuations induced by the electromagnetic field: the system response is determined entirely by the transverse current fluctuations. The equal transverse components of the dielectric tensor are given by

$$\epsilon_{\perp}(\mathbf{q}, \omega) = 1 - \frac{4\pi e^2}{\omega^2}\left\{\chi_{\perp}(\mathbf{q}, \omega) + \frac{N}{m}\right\}; \tag{4.166}$$

where $\chi_{\perp}(\mathbf{q}, \omega)$ is the transverse current-current correlation function,

$$\chi_{\perp}(\mathbf{q}, \omega) = \sum_n |(J_{\mathbf{q}\perp}^{+})_{no}|^2 \frac{2\omega_{no}}{(\omega + i\eta)^2 - \omega_{no}^2} \tag{4.167}$$

($J_{\mathbf{q}\perp}$ is an arbitrary component of $\mathbf{J_q}$ perpendicular to $\mathbf{q}$).

RESPONSE TO A LONGITUDINAL VECTOR POTENTIAL—GAUGE INVARIANCE

The preceding approach may be used as well in the case of a longitudinal vector potential $\mathcal{A}_{ext}(\mathbf{q}, \omega)$. According to Eq. (4.146), such a probe gives rise only to a longitudinal applied electric field, $\mathfrak{D}$, whose Fourier transform is

$$\mathfrak{D}(\mathbf{q}, \omega) = \frac{i\omega}{c}\mathcal{A}_{ext}(\mathbf{q}, \omega). \tag{4.168}$$

Clearly, one would obtain the same field $\mathfrak{D}(\mathbf{q}, \omega)$ by applying a suitable scalar potential $\varphi_{ext}(\mathbf{q}, \omega)$ to the system in place of the vector potential $\mathcal{A}_{ext}$. On using Poisson's equation, we see that the electrostatic potential "equivalent" to $\mathcal{A}_{ext}$ is given by

$$\varphi_{ext}(\mathbf{q}, \omega) = -\frac{\omega}{cq^2}\mathbf{q}\cdot\mathcal{A}_{ext}(\mathbf{q}, \omega). \tag{4.169}$$

The passage from $\mathcal{A}_{ext}$ to φ_{ext} corresponds to a *gauge transformation*, well known in electromagnetic theory. More generally, the potentials $\mathcal{A}_{ext}$ and φ_{ext} are determined within a transformation

$$\begin{aligned} \mathcal{A}_{ext} &\rightarrow \mathcal{A}_{ext} + \mathbf{grad}\,\lambda, \\ \varphi_{ext} &\rightarrow \varphi_{ext} - \frac{1}{c}\frac{\partial\lambda}{\partial t}, \end{aligned} \tag{4.170}$$

where $\lambda(\mathbf{r}, t)$ is an arbitrary function. In order for the theory to be "gauge invariant," all physical results must be independent of λ.

Since we are interested only in the *linear* response of the system to an external probe, it is sufficient to compare the pure longitudinal vector probe with the pure scalar probe. Let us consider the former first. The interaction of the electrons with the applied vector potential is described by the Hamiltonian

$$H_{ext} = -\frac{e}{c}\mathbf{J}_{\mathbf{q}} \cdot \mathcal{A}_{ext}(\mathbf{q}, \omega)e^{i(\mathbf{q}\cdot\mathbf{r}-\omega t)}e^{\eta t} \tag{4.171}$$

[note that Eq. (4.171) involves the *external* potential $\mathcal{A}_{ext}$, in contrast with (4.158)]. For an isotropic system, the induced electrical current is given by

$$\langle \mathcal{J} \rangle = -\frac{e^2}{c}\left[\chi_{\parallel}(\mathbf{q}, \omega) + \frac{N}{m}\right]\mathcal{A}_{ext}(\mathbf{q}, \omega) \tag{4.172}$$

[compare with Eq. (4.162)], where $\chi_{\parallel}(\mathbf{q}, \omega)$ is the longitudinal current-current correlation function, defined as

$$\chi_{\parallel}(\mathbf{q}, \omega) = \frac{1}{q^2}\sum_n |(\mathbf{q} \cdot \mathbf{J}_{\mathbf{q}}{}^{+})_{no}|^2 \frac{2\omega_{no}}{(\omega + i\eta)^2 - \omega_{no}^2}. \tag{4.173}$$

The induced density fluctuation, $\langle\rho(\mathbf{q}, \omega)\rangle$, may then be obtained from the continuity equation

$$\operatorname{div}\langle \mathcal{J} \rangle + e\left\langle\frac{\partial\rho}{\partial t}\right\rangle = 0. \tag{4.174}$$

On taking the Fourier transform of Eq. (4.174), and using (4.172), we find

$$\langle\rho(\mathbf{q}, \omega)\rangle = -\frac{e}{c}\left[\chi_{\parallel}(\mathbf{q}, \omega) + \frac{N}{m}\right]\frac{\mathbf{q} \cdot \mathcal{A}_{ext}(\mathbf{q}, \omega)}{\omega}. \tag{4.175}$$

Let us assume now that the electric field $\mathcal{D}$ is described in terms of the scalar electrostatic potential φ_{ext}. The induced density fluctuation is equal to

$$\langle\rho(\mathbf{q}, \omega)\rangle = \chi(\mathbf{q}, \omega)e\varphi_{ext}(\mathbf{q}, \omega), \tag{4.176}$$

where $\chi(\mathbf{q}, \omega)$ is the density-density response function. On replacing $\varphi_{ext}(\mathbf{q}, \omega)$ by its expression (4.169), we see that the results (4.175) and (4.176) will be equivalent if the following condition is satisfied:

$$\chi_{\parallel}(\mathbf{q}, \omega) + \frac{N}{m} = \frac{\omega^2}{q^2}\chi(\mathbf{q}, \omega). \tag{4.177}$$

The result (4.177) expresses the gauge invariance of the theory.

The explicit proof of Equation (4.177) is straightforward. We replace $\chi_{\parallel}(\mathbf{q}, \omega)$ and $\chi(\mathbf{q}, \omega)$ respectively by their expressions (4.173) and (2.57). We further note that

$$(\mathbf{q} \cdot \mathbf{J}_{\mathbf{q}}{}^{+})_{no} = \omega_{no}(\rho_{\mathbf{q}}{}^{+})_{no}, \tag{4.178}$$

a relation which follows from particle conservation [see Eq. (2.27)]. After some easy algebra, we obtain

$$\frac{\omega^2}{q^2}\chi(\mathbf{q},\omega) - \chi_{\parallel}(\mathbf{q},\omega) = \frac{1}{q^2}\sum_n 2\omega_{no}|(\rho_{\mathbf{q}}^{+})_{no}|^2. \tag{4.179}$$

Equation (4.179) reduces to (4.177) if we make use of the f-sum rule, (2.29). Gauge invariance is thus guaranteed by particle (charge) conservation, since only the latter is used to derive the f-sum rule.

A case of special interest is that of a static longitudinal potential, $\mathfrak{A}_{ext}(\mathbf{q}, 0)$. According to Eq. (4.168), such a potential has no physical effect: it cannot induce any current. It follows that

$$\chi_{\parallel}(\mathbf{q}, 0) = -\frac{N}{m}. \tag{4.180}$$

On using Eqs. (4.173) and (4.178), we see that Eq. (4.180) is directly equivalent to the f-sum rule.

We have seen in Section 4.3 that plasmons exhaust the f-sum rule in the long wavelength limit. Likewise, they provide the dominant contribution to $\chi(\mathbf{q}, \omega)$ and $\chi_{\parallel}(\mathbf{q}, \omega)$. In that sense, we may consider that the gauge invariance condition, (4.177), is a purely "plasmon affair" in the long wavelength limit. If the theory is to be explicitly gauge invariant, plasmons *must* be taken into account. If instead one wants to neglect collective oscillations, one must work in a *transverse* gauge [$\mathbf{q}\cdot\mathfrak{A}_{ext}(\mathbf{q}, 0) = 0$], in order to avoid any spurious longitudinal currents. As we shall see, these considerations apply to superconductors as well as to normal metals.

The electron response to a time varying longitudinal vector potential may also be obtained by regarding the induced polarization potential as an added applied field. The calculation is then identical to that given for the transverse response. The interaction Hamiltonian is given by Eq. (4.158), the induced electric current by (4.161). The only difference is that $\mathfrak{A}(\mathbf{q}, \omega)$ is now a *longitudinal* self-consistent vector potential. The induced currents may be written as

$$\begin{aligned} \langle \mathbf{J}(\mathbf{q},\omega)\rangle &= -\frac{e}{c}\chi_{\parallel}^{sc}(\mathbf{q},\omega)\mathfrak{A}(\mathbf{q},\omega), \\ \langle \mathcal{J}(\mathbf{q},\omega)\rangle &= -\frac{e^2}{c}\left[\chi_{\parallel}^{sc}(\mathbf{q},\omega) + \frac{N}{m}\right]\mathfrak{A}(\mathbf{q},\omega), \end{aligned} \tag{4.181}$$

where $\chi_{\parallel}^{sc}(\mathbf{q}, \omega)$ is the *screened* longitudinal current-current response function. In calculating $\chi_{\parallel}^{sc}$, we must neglect all polarization processes; these are already included in the definition of the self-consistent potential $\mathfrak{A}$.

The result, (4.181), is the longitudinal analog of Eq. (4.162). The screened quantity $\chi_{\parallel}^{sc}$ appears as the longitudinal component of the general response tensor $\chi_{\mu\nu}$ introduced in Eq. (4.163). The latter describes the system's response to an arbitrary *self-consistent* field $\mathfrak{A}$: for that reason, it involves $\chi_{\parallel}^{sc}$ rather than $\chi_{\parallel}$. The longitudinal dielectric function is given by

$$\epsilon_{\parallel}(\mathbf{q}, \omega) = 1 - \frac{4\pi e^2}{\omega^2}\left[\chi_{\parallel}^{sc}(\mathbf{q}, \omega) + \frac{N}{m}\right]. \tag{4.182}$$

Again, $\epsilon_{\parallel}$ is the longitudinal component of the general dielectric tensor $\epsilon_{\mu\nu}$.

Since a longitudinal vector potential is equivalent to a scalar potential, given by Eq. (4.169), it follows that

$$\mathfrak{A}(\mathbf{q}, \omega) = \frac{\mathfrak{A}_{ext}(\mathbf{q}, \omega)}{\epsilon(\mathbf{q}, \omega)}, \tag{4.183}$$

where $\epsilon(\mathbf{q}, \omega)$ is the scalar dielectric function. On comparing the expressions (4.172) and (4.181) of the electric current, we see that

$$\chi_{\parallel}^{sc}(\mathbf{q}, \omega) + \frac{N}{m} = \epsilon(\mathbf{q}, \omega)\left[\chi_{\parallel}(\mathbf{q}, \omega) + \frac{N}{m}\right]. \tag{4.184}$$

If we now combine Eq. (4.184) with Eqs. (4.177) and (4.12), we obtain the relation

$$\chi_{\parallel}^{sc}(\mathbf{q}, \omega) + \frac{N}{m} = \frac{\omega^2}{q^2}\chi_{sc}(\mathbf{q}, \omega). \tag{4.185}$$

Equation (4.185) is a "screened" version of the condition for gauge invariance. We shall see that it is somewhat more subtle than the "unscreened" condition (4.177) (even though the two relations are mathematically equivalent). By combining Eq. (4.185) with (4.182), and comparing with Eq. (4.13), we see that

$$\epsilon_{\parallel}(\mathbf{q}, \omega) = \epsilon(\mathbf{q}, \omega). \tag{4.186}$$

The longitudinal dielectric constant is thus uniquely defined.

In Eq. (4.185), let us write the screened density-density response function, $\chi_{sc}(\mathbf{q}, \omega)$, in the form

$$\chi_{sc}(\mathbf{q}, \omega) = \sum_n |(\rho_{\mathbf{q}}^{(1)})_{no}|^2 \frac{2\omega_{no}}{(\omega + i\eta)^2 - \omega_{no}^2}, \tag{4.187}$$

where the matrix element $(\rho_{\mathbf{q}}^{(1)})_{no}$ is given by Eq. (4.45). [The result (4.187) follows at once from Eq. (4.44) together with the Kramers–Kronig relation (2.78).] Let us further use the screened f-sum rule (4.41), which we can write in the form

$$\frac{N}{m} = \frac{1}{q^2}\sum_n 2\omega_{no}|(\rho_{\mathbf{q}}^{(1)})_{no}|^2. \tag{4.188}$$

With the help of Eq. (4.178), we may cast the function $\chi_{\parallel}^{sc}(\mathbf{q}, \omega)$ in the form

$$\chi_{\parallel}^{sc}(\mathbf{q}, \omega) = \frac{1}{q^2}\sum_n |(\mathbf{q}\cdot\mathbf{J}_{\mathbf{q}}^{(1)})_{no}|^2 \frac{2\omega_{no}}{(\omega + i\eta)^2 - \omega_{no}^2}, \tag{4.189}$$

where $(\mathbf{q}\cdot\mathbf{J}_{\mathbf{q}}^{(1)})_{no}$ is defined in analogy with Eq. (4.45):

$$(\mathbf{q}\cdot\mathbf{J}_{\mathbf{q}}^{(1)})_{no} = \omega_{no}(\rho_{\mathbf{q}}^{(1)})_{no} = \epsilon(\mathbf{q}, \omega_{no})(\mathbf{q}\cdot\mathbf{J}_{\mathbf{q}})_{no}. \tag{4.190}$$

The relation (4.189) is the screened analog of Eq. (4.173).

The relevant excited states in Eq. (4.189) are the same as in Eq. (4.187). They involve single pair and multipair excitations, as well as the "collective modes" appropriate to the equivalent neutral system (see Section 3.4). Plasmons do *not* contribute to Eq. (4.189), since the corresponding screened matrix element, (4.190), is equal to zero. The screened condition for gauge invariance, (4.185), is therefore different in nature from the unscreened one, (4.177). While the latter involves only plasmons in the long wavelength limit, the former involves the detailed screened excitation spectrum (pair excitations and screened collective mode). Equation (4.185) clearly represents a more subtle condition than (4.177).

In conclusion, let us emphasize that we are now able to treat on a symmetric basis the longitudinal and transverse dielectric functions, by considering the response to the self-consistent vector potential $\mathfrak{A}$. The response is characterized by a dielectric tensor $\epsilon_{\mu\nu}$, which is related to the screened response tensor $\chi_{\mu\nu}$ according to Eq. (4.165). One should, however, note that $\chi_{\mu\nu}$ is in a sense a construct, which does not describe the response to a well-defined external probe. In principle, one might set up a symmetric description of the response to an external probe by treating on an equal footing the Coulomb and magnetic interactions between electrons. Such a theory is rather involved, and not especially rewarding, since magnetic interactions are negligible in nonrelativistic systems; its derivation is sketched in Problem 4.4.

ASYMPTOTIC BEHAVIOR OF $\chi_\perp(\mathbf{q}, \omega)$; ELECTROMAGNETIC WAVE PROPAGATION, DIAMAGNETIC RESPONSE, AND A SUM RULE

As has been the case for all the correlation functions considered thus far, the behavior of $\chi_\perp(\mathbf{q}, \omega)$ in the limit $\mathbf{q} \to 0$, $\omega \to 0$ depends sensitively on which limit one takes first. In a translationally invariant system, the only contributions to Eq. (4.167) come from single-pair excitations, and (possibly) from a transverse collective mode (the longitudinal plasmon mode does not contribute for reasons of symmetry, while multipair excitations are negligible in the long wavelength limit). For such states, $\omega_{no} \sim qv_F$. It follows that

$$\chi_\perp(0, \omega) = 0. \tag{4.191a}$$

In the limit $\mathbf{q} \equiv 0$, the longitudinal and transverse response functions are thus equal. Such a result was perhaps obvious, since the direction of $\mathbf{q}$ is no longer defined for a uniform field (the response depends only on the polarization of the field). Indeed, one still has

$$\chi_\parallel(0, \omega) = \chi_\perp(0, \omega) \tag{4.191b}$$

in a real solid, as long as the polarization of the field remains the same.

According to Eq. (4.191b), we may apply the discussion of Section 4.3 to the optical properties of solids in the infrared region (where $\omega \gg qv_F$). A detailed comparison of theory and experiment will not be presented here; we refer the interested reader to Pines (1963) for a recent survey.

It follows from Eq. (4.191a) that

$$\lim_{\mathbf{q}\to 0} \epsilon_\perp(\mathbf{q}, \omega) = 1 - \frac{\omega_p^2}{\omega^2}, \tag{4.192}$$

a result we obtained earlier using the Landau–Silin theory. We find, from Eq. (4.155), that in this limit the frequency of propagation of electromagnetic waves is

$$\omega^2 = \omega_p^2 + c^2q^2. \tag{4.193}$$

This is the strong-coupling limit discussed in Chapter 3, in which the natural frequency of the waves is changed from cq to ω_p by the presence of the electron liquid. This change occurs as a result of the transverse polarization processes; the Coulomb interaction between the electrons appears only through $\chi_\perp$, and is here negligible.

In the opposite limit, that of a static, quasihomogeneous external field, the electron response is determined by $\chi_\perp(\mathbf{q}, 0)$. One can calculate $\chi_\perp(\mathbf{q}, 0)$ in the long wavelength limit from the Landau theory; the result is

$$\lim_{\mathbf{q}\to 0} \chi_\perp(\mathbf{q}, 0) = -\frac{N}{m}, \tag{4.194}$$

It is clear that corrections to this value will be of order $(N/m)q^2/p_F^2$; they are responsible for the weak diamagnetism first calculated by Landau. To see this, we use Eqs. (4.156) and (4.164) to write the relation between $\mathcal{A}(\mathbf{q}, 0)$ and $\mathcal{A}_{\text{ext}}(\mathbf{q}, 0)$ as

$$\begin{aligned}\lim_{\mathbf{q}\to 0} \mathcal{A}(\mathbf{q}, 0) &= \lim_{\mathbf{q}\to 0} \frac{\mathcal{A}_{\text{ext}}(\mathbf{q}, 0)}{1 + (4\pi e^2/c^2q^2)(\chi_\perp(\mathbf{q}, 0) + N/m)} \\ &\cong \mathcal{A}_{\text{ext}}(\mathbf{q}, 0)\left\{1 - \frac{\omega_p^2}{c^2p_F^2}\alpha\right\},\end{aligned} \tag{4.195}$$

where α is a positive coefficient of order unity, which cannot be obtained using the Fermi liquid theory. The *local* potential $\mathcal{A}$ determines the magnetic induction $\mathcal{B}$, while $\mathcal{A}_{\text{ext}}$ gives rise to the magnetic field $\mathcal{H}$. The result, (4.195), clearly describes a diamagnetic reduction in $\mathcal{H}$. We give a microscopic calculation of α in the following chapter. For the normal Fermi liquid, diamagnetic screening is a very weak effect, of order v^2/c^2.

Matters are quite different for a superconductor. We shall show that for a spatially homogeneous superconductor at $T = 0$,

$$\lim_{\mathbf{q}\to 0} \chi_\perp(\mathbf{q}, 0) = 0 \tag{4.196}$$

and hence

$$\lim_{\mathbf{q}\to 0} \mathcal{A}(\mathbf{q}, 0) = \frac{\mathcal{A}_{ext}(\mathbf{q}, 0)}{1 + \omega_p^2/c^2q^2}. \tag{4.197}$$

In the long wavelength limit, then, one has "perfect" diamagnetic screening, i.e. the Meissner effect. According to Eq. (4.197), a magnetic field penetrates only to a distance c/ω_p, the penetration depth predicted by London. We give a detailed discussion of these and related matters in Volume II; here we wish merely to emphasize the difference between normal metals and superconductors, and call attention to the usefulness of $\chi_\perp(\mathbf{q}, 0)$ as a measure of diamagnetic behavior.

In considering the question of a sum rule satisfied by $\epsilon_\perp(\mathbf{q}, \omega)$ or $\sigma_\perp(\mathbf{q}, \omega)$, we do not have available a transverse analog of the f-sum rule. We can proceed, however, to derive a sum rule provided we make the physically reasonable assumption that $\epsilon_\perp(\mathbf{q}, \omega)$ is analytic in the upper half of the complex ω-plane. [As was the case with the longitudinal dielectric function, since $\epsilon_\perp$ does not measure the response to an external field, causality is of no avail in determining its analytic properties.] We note that

$$\lim_{\omega\to\infty} \chi_\perp(\mathbf{q}, \omega) \cong O\left(\frac{1}{\omega^2}\right),$$
$$\lim_{\omega\to\infty} \epsilon_\perp(\mathbf{q}, \omega) = 1 - \frac{\omega_p^2}{\omega^2}. \tag{4.198}$$

We next remark that the integral

$$I = \int_C \omega \, d\omega \, \epsilon_\perp(\mathbf{q}, \omega), \tag{4.199}$$

evaluated along the contour C shown in Fig. 4.9, vanishes, in view of the analytic properties of $\epsilon_\perp$.

The contribution to I from the contour at infinity is simply $-\pi\omega_p^2$. We have therefore

$$\int_{-\infty}^{\infty} d\omega \, \omega\epsilon_\perp(\mathbf{q}, \omega) = \pi\omega_p^2, \tag{4.200}$$

where, in the integration over ω, it is necessary to pass above any singularities in $\epsilon_\perp(\mathbf{q}, \omega)$.

There are two possible sources of singularities in $\epsilon_\perp(\mathbf{q}, \omega)$. One is at the origin and is of strength $-4\pi e^2[\chi_\perp(\mathbf{q}, 0) + N/m]$; the second (present only at long wavelengths) is the transverse collective mode discussed in Chapter 3, which would appear as a collective pole in $\chi_\perp(\mathbf{q}, \omega)$. We can, if we like, take explicit account of the possible singularity at the origin; we then obtain

$$\fint_{-\infty}^{\infty} d\omega\, \omega\epsilon_\perp(\mathbf{q}, \omega) = -4\pi^2 e^2 \chi_\perp(\mathbf{q}, 0), \tag{4.201}$$

where the bar through the integral sign indicates that principal parts are to be taken near the origin. We see that in the long wavelength limit this sum rule will be very different for superconductors and normal electron liquids.

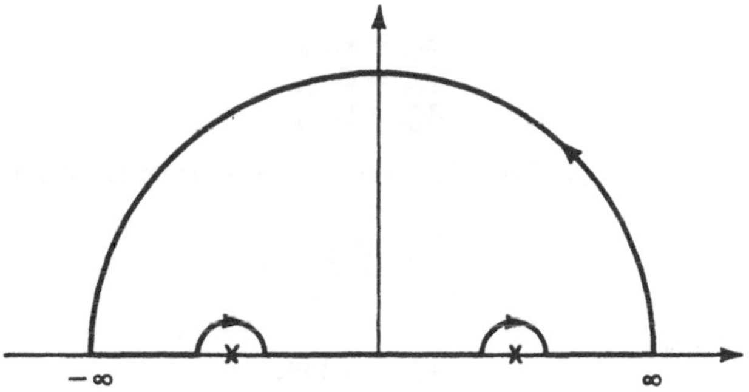

FIGURE 4.9. *Contour for establishing the sum rule obeyed by $\epsilon_\perp(\mathbf{q}, \omega)$. The contour passes above any possible singularity in $\epsilon_\perp$ (here denoted by a cross).*

QUANTIZATION OF THE PHOTON FIELD: RAMAN SCATTERING

We now consider briefly the quantum-mechanical treatment of the coupling between electrons and the electromagnetic field.* Such a treatment shows to what extent one is justified in using the self-consistent electron-field interaction Hamiltonian, (4.158), and also enables one to treat, in straightforward fashion, the Raman scattering of radiation by the quantum plasma. The Hamiltonian for the coupled system of electrons and photons may be written as follows:

$$H = \sum_i \frac{[\mathbf{p}_i + (e/c)\mathbf{\mathfrak{A}}(\mathbf{r}_i)]^2}{2m} + H_{\text{coul}} + H_{\text{field}}, \tag{4.202}$$

where H_{coul} represents the Coulomb interaction between the electrons, H_{field} is the Hamiltonian for the free electromagnetic wave field, and $\mathbf{\mathfrak{A}}(\mathbf{r})$ is the vector potential associated with that field. As usual, we work in a transverse gauge, in which div $\mathbf{\mathfrak{A}} = 0$.

As is well known, the free-field Hamiltonian may easily be expressed as a sum of harmonic oscillator terms. To do this, one expands the vector potential

* The approach used is essentially that of the authors [Nozières and Pines (1959)].

in a Fourier series:

$$\mathcal{A}(\mathbf{r}) = \sum_{\mathbf{q}\mu} (4\pi c^2)^{1/2} Q_{\mathbf{q}\mu} \mathbf{n}_{\mathbf{q}\mu} e^{i\mathbf{q}\cdot\mathbf{r}}. \tag{4.203}$$

$Q_{\mathbf{q}\mu}$ is the amplitude of the **q**th normal mode of the field, $\mathbf{n}_{\mathbf{q}\mu}$ is its polarization vector (μ takes the values 1 and 2 corresponding to the two possible directions of polarization perpendicular to **q**). The coordinate canonically conjugate to $Q_{\mathbf{q}\mu}$, $\Pi_{\mathbf{q}\mu}$, appears in the Fourier expansion of the electric field vector,

$$\mathcal{E}(\mathbf{r}) = -\frac{1}{c}\frac{\partial}{\partial t}\mathcal{A}(\mathbf{r}) = \sum_{\mathbf{q}\mu} (4\pi)^{1/2} \Pi^+_{\mathbf{q}\mu} \mathbf{n}_{\mathbf{q}\mu} e^{i\mathbf{q}\cdot\mathbf{r}}. \tag{4.204}$$

$\mathcal{A}$ and $\mathcal{E}$ are real, so that one has

$$\begin{aligned} \mathbf{n}_{\mathbf{q}\mu} &= \mathbf{n}_{-\mathbf{q}\mu}, \\ Q^+_{\mathbf{q}\mu} &= Q_{-\mathbf{q}\mu}, \\ \Pi^+_{\mathbf{q}\mu} &= \Pi_{-\mathbf{q}\mu}. \end{aligned} \tag{4.205}$$

The Hamiltonian for the free field is then obtained from the expression for the field energy,

$$\int d^3r \, \frac{\mathcal{E}^2 + \mathcal{H}^2}{8\pi}, \tag{4.206}$$

where, as usual,

$$\mathcal{H} = \text{curl}\, \mathcal{A}. \tag{4.207}$$

On making use of Eqs. (4.204) and (4.207), one finds easily that Eq. (4.206) may be written in the form

$$H_{\text{field}} = \frac{1}{2}\sum_{\mathbf{q}\mu} [\Pi^+_{\mathbf{q}\mu}\Pi_{\mathbf{q}\mu} + c^2q^2 Q^+_{\mathbf{q}\mu} Q_{\mathbf{q}\mu}]. \tag{4.208}$$

With the aid of the expansion (4.203) the Hamiltonian (4.202) may be rewritten as follows:

$$H = H_0 + H'_{\text{field}} + H_1 + H_2, \tag{4.209}$$

where

$$H_0 = \sum_i \frac{\mathbf{p}_i^2}{2m} + H_{\text{coul}}, \tag{4.210}$$

$$H'_{\text{field}} = \frac{1}{2}\sum_{\mathbf{q}\mu} [\Pi^+_{\mathbf{q}\mu}\Pi_{\mathbf{q}\mu} + (c^2q^2 + \omega_p^2) Q^+_{\mathbf{q}\mu} Q_{\mathbf{q}\mu}], \tag{4.211}$$

$$H_1 = \sum_{\mathbf{q}\mu} (4\pi e^2)^{1/2} (\mathbf{J}_{\mathbf{q}} \cdot \mathbf{n}_{\mathbf{q}\mu}) Q_{\mathbf{q}\mu}, \tag{4.212}$$

$$H_2 = \sum_{\mathbf{q}\neq\mathbf{k},\mu\nu} \frac{2\pi e^2}{m} (\mathbf{n}_{\mathbf{k}\nu} \cdot \mathbf{n}_{\mathbf{q}\mu}) \rho^+_{\mathbf{k}-\mathbf{q}} Q^+_{\mathbf{q}\mu} Q_{\mathbf{k}\nu}. \tag{4.213}$$

$\mathbf{J}_{\mathbf{q}}$ is the particle current density fluctuation as given by Eq. (2.22). We have included in Eq. (4.211) the diagonal contribution from the $\mathfrak{A}^2$ term of (4.202), which acts to shift the photon frequency from cq to $(c^2q^2 + \omega_p{}^2)^{1/2}$.

We have separated the photon-electron coupling into the distinct interaction terms, H_1 and H_2. H_1 gives rise to single photon emission or absorption: it also acts to shift the photon frequencies. H_2 gives rise to two-photon processes. It is not difficult to show, using perturbation-theoretic techniques, that to

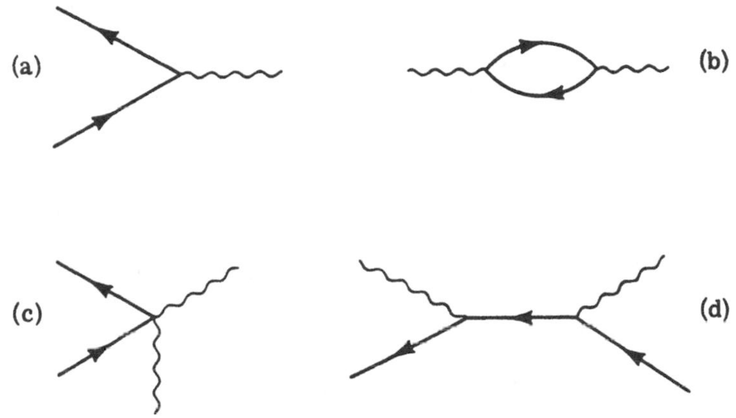

FIGURE 4.10. *Different electron photon interaction processes: (a) Photon decay into a particle-hole pair, (b) polarization correction to the photon propagation, (c) Raman scattering via the interaction H_2, (d) Raman scattering via the interaction H_1.*

lowest order in the electron-photon coupling, the two terms may be treated independently. Since that coupling constant is

$$\frac{e^2}{\hbar c} \approx \frac{1}{137},$$

this is quite a good approximation, and we shall follow it here. To this same order, the only processes of interest arising from H_2 correspond to Raman scattering of the photons by the density fluctuations. There is also a contribution to Raman scattering from higher-order terms arising from H_1, which, however, may be neglected to order v^2/c^2. A pictorial representation of these various terms is given in Fig. 4.10. In Fig. 4.10(a), we see photon decay into a particle-hole pair; 4.10(b) shows a polarization process which contributes to a photon energy shift. Figure 4.10(c) shows a Raman scattering process governed by H_2, while 4-10(d) shows the negligible (to order v^2/c^2) Raman scattering contribution coming from H_1.

The treatment of a system of electrons and photons coupled via H_1 proceeds along the same lines as that given in Section 4.5 for the coupled electron-phonon system. In order to carry out a self-consistent calculation of the photon frequencies and the induced current fluctuations, one assumes the existence of an electromagnetic wave of some frequency $\omega_{q\mu}$, and one then determines the induced current fluctuation that leads to a wave of that frequency. Analogous to the phonon case, one allows for all possible electronic polarization phenomena, but one neglects any contribution to the electronic response from virtual processes involving the photon field. The approximation is valid to order $e^2/\hbar c$, the electron-photon coupling constant. It is completely equivalent to the calculation of the electron response via the effective coupling Hamiltonian, (4.158), as the reader may easily verify. It is left as a problem to the reader to show that the shifted photon frequencies are given by the dispersion relation

$$\omega_{\mathbf{q}\mu}^2 = c^2q^2 + \omega_p{}^2 + 4\pi e^2 \sum_n |(\mathbf{J_q}^+ \cdot \mathbf{n}_{\mathbf{q}\mu})_{no}|^2 \left\{\frac{1}{\omega - \omega_{no} + i\eta} - \frac{1}{\omega + \omega_{no} + i\eta}\right\}. \tag{4.214}$$

This result could, in fact, have been obtained directly from Maxwell's equations and our calculation of $\chi_{\mu\nu}$, (4.163).

We consider now that part of H_2 which gives rise to Raman scattering of electromagnetic waves. Let us introduce the photon creation and annihilation operators, according to

$$Q_{\mathbf{q}\mu} = (A_{\mathbf{q}\mu} + A_{-\mathbf{q}\mu}^+)/(2\omega_{\mathbf{q}\mu})^{1/2}. \tag{4.215}$$

We then find

$$H_2 = \frac{2\pi e^2}{m} \sum_{\mathbf{qk},\mu\nu} \left(\frac{1}{4\omega_\mathbf{k}\omega_{\mathbf{k-q}}}\right)^{1/2} (\mathbf{n}_{\mathbf{k-q},\nu} \cdot \mathbf{n}_{\mathbf{k}\mu})(A_{\mathbf{k}\mu} + A_{-\mathbf{k}\mu}^+)(A_{\mathbf{k-q},\nu}^+ + A_{-\mathbf{k+q},\nu})\rho_\mathbf{q}^+. \tag{4.216}$$

H_2 gives rise to scattering processes in which a photon of momentum $\mathbf{k}$, energy $\omega_\mathbf{k}$, is scattered to a state $\mathbf{k} - \mathbf{q}$, energy $\omega_{\mathbf{k-q}}$, while transferring momentum $\mathbf{q}$, energy $\omega = \omega_\mathbf{k} - \omega_{\mathbf{k-q}}$, to the electron system. The matrix element for such a scattering is clearly proportional to the matrix element of the density fluctuation $(\rho_\mathbf{q}^+)_{no}$, taken between the ground state and the final electronic excited state of energy $\omega_{no} = \omega$. We see that the electromagnetic wave is acting as a probe of the particle density fluctuations. It is a straightforward calculation to show that the differential cross section for momentum transfer $\mathbf{q}$, energy transfer ω, is given by

$$\frac{d^2\sigma}{d\omega\, d\theta} = \left(\frac{e^2}{mc^2}\right)^2 \left[1 - \frac{\sin^2\theta}{2}\right] S(\mathbf{q}, \omega), \tag{4.217}$$

where θ is the angle between the incident and scattered waves; we have averaged over the polarizations of the incident wave, and summed over the polarizations of the scattered wave. Equation (4.217) furnishes the starting point for a discussion of the use of Raman scattering as a probe of density fluctuations in quantum plasmas. It may easily be extended to the case of electrons in a solid, or to the case of classical plasmas, both of which problems are of considerable experimental interest.

PROBLEMS

4.1. Show that if an external charge of density $\delta\rho(\mathbf{q}, 0)$ is introduced into the electron liquid, the change in energy of the system is

$$\delta E_{\mathbf{q}} = (2\pi e^2/q^2)|\delta\rho(\mathbf{q}, 0)|^2/\epsilon(\mathbf{q}, 0).$$

4.2. For electrons in solids, the f-sum rule continues to be valid. Consider the case of a metal, for which there is only one band of interest, the conduction band. In this case, in the long wavelength limit, it is convenient to distinguish between the intraband and interband oscillator strengths, and to define the total intraband oscillator strength by

$$\lim_{\mathbf{q}\to 0} \sum_n f_{on}^{\text{intra}} = \lim_{\mathbf{q}\to 0} \frac{2m}{q^2} \sum_n{}' \omega_{no}|(\rho_{\mathbf{q}}{}^+)_{no}|^2 = \frac{Nm}{m_c},$$

where the prime denotes the restriction to intraband transitions (for which $\lim_{\mathbf{q}\to 0} \omega_{no} = 0$) and N is the number of conduction electrons.

(i) Use the formal definition of $\epsilon(\mathbf{q}, \omega)$, Eq. (4.7), to prove that for ω less than any interband excitation frequency, one finds the exact result

$$\epsilon(0, \omega) \cong 1 - \frac{4\pi Ne^2}{m_c\omega^2}.$$

(ii) By comparing this result with that obtained using the Landau theory, [cf. Eq. (3.167)], show that the current carried by a single quasiparticle of momentum $\mathbf{p}$ is $\mathbf{p}/m_c$, where m_c is defined above.

Note: In this fashion, one obtains a microscopic definition of the crystalline mass, m_c, introduced through Eq. (3.157).

4.3. Obtain an explicit expression for the dielectric function of a system of electrons *and ions*, $\epsilon_{\text{tot}}(\mathbf{q}, \omega)$, by considering the response of the ions, as well as that of the electrons, to an external test charge. Show that in the jellium model, for frequencies $\omega \ll qv_F$,

$$\frac{1}{\epsilon_{\text{tot}}} = \frac{1}{\epsilon(\mathbf{q}, \omega)} + \frac{|v_{\mathbf{q}}{}^i|^2}{\epsilon(\mathbf{q}, 0)} \frac{1}{\omega^2 - \omega_{\mathbf{q}}{}^2},$$

but that otherwise the relation (4.143) is not valid.

4.4. Consider the coupling of an electron liquid to an external transverse disturbance, characterized by $\mathcal{A}_{\text{ext}}(\mathbf{q}, \omega)$.

(i) Show that the induced current is

$$\langle \mathcal{J} \rangle = \frac{e^2}{c} \left\{ \chi_{\perp}^{\text{ext}} + \frac{N}{m} \right\} \mathcal{A}_{\text{ext}},$$

where $\chi_{\perp}^{\text{ext}}$ is the transverse current-current response function calculated *including* the magnetic interactions between the electrons.

(ii) A transverse dielectric function, $\epsilon_{\perp}^{\text{ext}}(\mathbf{q}, \omega)$, may be defined through the relation

$$\mathcal{A} = \frac{\mathcal{A}_{\text{ext}}}{\epsilon_{\perp}^{\text{ext}}(\mathbf{q}, \omega)}$$

in direct analogy to the longitudinal case. (Here $\mathcal{A}$ is the *local* vector potential, $\mathcal{A}_{\text{pol}} + \mathcal{A}_{\text{ext}}$.) Show that

$$\frac{1}{\epsilon_{\perp}^{\text{ext}}(\mathbf{q}, \omega)} = 1 + \frac{4\pi e^2}{(\omega + i\eta)^2 - c^2q^2} \left\{\chi_{\perp}^{\text{ext}} + \frac{N}{m}\right\}.$$

(iii) Obtain the relationship between $\epsilon_{\perp}^{\text{ext}}(\mathbf{q}, \omega)$ and $\epsilon_{\perp}(\mathbf{q}, \omega)$, and use this result to discuss the limiting behavior of $\epsilon_{\perp}^{\text{ext}}(\mathbf{q}, \omega)$ and $\chi_{\perp}^{\text{ext}}(\mathbf{q}, \omega)$ in the following cases: (a) $\mathbf{q} = 0$, ω finite; (b) $\omega = 0$, $\mathbf{q} \to 0$.

4.5. Discuss the behavior, in the long wavelength limit, of the matrix elements $(\rho_{\mathbf{q}}^{+})_{no}$ and excitation frequencies ω_{no} for electrons in an external dc magnetic field (a) with $\mathbf{q}$ parallel to $\mathcal{H}_o$; (b) with $\mathbf{q}$ perpendicular to $\mathcal{H}_o$. Use the results to discuss the sum rules in the presence of a magnetic field, and the limiting behavior of $S(\mathbf{q}, \omega)$.

Hint: The results obtained using the Landau theory (see Problem 3.8) serve as a useful check on the above considerations.

4.6. In the presence of a weak static external magnetic field, calculate the phonon sound velocity in the jellium model for the case of a sound wave which propagates (a) parallel and (b) perpendicular to the external field.

Hint: Neglect the influence of the magnetic field on the ions, and make use of the results of Problem 3.8 to treat the electron reponse to the ionic motion.

REFERENCES

Bardeen, J. (1937), *Phys. Rev.* **52,** 688.

Bardeen, J. and Pines, D. (1955), *Phys. Rev.* **99,** 1140.

Bohm, D. and Pines, D. (1951), *Phys. Rev.* **82,** 625.

Bohm, D. and Staver, T. (1952), *Phys. Rev.* **84,** 836.

Bohr, A. (1948), *Kgl. Danske Videnskab, Selskab, Mat-fys. Medd.* **24,** No 19.

Born, M. and Oppenheimer, J. R. (1928), *Ann. Phys.* **84,** 457.

Fermi, E. (1940), *Phys. Rev.* **57,** 485.

Ferrell, R. A. (1957), *Phys. Rev.* **107,** 450.

Frölich, H. and Pelzer, H. (1955), *Proc. Phys. Soc. (London)* **68,** 525.

Glick, A. and Ferrell, R. A. (1960), *Ann. Phys.* **11,** 359.

Harrison, W. (1966), *Pseudopotentials in the Theory of Metals,* W. A. Benjamin, New York.

Hubbard, J. (1955), *Proc. Phys. Soc. (London)* **68,** 976.

Ichimaru, S. (1962), *Ann. Phys.* **20,** 78.

Ichimaru, S. (1965), *Phys. Rev.*, **140** B226.

Klemperer, O. and Shepherd, J. P. G. (1963), *Advan. Phys.* **12,** 355.

Kramers, H. A. (1947), *Physica* **13,** 401.

Marton, L., Simpson, J. A., Fowler, H. A., and Swanson, N. (1962), *Phys. Rev.* **126,** 182.

Migdal, A. B. (1958), *Sov. Phys. JETP* **7**, 996.
Nozières, P. and Pines, D. (1958), *Nuovo Cimento* **9**, 470.
Nozières, P. and Pines, D. (1959), *Phys. Rev.* **113**, 1254.
Pines, D. (1956), *Rev. Mod. Phys.* **28**, 184.
Pines, D. (1960), *Physica* **26**, S 103.
Pines, D. (1963), *Elementary Excitations in Solids*, W. A. Benjamin, New York.
Powell, C. J. (1960), *Proc. Phys. Soc. (London)* **76**, 593.
Powell, C. J. and Swan, J. B. (1960), *Phys. Rev.* **118**, 640.
Quinn, J. J. (1962), *Phys. Rev.* **126**, 1453.
Ritchie, R. H. (1957), *Phys. Rev.* **106**, 874.
Robins, J. L. and Swan, J. B. (1960), *Proc. Phys. Soc. (London)* **76**, 857.
Robins, J. L. (1961), *Proc. Phys. Soc. (London)* **78**, 1177.
Sham, L. J. and Ziman, J. (1964), *Solid State Phys.* **15**.
Stern, E. A. and Ferrell, R. A. (1960), *Phys. Rev.* **120**, 130.
Thompson, W. and Hubbard, J. (1960), *Rev. Mod. Phys.* **32**, 714.
Watanabe, H. (1956), *J. Phys. Soc. Japan* **11**, 12.

CHAPTER 5

MICROSCOPIC THEORIES OF THE ELECTRON LIQUID

The macroscopic theory of Fermi liquids we have developed in the preceding chapters possesses the advantage of being exact for any normal liquid, irrespective of its density or law of interaction, so long as the temperature is sufficiently low and the phenomena under discussion involve both long wavelengths and low frequencies. It is, of course, incomplete, since it can provide no account of microscopic phenomena involving short wavelengths and large excitation energies, or of equilibrium properties such as the ground state energy and pair distribution function. In contrast, microscopic theories are able to describe both macroscopic and microscopic phenomena. They are, however, subject to one of the following limitations:

(1) They are exact for only a limited range of particle density or interaction, one which rarely corresponds to physical reality.
(2) The extent to which they provide an approximate account of a real physical system can rarely be estimated with any precision.

Microscopic theories in the first category may be thought of as solutions to model problems, that is, solutions based on well-defined approximations which can be shown to be clearly valid for a given class of particle interactions and densities. For electron systems there exist two such models: the high-density electron gas, and the low-density electron solid. The behavior of the electron liquid in the limit of very high densities is simple because here the Coulomb interaction represents a relatively small perturbation; the average potential energy (measured by e^2/r_0, where r_0 is the interparticle spacing) is small compared to the kinetic energy (measured by $\hbar^2/mr_0{}^2$). The ratio of the two energies is

$$\frac{e^2mr_0{}^2}{r_0\hbar^2} = \frac{r_0}{a_0} = r_s,$$

where the interelectron spacing is measured in units of the Bohr radius. When $r_s \ll 1$, the system properties are well described in the so-called random phase approximation (RPA).

In the low-density limit ($r_s \gg 1$), the electron behavior is dominated by the Coulomb interaction. As Wigner (1938) first remarked, in this strong-coupling limit the electron "liquid" will not be a liquid at all, but rather a solid; the electrons form a stable lattice in the sea of uniform positive charge. The Coulomb repulsion keeps the electrons apart, and the kinetic energy for sufficiently large r_s is insufficient to keep the electrons from becoming localized at fixed lattice sites. The strong coupling limit is likely to be reached when $r_s \gg 10$.

The region of actual metallic densities ($1.8 \leq r_s \lesssim 5.6$) is essentially an intermediate coupling regime. As such it is far more difficult to treat; the kinetic energy and potential energy are comparable, and consequently no rigorous approximation schemes are applicable. Any theory of the electron liquid at metallic densities will perforce be a theory of the second category mentioned above, one whose accuracy cannot be determined with precision.

The present chapter is devoted to a consideration of the electron liquid in both the high-density and the intermediate-density regime. We do not consider the low-density electron solid here, primarily because its behavior is not especially relevant to the properties of metals.* For the same reason we do not consider at all the single model we possess for a neutral fermion system, a low-density neutral gas. Although its properties may be worked out in detail, they teach us little about the behavior of liquid ^{3}He. By contrast, as we shall see, one learns much about the behavior of electrons in metals by considering the RPA, despite the fact that it is valid only in the high-density limit.

We begin, in Section 5.1, with a consideration of the Hartree–Fock approximation for a many-body system. For uniform systems, the variational calculations of Hartree, Fock, and Slater, which were the first serious attempts to calculate quantum-mechanical many-body properties, reduce to the lowest-order perturbation-theoretic approach. The Hartree–Fock approximation thus represents a useful starting point for consideration of microscopic theories. Sections 5.2, 5.3, and 5.4 are

* For a brief summary of the properties of the low-density electron solid, see Pines (1963), Chapter 3.

devoted to an exposition of the RPA, and its application to the calculation of the ground state energy and quasiparticle properties of an electron liquid. We devote so much attention to the RPA in part because it offers a nontrivial solution for a model problem, one in which *all* the various effects of electron interaction may be taken into account; in part, because consideration of the RPA, and especially of its limitations, offers a useful guide to the construction of an approximate theory of the intermediate-density electron liquid.

In Section 5.5 we develop the "equation-of-motion" method in order to consider the RPA, and its possible generalizations, as part of a hierarchy of successive approximations to the electron liquid. In many respects, such considerations are most naturally carried out with the aid of diagrammatic techniques analogous to those developed by Feynman and Dyson for quantum electrodynamics; however, such sophisticated techniques are not necessary for the limited class of problems we consider here. In Section 5.6, we discuss two of the approximation methods which have been developed for the intermediate-density electron liquid; we then apply those to a calculation of a number of equilibrium properties of simple metals. It is, in fact, a somewhat long and tortuous path one must follow to go from an understanding of the electron liquid at metallic electron densities to the microscopic calculation of metallic properties; we attempt to chart the main obstacles, and indicate the nature of the results one obtains upon making the traversal.

5.1. HARTREE–FOCK APPROXIMATION

We present in this section a description of the electron liquid in which only the lowest-order (first-order in V) effects of particle interaction are taken into account. For a uniform system, this approximation is equivalent to the Hartree–Fock approximation. We calculate first the ground state energy, and the pair correlation function. The interaction between quasiparticles is determined from the ground state energy, and is subsequently applied in the calculation of quasiparticle properties.

GROUND STATE ENERGY

The basic Hamiltonian for a many-fermion system may be written as

$$H = \sum_{\mathbf{p}\sigma} \epsilon_{\mathbf{p}}{}^{0} c_{\mathbf{p}\sigma}^{+} c_{\mathbf{p}\sigma} + \sum_{\mathbf{p}\mathbf{p}'\mathbf{q}\sigma\sigma'} (V_{\mathbf{q}}/2) c_{\mathbf{p}+\mathbf{q}\sigma}^{+} c_{\mathbf{p}'-\mathbf{q}\sigma'}^{+} c_{\mathbf{p}'\sigma'} c_{\mathbf{p}\sigma}, \tag{5.1}$$

where $V_{\mathbf{q}}$ is the Fourier transform of the interaction potential. For the electron liquid, immersed in a uniform background of positive charge,

$$V_{\mathbf{q}} = \frac{4\pi e^2}{q^2}(1 - \delta_{\mathbf{q}o}). \tag{5.2}$$

There is no $\mathbf{q} = 0$ term in the electron interaction because this part of the potential is canceled by the field of the uniform positive charge.

The lowest-order approximation to the ground state energy is obtained if we assume that the ground state wave function, $|0\rangle$, is that of a noninteracting Fermi gas. It thus corresponds to filling all the plane wave states inside a Fermi sphere radius, p_F. We have

$$c^+_{\mathbf{p}\sigma}c_{\mathbf{p}\sigma}|0\rangle = n^{\circ}_{\mathbf{p}\sigma}|0\rangle, \tag{5.3}$$

where $n^{\circ}_{\mathbf{p}\sigma}$ is the ground state distribution given by

$$\begin{aligned} n^{\circ}_{\mathbf{p}\sigma} &= 1, \quad p < p_F, \\ &= 0, \quad p > p_F. \end{aligned} \tag{5.4}$$

When the potential energy term in Eq. (5.1) operates on $|0\rangle$, it acts to "destroy" a pair of particles $\mathbf{p}\sigma$, $\mathbf{p}'\sigma'$ inside the Fermi sphere. Diagonal terms are obtained only if these particles are in turn "created" by the operators $c^+_{\mathbf{p}+\mathbf{q}\sigma}c^+_{\mathbf{p}'-\mathbf{q}\sigma'}$. This will be the case either if $\mathbf{q} = 0$, or if $\mathbf{p}' - \mathbf{q} = \mathbf{p}$ and $\sigma' = \sigma$. The ground state energy is therefore

$$\begin{aligned} E_o = \langle 0|H|0\rangle = \sum_{\mathbf{p}\sigma} \epsilon_{\mathbf{p}}{}^{\circ} n^{\circ}_{\mathbf{p}\sigma} &+ \frac{1}{2}\sum_{\mathbf{p}\mathbf{p}'\sigma\sigma'} V_o \langle 0|c^+_{\mathbf{p}\sigma}c^+_{\mathbf{p}'\sigma'}c_{\mathbf{p}'\sigma'}c_{\mathbf{p}\sigma}|0\rangle \\ &+ \frac{1}{2}\sum_{\mathbf{p}\mathbf{q}\sigma,\mathbf{q}\neq 0} V_{\mathbf{q}} \langle 0|c^+_{\mathbf{p}+\mathbf{q}\sigma}c^+_{\mathbf{p}\sigma}c_{\mathbf{p}+\mathbf{q}\sigma}c_{\mathbf{p}\sigma}|0\rangle. \end{aligned} \tag{5.5}$$

On commuting through the relevant operators, so as to obtain an expression in terms of the distribution function, we find

$$E_o = \sum_{\mathbf{p}\sigma} \epsilon_{\mathbf{p}}{}^{\circ} n^{\circ}_{\mathbf{p}\sigma} + \frac{1}{2}\sum_{\mathbf{p}\mathbf{p}'\sigma\sigma'} V_o n^{\circ}_{\mathbf{p}\sigma}(n^{\circ}_{\mathbf{p}'\sigma'} - \delta_{\mathbf{p}'\mathbf{p}}\delta_{\sigma'\sigma}) - \frac{1}{2}\sum_{\mathbf{p}\mathbf{q}\sigma} V_{\mathbf{q}} n^{\circ}_{\mathbf{p}+\mathbf{q}\sigma} n^{\circ}_{\mathbf{p}\sigma}. \tag{5.6}$$

The first term in Eq. (5.6) is a "zeroth-order" approximation to the ground state energy; it is the kinetic energy for a noninteracting particle system. The second and third terms represent the first-order (in the potential) contributions to the ground state energy. The second term is just what would have been obtained if we had carried out a Hartree (1928) self-consistent procedure, in which we attempted to find the best many-body wave function which was a simple product of one-electron wave functions. For a uniform many-body system, those functions are simply plane waves. The third term is known as the exchange energy; it is the additional term one finds if one carries out the Fock (1930) procedure, in which one seeks a properly antisymmetric wave function, and so attempts to find the best many-body wave function which is a simple determinant of one-electron wave functions. For a uniform system, the self-consistent wave functions are again plane waves, so that the ground state in the Hartree–Fock approximation is that appropriate to the noninteracting particle system.* For that reason, we shall consider Eq. (5.6) as the HFA (Hartree–Fock approximation) calculation of the ground state energy. Let us emphasize that the HFA reduces to a first-order calculation *only* in uniform systems. In the nonuniform case, the HFA requires a *self-consistent* calculation of the average potential acting on the particles: it is then far more subtle than a plain perturbation expansion.

For the electron liquid, the Hartree term in Eq. (5.6) is missing; the uniform background of positive charge cancels out the influence of the averaged electron charge density. The ground state energy per particle is therefore

$$\frac{E_0^{\text{HFA}}}{N} = \left(\frac{3}{5}\right)\frac{p_F^2}{2m} - \left(\frac{1}{N}\right) \sum_{p<p_F,\,p'<p_F} \frac{2\pi e^2}{|\mathbf{p} - \mathbf{p}'|^2}. \tag{5.7}$$

It is convenient to measure the energy in rydbergs, and to measure lengths in units of the Bohr radius a_0. We define r_s, the dimensionless interparticle spacing, by the equation

$$\frac{4\pi r_s^3 a_0^3}{3} = \frac{1}{N}. \tag{5.8}$$

The Fermi momentum, p_F, is related to r_s by

$$p_F = \frac{1}{\alpha r_s a_0}, \tag{5.9}$$

where

$$\alpha = \left(\frac{4}{9\pi}\right)^{1/3} \cong 0.521. \tag{5.10}$$

On carrying out the integration in Eq. (5.7), and making use of the above definitions, we find

$$\frac{E_0^{\text{HFA}}}{N} = \frac{3}{5}\frac{1}{\alpha^2 r_s^2} - \left(\frac{3}{2\pi}\right)\frac{1}{\alpha r_s} = \frac{2.21}{r_s^2} - \frac{0.916}{r_s} \text{ ryd.} \tag{5.11}$$

* For an excellent discussion of the Hartree–Fock approximation, see, for example, Seitz (1940).

The exchange energy is seen to be negative; it is comparable in size to the kinetic energy for $r_s \cong 1$.

PAIR CORRELATION FUNCTION AND DENSITY-FLUCTUATION EXCITATION SPECTRUM

Further insight into the microscopic behavior of the electron liquid, as calculated in the HFA, is obtained by considering the static pair correlation function. This quantity is simply related to the static form factor, $S_{\mathbf{q}}$, introduced in Chapter 2. We consider first the instantaneous density-density correlation function, $S(\mathbf{r})$, defined by

$$S(\mathbf{r}) = (1/N)\langle\Psi_0|\rho(\mathbf{r}+\mathbf{r}')\rho(\mathbf{r}')|\Psi_0\rangle, \tag{5.12}$$

where Ψ_0 is the exact ground state wave function (the particle density operators are evaluated at the same time). For a translationally invariant system, the correlation function on the right-hand side of Eq. (5.12) is independent of $\mathbf{r}'$, a fact we have assumed in writing S as a function of $\mathbf{r}$ alone. For a system of point particles, we have

$$\rho(\mathbf{r}) = \sum_i \delta(\mathbf{r}-\mathbf{r}_i). \tag{5.13}$$

Assuming that the system is enclosed in a unit volume, we may write

$$\begin{aligned} S(\mathbf{r}) &= (1/N)\left\langle\Psi_0\left|\sum_{ij}\delta(\mathbf{r}+\mathbf{r}'-\mathbf{r}_i)\delta(\mathbf{r}'-\mathbf{r}_j)\right|\Psi_0\right\rangle \\ &= (1/N)\left\langle\Psi_0\left|\sum_{ij}\delta(\mathbf{r}+\mathbf{r}_i-\mathbf{r}_j)\right|\Psi_0\right\rangle \\ &= \delta(\mathbf{r}) + (1/N)\left\langle\Psi_0\left|\sum_{i\neq j}\delta(\mathbf{r}+\mathbf{r}_i-\mathbf{r}_j)\right|\Psi_0\right\rangle \\ &= \delta(\mathbf{r}) + (N-1)g(\mathbf{r}), \end{aligned} \tag{5.14}$$

where

$$g(\mathbf{r}) = \frac{1}{N(N-1)}\left\langle\Psi_0\left|\sum_{i\neq j}\delta(\mathbf{r}+\mathbf{r}_i-\mathbf{r}_j)\right|\Psi_0\right\rangle \tag{5.15}$$

is the *pair distribution function*. $g(\mathbf{r})$ gives the probability that if one particle is observed at some point $\mathbf{R}$, another particle will be found at point $\mathbf{R}+\mathbf{r}$. What has been done in Eq. (5.14) is to separate out of $S(\mathbf{r})$ the "auto-correlations" (that part associated with the correlation of a particle with itself); what is left offers a direct measure of the correlation between pairs of particles.

It is straightforward to show that $S(\mathbf{r})$ is the Fourier transform (in space) of the static form factor, $S_{\mathbf{q}}$;

$$S_{\mathbf{q}} = \int d\mathbf{r}\, S(\mathbf{r})e^{-i\mathbf{q}\cdot\mathbf{r}} = (1/N)\langle\Psi_0|\rho_{\mathbf{q}}^{+}\rho_{\mathbf{q}}|\Psi_0\rangle. \tag{5.16}$$

It suffices therefore to calculate $S_{\mathbf{q}}$, the mean square density fluctuation, in order to determine the pair distribution function, $g(\mathbf{r})$. It is frequently convenient to calculate first the dynamic form factor, $S(\mathbf{q}, \omega)$, and then make use of the relation

$$S_{\mathbf{q}} = (1/N) \int_0^\infty d\omega\, S(\mathbf{q}, \omega). \tag{5.17}$$

In the HFA, $S(\mathbf{q}, \omega)$ is given by its value for a noninteracting fermion system, $S^\circ(\mathbf{q}, \omega)$. We have discussed $S^\circ(\mathbf{q}, \omega)$ in some detail in Chapter 2, where we found that it consisted entirely in single-pair excitations, of energy

$$\omega_{\mathbf{pq}}^\circ = \epsilon_{\mathbf{p}+\mathbf{q}}^\circ - \epsilon_{\mathbf{p}}^\circ. \tag{5.18}$$

Thus

$$S^{\mathrm{HFA}}(\mathbf{q}, \omega) = S^\circ(\mathbf{q}, \omega) = \sum_{\mathbf{p}\sigma} n_{\mathbf{p}\sigma}^\circ(1 - n_{\mathbf{p}+\mathbf{q}\sigma}^\circ)\delta(\omega - \omega_{\mathbf{pq}}^\circ). \tag{5.19}$$

From Eq. (5.19), we obtain

$$S_{\mathbf{q}}^{\mathrm{HFA}} = S_{\mathbf{q}}^\circ = \begin{cases} N, & \mathbf{q} \equiv 0, \\ (3q/4p_F) - q^3/16p_F^3, & 0 < q \leqq 2p_F, \\ 1, & q \geqq 2p_F \end{cases} \tag{5.20}$$

[see Eq. (2.96)]. On making use of Eqs. (5.20), (5.16), and (5.14), one readily finds that the pair distribution function is given by

$$g^{\mathrm{HFA}}(\mathbf{r}) = 1 - \frac{9}{2}\left[\frac{\sin p_F r - p_F r \cos p_F r}{p_F^3 r^3}\right]^2. \tag{5.21}$$

A plot of $g^{\mathrm{HFA}}(\mathbf{r})$ is given in Fig. 5.1.

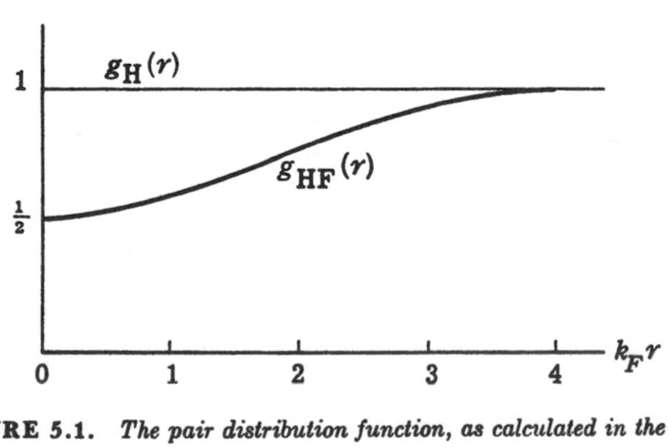

FIGURE 5.1. *The pair distribution function, as calculated in the HFA.*

If there were *no* correlations between particle positions, $g(\mathbf{r})$ would be unity. In the present case, the departure of $g^{\mathrm{HFA}}(\mathbf{r})$ from unity may be traced directly to the Pauli principle correlations which are built into the ground state wave function, $|0\rangle$. A glance at Eq. (5.19) shows that there are no correlations between particles of antiparallel spin; the correlations present in $g(\mathbf{r})$ are those for particles of parallel spin. Let us write $g(\mathbf{r})$ in the form

$$g(\mathbf{r}) = \tfrac{1}{2}[g_{\uparrow\uparrow}(\mathbf{r}) + g_{\uparrow\downarrow}(\mathbf{r})],$$

where $g_{\uparrow\uparrow}$ and $g_{\uparrow\downarrow}$ are the correlation functions for pairs of respectively parallel and antiparallel spins. In the HFA, we have $g_{\uparrow\downarrow}^{\text{HFA}}(\mathbf{r}) = 1$. It is easily verified that $g_{\uparrow\uparrow}^{\text{HFA}}(0) = 0$: because of the Pauli principle, two particles with the same spin cannot be found at the same point. One often speaks of an "exchange hole" around a given electron, corresponding to that region in space in which one is not likely to find another electron with the same spin. From Eq. (5.21) and Fig. 5.1 we see that the exchange hole is roughly of radius $p_F^{-1} = \alpha r_s a_0$.

It is the Pauli principle correlations that are responsible for the exchange energy in Eq. (5.11). Since the interaction between electrons is repulsive, any correlations that tend to keep the electrons apart will reduce the energy of the system. It is instructive to write the expectation value of the potential energy in the ground state in terms of $S_\mathbf{q}$; one sees readily that

$$\left\langle \Psi_0 \middle| \sum_{\mathbf{q}\mathbf{p}\mathbf{p}'\sigma\sigma'} (V_\mathbf{q}/2) c^+_{\mathbf{p}+\mathbf{q}\sigma} c^+_{\mathbf{p}'-\mathbf{q}\sigma'} c_{\mathbf{p}'\sigma'} c_{\mathbf{p}\sigma} \middle| \Psi_0 \right\rangle = \sum_\mathbf{q} V_\mathbf{q} N(S_\mathbf{q} - 1)/2. \quad (5.22)$$

The reader may verify that upon substituting Eq. (5.20) into (5.22), one obtains the exchange energy given in Eq. (5.11).

QUASIPARTICLE SPECTRUM

We may readily calculate quasiparticle properties in the HFA by using Eq. (5.6) to determine the appropriate Landau theory parameters. We have

$$\epsilon_{\mathbf{p}\sigma}^{\text{HFA}} = \frac{\partial E_0^{\text{HFA}}}{\partial n_{\mathbf{p}\sigma}} = \epsilon^0_{\mathbf{p}\sigma} + \sum_{\mathbf{p}'\sigma'} n^0_{\mathbf{p}'\sigma'} V_0 - \sum_{\mathbf{p}'\sigma'} n^0_{\mathbf{p}'\sigma'} V_{\mathbf{p}'-\mathbf{p}} \delta_{\sigma\sigma'}, \quad (5.23)$$

$$f^{\text{HFA}}_{\mathbf{p}\sigma,\mathbf{p}'\sigma'} = \frac{\partial^2 E_0^{\text{HFA}}}{\partial n_{\mathbf{p}\sigma} \partial n_{\mathbf{p}'\sigma'}} = V_0 - V_{\mathbf{p}'-\mathbf{p}} \delta_{\sigma,\sigma'}. \quad (5.24)$$

As an example, let us consider the electron liquid, for which the "Hartree" term V_0 is missing. For quasiparticles on the Fermi surface, $f_{\mathbf{p}\mathbf{p}'}$ is a function only of $x = \mathbf{p} \cdot \mathbf{p}'/p_F^2$. We have

$$f^{\text{HFA}}_{\mathbf{p}\sigma,\mathbf{p}'\sigma'}(x) = \frac{-2\pi e^2}{p_F^2(1 - x)} \delta_{\sigma\sigma'} = f_{\text{HFA}}(x). \quad (5.25)$$

The spin symmetric and antisymmetric parts of $f_{\rm HFA}(x)$ are thus equal, being given by [cf. Eq. (1.20)]

$$f^s_{\rm HFA}(x) = f^a_{\rm HFA}(x) = \frac{f_{\rm HFA}}{2} = \frac{-\pi e^2}{p_F{}^2(1-x)}. \qquad (5.26)$$

From Eqs. (1.22), (1.43), (1.58), (1.67), and (1.100) it is straightforward to derive the following exact expressions for the compressibility, spin susceptibility, and specific heat (see Problem 5.1):

$$\frac{\kappa^\circ}{\kappa} = \frac{s^2}{s_0{}^2} = \frac{1+F_0{}^s}{1+F_1{}^s/3} = 1 + N^\circ(0)\int_{-1}^{1} dx\, f^s(x)(1-x), \qquad (5.27)$$

$$\frac{\chi_P{}^\circ}{\chi_P} = \frac{1+F_0{}^a}{1+F_1{}^s/3} = 1 + N^\circ(0)\int_{-1}^{1} dx\, [f^a(x) - xf^s(x)], \qquad (5.28)$$

$$\frac{C_0}{C} = \frac{m}{m^*} = \frac{1}{1+F_1{}^s/3} = 1 - N^\circ(0)\int_{-1}^{1} dx\, xf^s(x), \qquad (5.29)$$

where $N^\circ(0)$ is the density of states for *noninteracting* particles of *one* kind of spin,

$$N^\circ(0) = \frac{3}{4}\frac{N}{\epsilon_F{}^\circ} = \frac{p_F m}{2\pi^2}. \qquad (5.30)$$

On making use of Eqs. (5.26) and (5.30), we find from Eqs. (5.27) and (5.28) that

$$\frac{s^2_{\rm HFA}}{s_0{}^2} = \frac{\chi_P{}^\circ}{\chi_P^{\rm HFA}} = 1 - \frac{\alpha r_s}{\pi}, \qquad (5.31)$$

while the specific heat displays a *logarithmic divergence*.

The source of the divergence is the long range of Coulomb interactions, which leads to a vanishing density of states in the HFA. This is the result we have mentioned in Section 3.1; it represents a failure of the HFA, and shows as well that one must take care in applying the Landau theory in the case of Coulomb interactions. $F_0{}^a$ and $F_0{}^s$ are likewise logarithmically divergent; the divergence does not, however, appear when one forms the combinations appropriate to the compressibility and spin susceptibility. This is as it should be, since these latter quantities may in fact be obtained directly from the ground state energy, which is itself well behaved.

Were the divergence of $F_1{}^s$ real, it would lead to a specific heat which varies as $T/\ln T$, rather than T, as Bardeen (1936) first showed. It is, of course, not real; when the higher-order terms in the perturbation series expansion are properly taken into account, they act to screen the quasiparticle interaction: the divergences are thereby removed.

It is the long range of the Coulomb interaction that renders subtle the systematic perturbation-theoretic description of the electron liquid. One cannot simply write down the second-order (in V_q) terms for the ground state energy and quasiparticle scattering amplitude. Both exhibit divergences which show clearly that second-order terms can in no sense be regarded as a "small" correction to the first. We shall see that the lowest-order, nontrivial, consistent approximation for the electron gas is obtained in the random phase approximation, hereafter known as the RPA.

5.2. RANDOM PHASE APPROXIMATION

The development, frequent independent rediscovery, and gradual appreciation of the RPA for the electron gas over a ten-year period (1950-1960) offers a useful object lesson to the theoretical physicist. First, it illustrates the splendid variety of ways that can be developed for saying the same thing. Second, it suggests the usefulness of learning more than one "language" of theoretical physics, and of attempting the reconciliation of seemingly different, but obviously related, results.

The RPA was developed in the course of studies of screening and collective behavior of an electron liquid. In the early work [Pines (1950), Pines and Bohm (1952), Bohm and Pines (1953)], primary emphasis was placed on the explicit introduction of collective coordinates (which described the plasmons) for this purpose.* The approximation acquired its name on the basis of the physical argument that under suitable circumstances a sum of exponential terms with randomly varying phases could be neglected compared to N. Thus the approximation

$$\rho_{\mathbf{k}-\mathbf{q}} = \sum_i e^{i(\mathbf{q}-\mathbf{k})\cdot\mathbf{x}_i} \cong N\delta_{\mathbf{k},\mathbf{q}}$$

was frequently made when $\rho_{\mathbf{k}-\mathbf{q}}$ appeared as a multiplicative factor in the basic equations of the theory. The resulting calculations of the ground state energy and quasiparticle properties led to a considerable improvement over the HFA [(Pines (1953)]; however, the theory was somewhat inelegant, since the introduction of extra collective coordinates meant that collective modes were treated on a quite different basis from quasiparticle excitations. As a result, a number of people questioned the validity of the RPA.

* For a review of the collective coordinate theory, see Pines (1955); a brief summary may be found in Pines (1963), Chapter 3.

During this time, several alternative derivations of the basic plasmon dispersion relation were developed, including the use of a time-dependent Hartree approximation [(Pines (1950)] and the approximate solution of the Heisenberg equations of motion for particle-hole pairs [Bohm and Pines (1953)]. A derivation based on a quantum kinetic equation was given by Klimontovitch and Silin (1952), while Lindhard (1954) calculated the dielectric function in the RPA using a self-consistent field method. These approaches appeared interesting and suggestive; however, it was not clear how one could apply them to other than a discussion of plasmon behavior.

The next major development came when Gell-Mann and Brueckner carried out a detailed calculation of the ground state energy and specific heat in an approximation which proved to be equivalent to the RPA [Gell-Mann and Brueckner (1957), Gell-Mann (1957)]. They showed that a summation of the most divergent terms (one from each order of V) in the perturbation-theoretic expansion led to a convergent result, which was valid in the high-density limit. [A similar calculation had earlier been carried out by Macke (1950), with less conclusive results.] Shortly thereafter Sawada, Brout, and their collaborators developed anew the equation-of-motion approach and showed that in the RPA it led to both plasma oscillations (which were nowhere evident in the Gell-Mann, Brueckner theory) and the Gell-Mann, Brueckner result for the ground state energy [Sawada, Brueckner, Fukuda, and Brout (1957)]. Independently, Hubbard (1957) used field-theoretic methods to establish the desired connection between the collective coordinate approach and the perturbation-theoretic summation.* Subsequently, the relationship between the equation-of-motion approach, the quantum kinetic equation, the calculation of the dielectric function, and the Landau Fermi liquid theory was established by Ehrenreich and Cohen (1959) and Goldstone and Gottfried (1959).

We shall not attempt to enter on the details of all these different derivations, or the relation between them. Our approach will rather be to derive the RPA expression for the dielectric function, and then show how it may be used to determine all the physical properties of the electron liquid within the RPA. After the preceding two chapters, it is clear that the dielectric function offers a natural vehicle for both the development of the theory and a discussion of the relationship between some of the many "apparently" different approaches to the problem.

* The connection was not obvious because Gell-Mann and Brueckner included certain terms which had been neglected in the work of Bohm and Pines. An explicit proof that when all appropriate terms are included both methods lead to the same result was given by the authors [Nozières and Pines (1958a)].

DERIVATIONS AND NATURE OF THE RPA

We give first a simple heuristic derivation of the dielectric function in the RPA, one which shows clearly the physical nature of the approximation. We have defined, in Section 4.1, a screened density-density response function, which measures the density fluctuations induced by a *screened* external field. We have further shown that the dielectric function may be expressed simply in terms of this response function, according to

$$\epsilon(\mathbf{q}, \omega) = 1 - \frac{4\pi e^2}{q^2} \chi_{sc}(\mathbf{q}, \omega). \tag{5.32}$$

As we have remarked, $\chi_{sc}(\mathbf{q}, \omega)$ must be calculated with some care, in order to avoid counting interactions twice. However, this scarcely need trouble us if we approximate $\chi_{sc}(\mathbf{q}, \omega)$ by its value for a noninteracting electron liquid, $\chi^{\circ}(\mathbf{q}, \omega)$. Doing so, we obtain the dielectric function in the RPA,

$$\begin{aligned} \epsilon_{RPA}(\mathbf{q}, \omega) &= 1 - \frac{4\pi e^2}{q^2} \chi^{\circ}(\mathbf{q}, \omega) \\ &= 1 - \frac{4\pi e^2}{q^2} \sum_{\mathbf{p}\sigma} \frac{n^{\circ}_{\mathbf{p}\sigma}(1 - n^{\circ}_{\mathbf{p}+\mathbf{q}\sigma})2\omega^{\circ}_{\mathbf{pq}}}{(\omega + i\eta)^2 - (\omega^{\circ}_{\mathbf{pq}})^2}. \end{aligned} \tag{5.33}$$

The RPA thus amounts to treating the long-range part of the Coulomb interaction as an "external" polarization field, in much the same way as in the Landau–Silin approach. This is done for all wavelengths; the response to the screened field, that is, the polarizability, is assumed to be that appropriate to a gas of noninteracting particles.

It is instructive to compare the RPA and HFA calculations of $\epsilon(\mathbf{q}, \omega)$. The latter is obtained by substituting for $\chi(\mathbf{q}, \omega)$, in Eq. (4.6), its value for the noninteracting system,

$$\frac{1}{\epsilon_{HFA}(\mathbf{q}, \omega)} = 1 + \frac{4\pi e^2}{q^2} \chi^{\circ}(\mathbf{q}, \omega). \tag{5.34}$$

If we formally regard $\chi^{\circ}(\mathbf{q}, \omega)$ as small, the RPA result may be written as

$$\begin{aligned} \frac{1}{\epsilon_{RPA}(\mathbf{q}, \omega)} = \frac{1}{1 - (4\pi e^2/q^2)\chi^{\circ}(\mathbf{q}, \omega)} &= 1 + \frac{4\pi e^2}{q^2} \chi^{\circ}(\mathbf{q}, \omega) \\ &+ \left\{\frac{4\pi e^2}{q^2} \chi^{\circ}(\mathbf{q}, \omega)\right\}^2 + \cdots. \end{aligned} \tag{5.35}$$

In this form, we may easily suggest the relationship between the above approach and the perturbation-theoretic summation carried out by Gell-Mann and Brueckner. The first term in the series on the right-hand side of Eq. (5.35) yields the HFA expression, (5.34); the higher-order terms are just those (most divergent) terms kept by Gell-Mann and Brueckner in their calculations of the ground state energy and specific heat. If one regards the series as summable, one finds a well-behaved, convergent result.

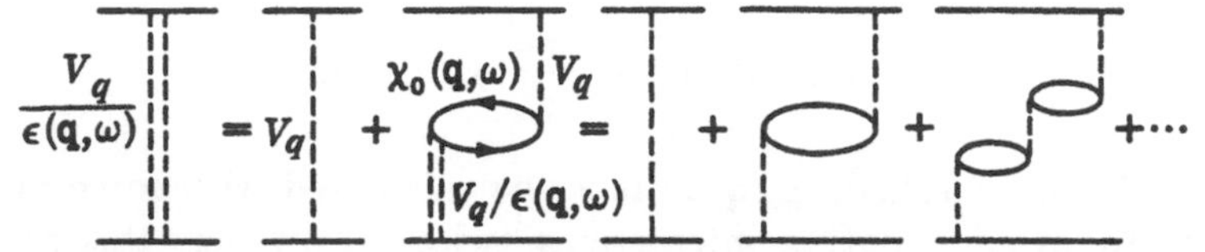

FIGURE 5.2. *Summation of polarization diagrams in the RPA.*

The same series summation is easily carried out with the aid of Feynman diagrams in time-dependent perturbation theory [Hubbard (1957), DuBois (1959)]. The appropriate pictures are shown in Fig. 5.2. The dotted line is the bare Coulomb interaction, $V_{\mathbf{q}}$. The double-dotted line is that interaction screened by the polarization of the medium, $V_{\mathbf{q}}/\epsilon(\mathbf{q}, \omega)$. The bubble represents the lowest-order polarization process, involving the excitation and subsequent de-excitation of a particle-hole pair. What is represented in Fig. 5.2 is the algebraic Dyson equation,

$$\frac{V_{\mathbf{q}}}{\epsilon(\mathbf{q}, \omega)} = V_{\mathbf{q}} + \frac{4\pi e^2}{q^2} \frac{\chi^0(\mathbf{q}, \omega)}{\epsilon(\mathbf{q}, \omega)} V_{\mathbf{q}}, \tag{5.36}$$

which sums the series (5.35). We note that the Dyson equation serves to sum automatically a given set of polarization diagrams; it is equivalent to replacing the external potential, $\varphi_{\text{ext}}(\mathbf{q}, \omega)$, by the screened potential, $\varphi = \varphi_{\text{ext}}/\epsilon$. For the benefit of those familiar with field-theoretic techniques, we further remark that the sum of *all* "irreducible" polarization diagrams (those diagrams that cannot be split into two polarization diagrams by cutting an interaction line) simply yields the exact "screened" polarizability, $(4\pi e^2/q^2)\chi_{\text{sc}}(\mathbf{q}, \omega)$.

We consider next an explicit microscopic derivation of $\epsilon_{\text{RPA}}(\mathbf{q}, \omega)$. The derivation we present is essentially that of Ehrenreich and Cohen (1959), and represents an application of the equation-of-motion method [Bohm and Pines (1953)] to the problem at hand. We study the motion of an electron-hole pair of momentum $-\mathbf{q}$,

$$\rho_{\mathbf{pq}\sigma} = c^{+}_{\mathbf{p}\sigma} c_{\mathbf{p}+\mathbf{q}\sigma}, \tag{5.37}$$

under the influence of the Hamiltonian (5.1) plus an external test field, of wave vector $\mathbf{q}$, frequency ω, specified by the forcing term,

$$\lim_{\eta\to 0} \{e\rho_{\mathbf{q}}{}^{+}\varphi_{\text{ext}}(\mathbf{q}, \omega)e^{-i\omega t} + \text{c.c.}\}e^{\eta t}. \tag{5.38}$$

We wish to determine the mean value of

$$\rho_{\mathbf{q}} = \sum_{\mathbf{p}\sigma} \rho_{\mathbf{pq}\sigma} \tag{5.39}$$

in the presence of the external field. We do so by writing down the explicit quantum-mechanical equation of motion,

$$i\dot{\rho}_{\mathbf{pq}} = [\rho_{\mathbf{pq}}, H], \tag{5.40}$$

and then keeping only those terms which will correspond to making the RPA. [We suppress the spin index in Eq. (5.40) and the equations which follow.]

Let us first neglect the electron-electron interaction; by so doing, we expect to find $\epsilon_{\text{HFA}}(\mathbf{q}, \omega)$. On evaluating the commutator of $\rho_{\mathbf{pq}}$ with the kinetic energy, and with the external field, we obtain

$$\begin{aligned} i\dot{\rho}_{\mathbf{pq}} = \omega^{\circ}_{\mathbf{pq}}\rho_{\mathbf{pq}} &+ \{c_{\mathbf{p}}{}^{+}c_{\mathbf{p}} - c^{+}_{\mathbf{p}+\mathbf{q}}c_{\mathbf{p}+\mathbf{q}}\}e\varphi_{\text{ext}}e^{-i\omega t}e^{\eta t} \\ &+ \{c_{\mathbf{p}}{}^{+}c_{\mathbf{p}+2\mathbf{q}} - c^{+}_{\mathbf{p}-\mathbf{q}}c_{\mathbf{p}+\mathbf{q}}\}e\varphi^{+}_{\text{ext}}e^{\eta t}e^{i\omega t}. \end{aligned} \tag{5.41}$$

We solve this equation by taking the expectation value in the ground state of both sides; since we are interested only in first-order terms in $\varphi_{\text{ext}}(\mathbf{q}, \omega)$, we make the replacement

$$\langle c_{\mathbf{p}}{}^{+}c_{\mathbf{p}}\rangle = n_{\mathbf{p}}{}^{\circ}, \qquad \langle c_{\mathbf{p}}{}^{+}c_{\mathbf{p}+2\mathbf{q}}\rangle = 0, \tag{5.42}$$

on the right-hand side of Eq. (5.41). The term proportional to φ^{+} drops out, and we find

$$\langle\rho_{\mathbf{q}}\rangle = \sum_{\mathbf{p}} \langle\rho_{\mathbf{pq}}\rangle = \sum_{\mathbf{p}} \frac{n_{\mathbf{p}}{}^{\circ} - n^{\circ}_{\mathbf{p}+\mathbf{q}}}{\omega - \omega^{\circ}_{\mathbf{pq}} + i\eta} e\varphi_{\text{ext}}. \tag{5.43}$$

The density-density response function is accordingly

$$\chi^{\circ}(\mathbf{q}, \omega) = \sum_{\mathbf{p}} \frac{n_{\mathbf{p}}{}^{\circ} - n^{\circ}_{\mathbf{p}+\mathbf{q}}}{\omega - \omega^{\circ}_{\mathbf{pq}} + i\eta}, \tag{5.44a}$$

a result which is equivalent to the standard HFA expression, obtained from Eq. (5.33):

$$\chi^{\circ}(\mathbf{q}, \omega) = \sum_{\mathbf{p}} n_{\mathbf{p}}{}^{\circ}(1 - n^{\circ}_{\mathbf{p}+\mathbf{q}}) \left\{\frac{1}{\omega - \omega^{\circ}_{\mathbf{pq}} + i\eta} - \frac{1}{\omega + \omega^{\circ}_{\mathbf{pq}} + i\eta}\right\} \tag{5.44b}$$

[one passes from Eq. (5.44a) to (5.44b) by making the dummy index transformation $\mathbf{p} \to -\mathbf{p} - \mathbf{q}$, $\omega^{\circ}_{\mathbf{pq}} \to -\omega^{\circ}_{\mathbf{pq}}$, on the last term of Eq. (5.44b)].

We next take into account electron interaction, by considering as well the commutator of $\rho_{\mathbf{pq}}$ with the potential energy term in Eq. (5.1). The following term is added on the right-hand side of Eq. (5.41):

$$\sum_{\mathbf{k}} (V_{\mathbf{k}}/2)\{\rho_{\mathbf{k}}(c_{\mathbf{p}}^{+}c_{\mathbf{p}+\mathbf{q}-\mathbf{k}} - c^{+}_{\mathbf{p}+\mathbf{k}}c_{\mathbf{p}+\mathbf{q}}) + (c_{\mathbf{p}}^{+}c_{\mathbf{p}+\mathbf{q}-\mathbf{k}} - c^{+}_{\mathbf{p}+\mathbf{k}}c_{\mathbf{p}+\mathbf{q}})\rho_{\mathbf{k}}\}. \tag{5.45}$$

The RPA consists, first of all, in keeping only a single term in this summation, that with $\mathbf{k} = \mathbf{q}$. On retaining this term, and taking expectation values in the ground state, the added term may, with the aid of Eq. (5.42), be factored as

$$V_{\mathbf{q}}\langle\rho_{\mathbf{q}}(c_{\mathbf{p}}^{+}c_{\mathbf{p}} - c^{+}_{\mathbf{p}+\mathbf{q}}c_{\mathbf{p}+\mathbf{q}})\rangle = V_{\mathbf{q}}\langle\rho_{\mathbf{q}}\rangle(n_{\mathbf{p}}^{\circ} - n^{\circ}_{\mathbf{p}+\mathbf{q}}). \tag{5.46}$$

Equation (5.43) is then replaced by

$$\langle\rho_{\mathbf{q}}\rangle = \chi^{\circ}(\mathbf{q}, \omega)\left\{e\varphi_{\text{ext}} + \frac{4\pi e^2}{q^2}\langle\rho_{\mathbf{q}}\rangle\right\} = \chi^{\circ}(\mathbf{q}, \omega)e\varphi(\mathbf{q}, \omega). \tag{5.47}$$

From Eq. (5.47) one finds at once the RPA result (5.33) for the dielectric function.

The RPA procedure takes into account electron interaction only to the extent required to produce the screening field, $\langle\rho_{\mathbf{q}}\rangle$. The response to the screened field is measured by χ°. We have thus justified the heuristic derivation starting from Eq. (5.32). We note that the RPA represents only an approximate linearization of the equation of motion for $\rho_{\mathbf{pq}}$; the results of a more systematic linearization of the terms present in Eq. (5.45) will be given in Section 5.5.

The equation-of-motion procedure we have sketched above is closely related to the Landau–Silin kinetic equation considered in Chapter 3. To establish the connection, we make the identification

$$\langle\rho_{\mathbf{pq}}\rangle = \delta n_{\mathbf{p}}(\mathbf{q}, \omega). \tag{5.48}$$

The *microscopic* equation of motion in the RPA, which is obtained from Eqs. (5.41) and (5.46), may be written as

$$(\omega - \omega^{\circ}_{\mathbf{pq}})\delta n_{\mathbf{p}} + (n^{\circ}_{\mathbf{p}+\mathbf{q}} - n_{\mathbf{p}}^{\circ})\{e\varphi_{\text{ext}} + (4\pi e^2/q^2)\langle\rho_{\mathbf{q}}\rangle\} = 0. \tag{5.49}$$

If we pass to the long wavelength limit, we have

$$\lim_{\mathbf{q}\to 0} (n^{\circ}_{\mathbf{p}+\mathbf{q}} - n_{\mathbf{p}}^{\circ}) = \mathbf{q}\cdot\frac{\partial n_{\mathbf{p}}^{\circ}}{\partial \mathbf{p}} = \mathbf{q}\cdot\mathbf{v}_{\mathbf{p}}^{\circ}\frac{\partial n^{\circ}}{\partial \epsilon_{\mathbf{p}}^{\circ}} = -\mathbf{q}\cdot\mathbf{v}_{\mathbf{p}}^{\circ}\delta(\epsilon_{\mathbf{p}}^{\circ} - \mu) \tag{5.50a}$$

and

$$\lim_{\mathbf{q}\to 0} \omega_{\mathbf{pq}}^{\circ} = \epsilon_{\mathbf{p}+\mathbf{q}}^{\circ} - \epsilon_{\mathbf{p}}{}^{\circ} = \mathbf{q}\cdot\mathbf{v}_{\mathbf{p}}{}^{\circ}, \tag{5.50b}$$

so that Eq. (5.49) takes the form

$$(\omega - \mathbf{q}\cdot\mathbf{v}_{\mathbf{p}}{}^{\circ})\delta n_{\mathbf{p}} - (\mathbf{q}\cdot\mathbf{v}_{\mathbf{p}}{}^{\circ})e\varphi(\mathbf{q},\omega)\delta(\epsilon_{\mathbf{p}}{}^{\circ} - \mu) = 0. \tag{5.51}$$

This result is to be compared with the Landau–Silin transport equation, (3.45), which may be written

$$(\omega - \mathbf{q}\cdot\mathbf{v}_{\mathbf{p}})\delta n_{\mathbf{p}} - \mathbf{q}\cdot\mathbf{v}_{\mathbf{p}}\{e\varphi(\mathbf{q},\omega) + \sum_{\mathbf{p}'} f_{\mathbf{pp}'}\delta n_{\mathbf{p}'}\}\delta(\epsilon_{\mathbf{p}} - \mu) = 0. \tag{5.52}$$

One passes from Eq. (5.52) to (5.51) by letting $\epsilon_{\mathbf{p}} \to \epsilon_{\mathbf{p}}{}^{\circ}$, $f_{\mathbf{pp}'} \to 0$. Insofar as the long wavelength kinetic equation is concerned, the RPA amounts to taking into account the average polarization field, while neglecting altogether the short-range interaction between quasiparticles.

PLASMON ENERGY AND PLASMON DAMPING

As long as there is no damping, the plasmon frequency, $\omega_{\mathbf{q}}$, is determined by the dispersion relation

$$\epsilon(\mathbf{q}, \omega_{\mathbf{q}}) = 0. \tag{5.53}$$

In order to evaluate $\omega_{\mathbf{q}}$, we note that in Eq. (5.33) the Pauli principle restriction can be neglected (to see this, interchange indices according to $\mathbf{p} \to -\mathbf{p} - \mathbf{q}$); we may thus write the dispersion relation as

$$1 = \frac{4\pi e^2}{q^2} \sum_{p<p_F} \left\{ \frac{1}{\omega_{\mathbf{q}} - (\mathbf{q}\cdot\mathbf{p}/m) - (q^2/2m)} - \frac{1}{\omega_{\mathbf{q}} + (\mathbf{q}\cdot\mathbf{p}/m) + (q^2/2m)} \right\} \tag{5.54}$$

or, to use the form first derived by Bohm and Pines,

$$1 = \frac{4\pi e^2}{m} \sum_{p<p_F} \frac{1}{(\omega_{\mathbf{q}} - \mathbf{q}\cdot\mathbf{p}/m)^2 - q^4/4m^2}. \tag{5.55}$$

[To obtain Eq. (5.55) change $\mathbf{p}$ to $-\mathbf{p}$ in the last term of Eq. (5.54).] One finds readily that the plasmon energy is

$$\omega_{\mathbf{q}} \cong \omega_p \left\{ 1 + \frac{3}{10}\left(\frac{qv_F{}^{\circ}}{\omega_p}\right)^2 + \cdots \right\}. \tag{5.56}$$

As one goes to larger values of **q**, the plasmon energy increases rather slowly; the maximum single-pair energy,

$$qv_F{}^\circ + q^2/2m,$$

increases more rapidly. At a critical wave vector, q_c, it first becomes possible for a plasmon to decay into a single-pair excitation. One finds q_c by solving the equations [Ferrell (1957), Sawada *et al.* (1957)]

$$\begin{aligned} \epsilon(\mathbf{q}, \omega_{\mathbf{q}_c}) &= 0, \\ \omega_{\mathbf{q}_c} &= q_c v_F{}^\circ + q_c{}^2/2m. \end{aligned} \tag{5.57}$$

It is not overly difficult to solve for q_c since at $\omega_{\mathbf{q}_c}$ the analytic expression for ϵ_1 simplifies considerably. Detailed calculation of q_c is left as a problem for the reader; it may be verified that at metallic electron densities ($2 \lesssim r_s \lesssim 5.5$), the simple long wavelength criterion

$$q_c \cong \omega_p/v_F{}^\circ = \left(\frac{4\alpha r_s}{3\pi}\right)^{1/2} p_F \tag{5.58}$$

furnishes a satisfactory order-of-magnitude estimate. In Fig. 5.3, we give a plot of the RPA predictions for the plasmon and pair excitation spectrum for a metal of density $r_s = 4$ (sodium).

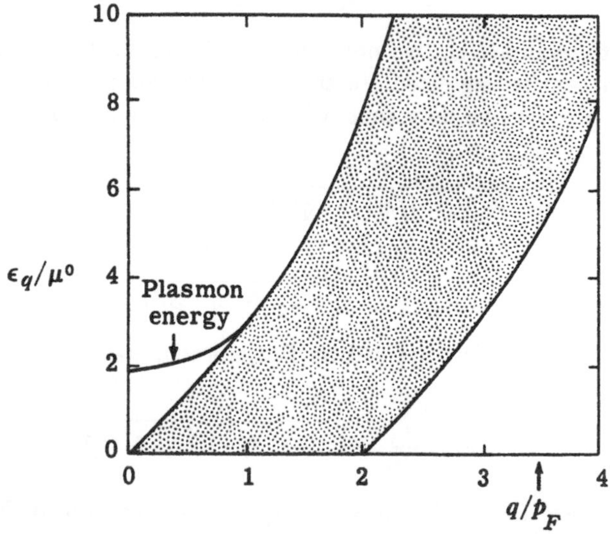

FIGURE 5.3. *E vs. p curves for density-fluctuation excitations, calculated in the RPA for an electron liquid of density $r_s = 4$. The pair excitations fall within the shaded area.*

DYNAMIC FORM FACTOR AND DIELECTRIC RESPONSE

According to Eq. (4.36), the dynamic form factor in the RPA is given by

$$S_{\mathrm{RPA}}(\mathbf{q}, \omega) = \frac{q^2}{4\pi^2 e^2} \frac{\epsilon_2(\mathbf{q}, \omega)}{\epsilon_1^2(\mathbf{q}, \omega) + \epsilon_2^2(\mathbf{q}, \omega)}, \tag{5.59}$$

where ϵ_1 and ϵ_2 are here defined as the real and imaginary parts of the RPA dielectric function,

$$\epsilon_{\mathrm{RPA}}(\mathbf{q}, \omega) = \epsilon_1(\mathbf{q}, \omega) + i\epsilon_2(\mathbf{q}, \omega). \tag{5.60}$$

S_{RPA} may also be written in the form

$$S_{\mathrm{RPA}}(\mathbf{q}, \omega) = \frac{S^{\circ}(\mathbf{q}, \omega)}{|\epsilon_{\mathrm{RPA}}(\mathbf{q}, \omega)|^2} = \sum_{\mathbf{p}} \frac{n_{\mathbf{p}}^{\circ}(1 - n_{\mathbf{p}+\mathbf{q}}^{\circ})}{|\epsilon_{\mathrm{RPA}}(\mathbf{q}, \omega_{\mathbf{pq}}^{\circ})|^2}\, \delta(\omega - \omega_{\mathbf{pq}}^{\circ}), \tag{5.61}$$

where $S^{\circ}(\mathbf{q}, \omega)$ is the dynamic form factor for the noninteracting particle system. Equation (5.61) clearly shows the dynamic screening of matrix elements in the RPA. We note that Eq. (5.61) contains only single-pair and plasmon contributions; in the RPA, the dynamic form factor and linear response functions do *not* contain multipair excitations.

Explicit expressions for ϵ_1 and ϵ_2 are found by carrying out the sum over $\mathbf{p}$ in Eq. (5.44a) or (5.44b); the calculation was first carried out by Lindhard (1954), who found

$$\begin{aligned} \epsilon_1(\mathbf{q}, \omega) = 1 + \frac{q_{FT}^2}{q^2} \Bigg\{ \frac{1}{2} + \frac{p_F}{4q} \Bigg[\left(\frac{(\omega + q^2/2m)^2}{(qv_F^{\circ})^2} - 1 \right) \ln \left| \frac{\omega - qv_F^{\circ} + q^2/2m}{\omega + qv_F^{\circ} + q^2/2m} \right| \\ - \left(\frac{(\omega - q^2/2m)^2}{(qv_F^{\circ})^2} - 1 \right) \ln \left| \frac{\omega - qv_F^{\circ} - q^2/2m}{\omega + qv_F^{\circ} - q^2/2m} \right| \Bigg] \Bigg\}, \end{aligned} \tag{5.62}$$

$$\begin{aligned} \epsilon_2(\mathbf{q}, \omega) &= \frac{\pi}{2} \frac{q_{FT}^2}{q^2} \frac{\omega}{qv_F^{\circ}}, \qquad 0 \le \omega \le qv_F^{\circ} - q^2/2m, \\ &= \frac{\pi}{2} \frac{q_{FT}^2}{q^2} \frac{p_F}{q} \left\{ 1 - \frac{(\omega - q^2/2m)^2}{(qv_F^{\circ})^2} \right\}, \qquad -\frac{q^2}{2m} \le \omega - qv_F^{\circ} \le \frac{q^2}{2m}, \\ &= 0, \qquad \omega \ge qv_F^{\circ} + q^2/2m. \end{aligned} \tag{5.63}$$

where q_{FT} is the Fermi–Thomas screening wave vector, given by

$$q_{FT}^2 = 3\omega_p^2/(v_F^{\circ})^2.$$

These expressions are valid for arbitrary wave vector and frequency; they completely determine $S_{\mathrm{RPA}}(\mathbf{q}, \omega)$.

In the long wavelength limit ϵ_1 takes on the expected values:

$$\epsilon_1(0, \omega) = 1 - \omega_p^2/\omega^2, \tag{5.64}$$

$$\lim_{\mathbf{q}\to 0} \epsilon_1(\mathbf{q}, 0) = 1 + q_{FT}^2/q^2 = 1 + 3\omega_p^2/(qv_F^{\circ})^2. \tag{5.65}$$

Equation (5.64) is, as we have seen, an exact property of the electron liquid. The result, (5.65), is that found in the Fermi–Thomas approximation, and is to be contrasted with the exact result, (4.27):

$$\lim_{\mathbf{q}\to 0} \epsilon_1(\mathbf{q}, 0) = 1 + \frac{\omega_p^2}{q^2 s^2}. \tag{5.66}$$

We note that the isothermal sound velocity which appears in Eq. (5.65) represents a lower order of approximation to s than that calculated by differentiating the ground state energy. Such an apparent "inconsistency," which is characteristic of all perturbation-theoretic treatments, arises because the ground state energy contains an extra $V_{\mathbf{q}}$. A similar "inconsistency" occurs in the HFA for a neutral system, where one finds

$$\lim_{\mathbf{q}\to 0} \chi_{\mathrm{HFA}}(\mathbf{q}, 0) = \chi^{\circ}(\mathbf{q}, 0) = -\frac{N}{m s_{\circ}^2} = -\frac{3N}{m(v_F^{\circ})^2}, \tag{5.67}$$

while

$$s_{\mathrm{HFA}}^2 = \frac{(v_F^{\circ})^2}{3}\{1 + (F_{\circ}^{s})_{\mathrm{HFA}}\}\frac{m}{m_{\mathrm{HFA}}^*}. \tag{5.68}$$

We consider now the microscopic behavior of $S_{\mathrm{RPA}}(\mathbf{q}, \omega)$ and $\epsilon_{\mathrm{RPA}}(\mathbf{q}, \omega)$. For calculational purposes it is useful to express ϵ_1 and ϵ_2 in terms of the dimensionless quantities x, y, F_1, and F_2, according to

$$x = \frac{q}{p_F}, \qquad y = \frac{\omega}{\epsilon_F^{\circ}}, \tag{5.69}$$

$$\epsilon_1(\mathbf{q}, \omega) = 1 + \xi F_1(x, y), \tag{5.70}$$

$$\epsilon_2(\mathbf{q}, \omega) = \xi F_2(x, y), \tag{5.71}$$

$$\xi = \frac{e^2 p_F}{\pi \epsilon_F^{\circ}} = \frac{2\alpha r_s}{\pi} \cong 0.33 r_s. \tag{5.72}$$

F_1 and F_2 play the role of "universal" dielectric functions. In Fig. 5.4, we give plots of F_1 and F_2 for the three representative wave vectors $q = p_F/2$, $q = p_F$, and $q = 2p_F$. We note that in these units

$$S_{\mathrm{RPA}}(\mathbf{q}, \omega) = S_{\mathrm{RPA}}(x, y) = \frac{N^{\circ}(0)x^2}{2\alpha r_s}\frac{\epsilon_2}{\epsilon_1^2 + \epsilon_2^2}. \tag{5.73}$$

In order to carry out a more detailed investigation, we consider an electron liquid of density $r_s = 3$ (i.e., $\xi \approx 1$). In that case, plasmon damping begins at about

$$x_c = q_c/p_F \cong 0.82 \tag{5.74}$$

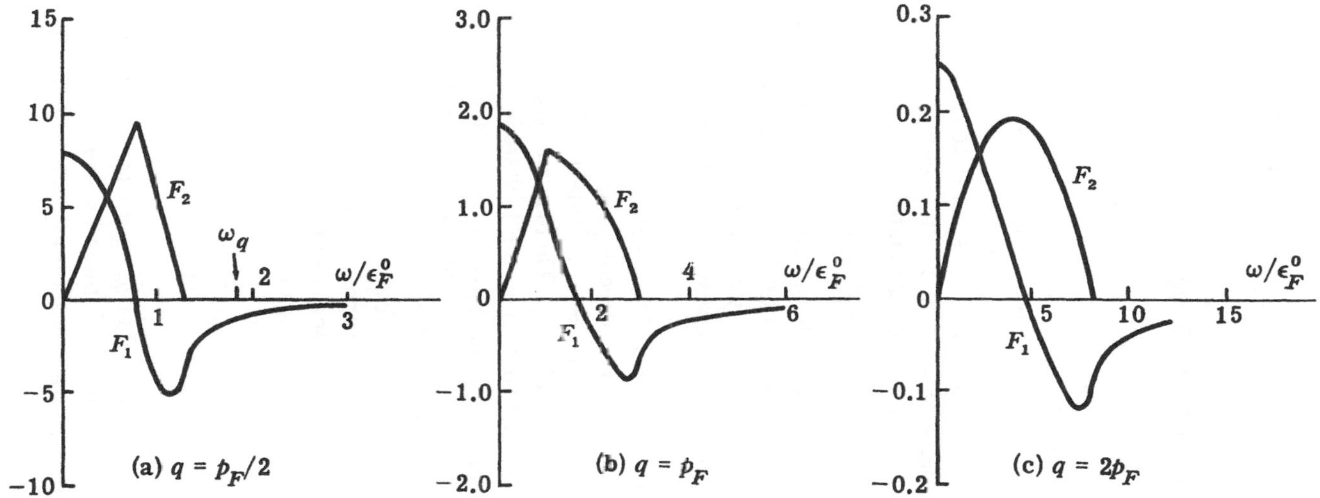

FIGURE 5.4. *Universal dielectric functions for three different wave vectors.*

and

$$x_{FT} = q_{FT}/p_F \cong 2. \tag{5.75}$$

In Fig. 5.5 we give a plot of $\epsilon_1/|\epsilon_{\text{RPA}}|^2$ and $\epsilon_2/|\epsilon_{\text{RPA}}|^2$ for the above three wave vectors. We remark on the following general features of the results shown there.

(*i*) $q = p_F/2$ Although the wavelength in question is no longer macroscopic, the behavior of the dielectric function and the dynamic form factor is little changed from the macroscopic regime. One sees clearly the almost complete screening of a low-frequency disturbance, corresponding to $\omega \lesssim qv_F{}^\circ$, $y \lesssim 1$. (Compare $\epsilon_2/|\epsilon_{\text{RPA}}|^2$ with $\epsilon_2 = F_2$.) At high frequencies such that $qv_F{}^\circ \lesssim \omega$, the response function is essentially that of a classical oscillator of frequency ω_p. One has a reversal of sign ("antiscreening") when

$$qv_F{}^\circ \lesssim \omega \lesssim \omega_p, \tag{5.76}$$

while for $\omega \gtrsim \omega_p$, one has a field enhancement. For $\omega \gg \omega_p$, the effect of particle interaction is negligible. We note that plasmons are a well-defined excitation, and that in view of the smallness of $\epsilon_2/|\epsilon_{\text{RPA}}|^2$, the plasmon pole continues to dominate the density-fluctuation excitation spectrum.

(*ii*) $q = p_F$ The dramatic low-frequency screening action has disappeared. For $\omega \lesssim qv_F{}^\circ$ the screening, as represented by $(\epsilon_1/|\epsilon_{\text{RPA}}|^2)$, is nearly independent of frequency; an external field is reduced in magnitude by a factor of roughly $\frac{1}{3}$. There is no longer "antiscreening" of a high-frequency disturbance ($\omega \gtrsim qv_F{}^\circ$); there is some enhancement in the region $\omega \cong \frac{3}{2}qv_F{}^\circ$. For the purpose of comparison, we have plotted ϵ_2 as a function of frequency. Although plasmons are now damped, the Coulomb interaction acts to shift the excitation spectrum toward high frequencies; the dynamic form factor resembles that for a screened pair spectrum on which there is superimposed a "plasmon" at frequency $\omega_q \cong \frac{3}{2}qv_F{}^\circ$. Whether one wishes to interpret the spectrum in this way, or as a modified pair excitation spectrum, is essentially a matter of taste.

(*iii*) *Very short wavelengths* $[q \cong 2p_F]$ Essentially all effects of electron interaction are unimportant; both the dynamic form factor and the dielectric response are nearly those of a noninteracting particle system.

The above results suggest that q_c represents the effective dividing line above which electron-electron interaction begins to be less important. For $q \lesssim q_c$, plasmons are not damped, and dominate the excitation spectrum. For $q \gtrsim q_c$, dielectric screening is no longer an essential feature of system behavior (although electron interaction may well affect the dynamic form factor).

It should be emphasized that the above results are not quantitatively

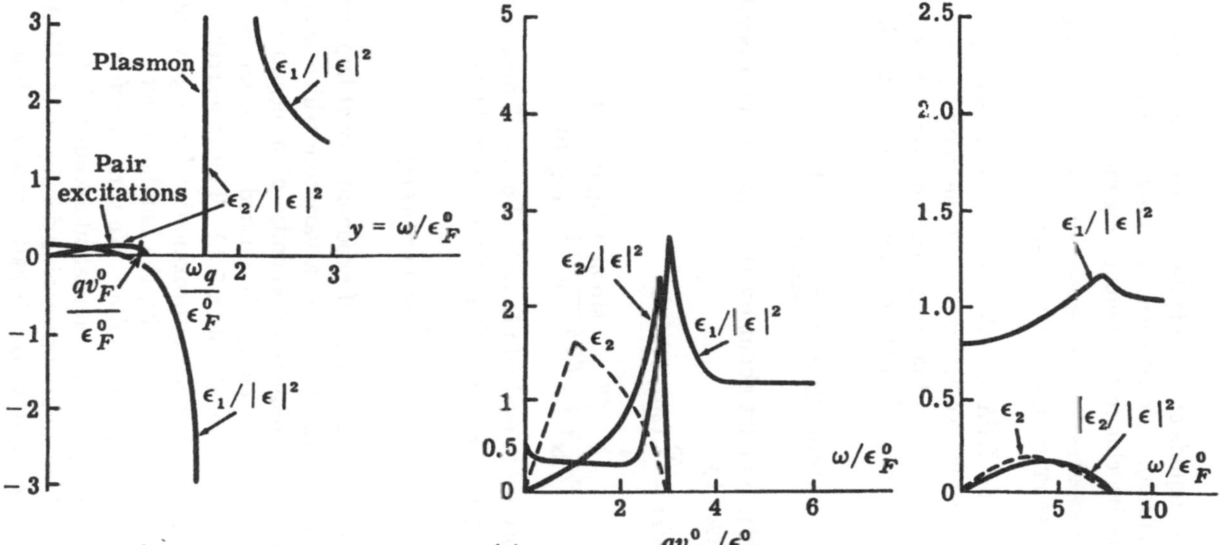

(a) $q = p_F/2$ (b) $q = p_F$ (c) $q = 2p_F$

FIGURE 5.5. *$\epsilon_1/|\epsilon|^2$ and $\epsilon_2/|\epsilon|^2$ for three different wave vectors, as calculated in the RPA.*

correct for electron liquids of metallic density (for which the RPA is not a valid approximation). Nonetheless, we may expect them to furnish a valuable qualitative clue in this density regime.

STATIC DIELECTRIC SCREENING

We consider now the RPA calculation of the screening of a static charged impurity in the electron gas. Let the impurity, of charge $-e$, be located at the origin. In that case, we find from Eq. (4.3) that the induced density fluctuation in the electron system is given by

$$\langle \rho(\mathbf{q}, 0)\rangle = -\frac{1}{\epsilon(\mathbf{q}, 0)} + 1. \tag{5.77}$$

The density change brought about by the impurity at a point $\mathbf{r}$ in the electron gas is given by

$$\begin{aligned} \delta\rho(\mathbf{r}) &= \sum_{\mathbf{q}} \langle \rho(\mathbf{q}, 0)\rangle e^{i\mathbf{q}\cdot\mathbf{r}} \\ &= \frac{1}{2\pi^2}\int_0^\infty dq\, q\,\frac{\sin qr}{r}\left\{\frac{\epsilon(\mathbf{q}, 0) - 1}{\epsilon(\mathbf{q}, 0)}\right\}. \end{aligned} \tag{5.78}$$

It is straightforward to evaluate $\delta\rho$ in the Fermi–Thomas approximation, for which $\epsilon(\mathbf{q}, 0)$ is given by Eq. (5.65). One finds

$$\delta\rho_{FT}(\mathbf{r}) = \frac{q_{FT}^2}{4\pi r}\exp\left(-q_{FT}r\right). \tag{5.79}$$

The induced screening density falls off exponentially at large values of r; for small values of r it varies as $1/r$, a variation which leads to nonphysical results in the immediate vicinity of a negatively charged impurity. (More negative charge is pushed out than exists in the vicinity of the impurity—see Problem 5.3.) This is scarcely surprising; we have used the Fermi–Thomas approximation for large values of $\mathbf{q}$ for which it is unapplicable. Obviously, the Landau theory result, (5.66), will yield no more satisfactory results. We turn therefore to the RPA, which has the virtue of yielding a consistent account of static screening.

The calculation of the induced screening density in the RPA has been carried out by Langer and Vosko (1960). One has from Eq. (5.62),

$$\epsilon_1(q, 0) = 1 + \frac{q_{FT}^2}{q^2}\left\{\frac{1}{2} + \frac{p_F}{2q}\left(1 - \frac{q^2}{4p_F^2}\right)\ln\left(\frac{q + 2p_F}{q - 2p_F}\right)\right\}. \tag{5.80}$$

On substituting Eq. (5.80) into (5.78), one finds that the resulting integral is too difficult to carry out analytically, and must be done numerically. The results obtained for an electron gas of density $r_s = 3$ are shown in Fig. 5.6. There one sees two striking differences between the RPA and the Fermi–Thomas calculation of screening: (i) The screening density at the origin is finite in the RPA. (ii) In the RPA, the screening density does not go to zero exponentially, but rather oscillates at large values of r.

The first result is gratifying. The second feature is, at first sight, quite surprising, although it had appeared in earlier calculations of Friedel (1958).

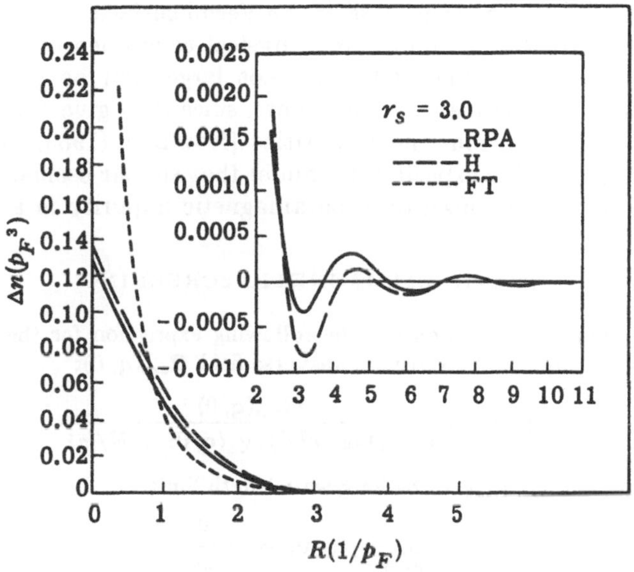

FIGURE 5.6. *Density change for $r_s = 3$ vs. distance from impurity [after Langer and Vosko (1960)]. The screening density is shown in the Fermi–Thomas (FT), RPA, and Hubbard (H) approximations.*

To see how these oscillations come about, we remark that at $q = 2p_F$, $\epsilon_{\text{RPA}}(\mathbf{q}, 0)$ is finite and continuous, but its derivative is logarithmically divergent:

$$\left(\frac{\partial\epsilon(\mathbf{q}, 0)}{\partial q}\right)_{q=2p_F} = \infty. \tag{5.81}$$

As a consequence of the singularity (5.81), $\delta\rho(\mathbf{r})$ behaves for large $\mathbf{r}$ like

$$\frac{\delta\rho_{\text{RPA}}(\mathbf{r})}{p_F{}^3} = \frac{-2\xi}{\pi(4+\xi)^2}\frac{\cos 2p_F r}{(p_F r)^3}, \tag{5.82}$$

where ξ is the dimensionless quantity (5.72). The physical origin of the singularity (5.81) is not difficult to trace. When $q < 2p_F$, one may excite an electron from one part of the Fermi surface to another; the electron-hole excitation spectrum begins at zero energy. When $q > 2p_F$, one must instead supply energy to excite an electron; the corresponding excitation spectrum begins at some finite energy, as shown in Fig. 5.3.

Such a property is in fact quite general, and not specific to the RPA. The single-pair contributions to the screened dielectric response function will always possess the singularity (5.81); moreover, we have seen in Section 2.4 that multipair excitations are negligible for low frequencies. It follows that oscillations like those in Eq. (5.82) are an exact microscopic property of normal electron liquids; they may be considered as a direct reflection of the sharpness of the Fermi surface.

The result (5.82) has important physical consequences. The long-range density oscillations, although not large, can be picked up in nuclear magnetic resonance experiments, since they give rise to electric field gradients far from the impurities [Rowland (1960); Kohn and Vosko (1960)]. One expects (and finds) that similar oscillations in the spin density are produced by a paramagnetic impurity in a metal.

STATIC DIAMAGNETIC SCREENING

We have obtained in Section 4.7 the following expression for the static diamagnetic screening of an external transverse field $\mathcal{A}_{ext}(\mathbf{q}, 0)$:

$$\mathcal{A}(\mathbf{q}, 0) = \frac{\mathcal{A}_{ext}(\mathbf{q}, 0)}{1 + (4\pi e^2/c^2q^2)\{\chi_\perp(\mathbf{q}, 0) + N/m\}}. \tag{4.195}$$

We remarked there that in the long wavelength limit

$$\lim_{\mathbf{q}\to 0} \chi_\perp(\mathbf{q}, 0) = -\frac{N}{m},$$

a result which can be derived using the Landau theory. We may now use the RPA to evaluate $\chi_\perp(\mathbf{q}, 0)$ for microscopic values of q. Since there is no dielectric screening of a transverse field, and since in the RPA the single-particle excitations are unchanged from their free particle values, $\chi_\perp(\mathbf{q}, 0)$ is the same as for a noninteracting fermion system. According to Eq. (4.167), we have

$$\chi_\perp^{\circ}(\mathbf{q}, 0) = -\sum_{\mathbf{p}} \frac{2n_{\mathbf{p}}^{\circ}(1 - n_{\mathbf{p}+\mathbf{q}}^{\circ})(\mathbf{p}\cdot\mathbf{n}_\perp)^2}{m^2\omega_{\mathbf{pq}}^{\circ}}. \tag{5.83}$$

It is straightforward to show, by a suitable interchange of indices, that Pauli principle restrictions play no role in the summation over $\mathbf{p}$ in Eq. (5.83). We may therefore write

$$\chi_\perp^{\circ}(\mathbf{q}, 0) = -\frac{2}{m}\sum_{p<p_F} \frac{(\mathbf{p}\cdot\mathbf{n}_\perp)^2}{\mathbf{p}\cdot\mathbf{q} + q^2/2}. \tag{5.84}$$

On changing the sum to an integral and carrying out the angular integrations, we obtain

$$\chi_\perp(\mathbf{q}, 0) = -\frac{N}{2m} + \frac{1}{2\pi^2 mq}\int_0^{p_F} dp\, p(p^2 - q^2/4)\ln\left(\frac{p - q/2}{p + q/2}\right). \tag{5.85}$$

The integration over p is straightforward, but lengthy: actually we are not interested so much in the exact expression as in the corrections of order q^2 to $(\chi_\perp + N/m)$. If we expand the logarithm in powers of $q/2p$, we easily find the result

$$\lim_{\mathbf{q}\to 0} \chi_\perp(\mathbf{q}, 0) = -\frac{N}{m} + \frac{N}{m}\frac{q^2}{4p_F^2}. \tag{5.86}$$

The term of order q^2 is responsible for the weak diamagnetism first calculated by Landau; on substituting Eq. (5.86) into (4.195), we obtain the following expression for the magnetic permeability, μ_M:

$$\mu_M = \lim_{\mathbf{q}\to 0} \frac{\mathcal{A}(\mathbf{q}, 0)}{\mathcal{A}_{\text{ext}}(\mathbf{q}, 0)} = \frac{1}{1 + \omega_p^2/4c^2p_F^2}.$$

The corresponding diamagnetic susceptibility is

$$\chi_D = (\mu_M - 1)/4\pi \cong -\frac{\omega_p^2}{16\pi c^2 p_F^2} = -\tfrac{1}{3}\chi_P^{\circ}, \tag{5.87}$$

where χ_P° is the paramagnetic spin susceptibility for the noninteracting Fermi gas. The simple relation (5.87) between χ_P and χ_D is only valid in the framework of the RPA. We note that diamagnetic screening is of order v^2/c^2, in accord with our dimensional argument of Section 4.7.

5.3. GROUND STATE ENERGY IN THE RPA

The calculation of the ground state energy of an electron liquid has been the object of considerable theoretical interest ever since the pioneering studies of Wigner and Seitz on the cohesive energy of metals. [Wigner and Seitz (1933, 1934), Wigner (1934)]. Wigner and Seitz showed that electron interaction plays an important role in metallic cohesion, and that a satisfactory calculation from "first principles" of the cohesive energy requires accurate knowledge of the ground state energy of an electron liquid at the appropriate metallic density. They defined the *correlation energy* of the electron liquid as the difference between the HFA ground state energy, (5.11), and any better calculation. The name is appropriate because in the HFA the only particle correlations which are taken into account are the kinematic correlations associated with the Pauli principle. Dynamic correlations arising from the fact that the electrons possess a charge are taken into account only in better calculations, such as the RPA.

It is clear that such "charge-induced" correlations will tend to keep electrons apart (since the electrons repel one another), and so give rise to a further lowering of the system energy. The correlation energy is therefore negative. The effect of charge-induced correlations will be especially important for electrons of antiparallel spin, their relative motion being entirely uncorrelated in the HFA. Such correlations were first considered by Wigner (1934, 1938), who introduced correlations between electrons of antiparallel spin into his approximate variational wave functions, and carried out a numerical evaluation of the resulting ground state energy. For an electron liquid of density $r_s \cong 1$, he estimated the correlation energy per electron to be of the order of 0.1 ryd. Wigner calculated as well the correlation energy for a low-density electron system, one in which the potential energy dominates to such an extent that the electrons oscillate about fixed lattice sites. (This will be the case for $r_s \gtrsim 100$.) He suggested that one could approximate the correlation energy for metallic electron densities by interpolating between the high-density and low-density regimes; the approximate interpolation formula is*

$$E_{\text{corr}} = -\frac{0.88}{r_s + 7.8} \text{ ryd.} \tag{5.88}$$

We shall see in Section 5.6 that this expression continues to offer a good estimate of the correlation energy at metallic densities.

In the present section we are concerned with the evaluation of the correlation energy in the RPA, an approximation which will subsequently be shown to be valid in the high-density limit ($r_s \lesssim 1$). We derive a useful theorem for the calculation of the ground state energy from the dynamic form factor or dielectric function, and apply it to the desired RPA calculation. We then elaborate on the relationship between the RPA and a formal perturbation-theoretic treatment, and on the behavior of the long-range contributions to the correlation energy. We conclude with a brief discussion of the pair correlations as calculated in the RPA.

GROUND STATE ENERGY THEOREM

The calculation of the ground state energy is very much simplified by application of a theorem, apparently first discovered by Pauli (unpublished), which enables one to determine E_0 once the dielectric function, $\epsilon(\mathbf{q}, \omega)$, is known for all wave vectors, frequencies, and coupling constants [Nozières and Pines (1958b)]. We define the *interaction energy* in

* The early published values of Wigner's interpolation formula give it as $-0.58/(r_s + 5.1)$ ryd. This expression was based on an incorrect low-density limit for the correlation energy, as discussed in a footnote in Wigner (1938).

the ground state as the expectation value of the potential energy, (5.22):

$$E_{\text{int}} = \left\langle \Psi_0 \left| \sum_{\mathbf{q}} \frac{2\pi e^2}{q^2} (\rho_{\mathbf{q}}^+ \rho_{\mathbf{q}} - N) \right| \Psi_0 \right\rangle = N \sum_{\mathbf{q}} \frac{2\pi e^2}{q^2} (S_{\mathbf{q}} - 1).$$

On making use of Eqs. (5.17) and (4.36), we may write

$$E_{\text{int}} = \sum_{\mathbf{q}} E_{\text{int}}(\mathbf{q}) = -\sum_{\mathbf{q}} \int_0^\infty \frac{d\omega}{2\pi} \operatorname{Im} \frac{1}{\epsilon(\mathbf{q}, \omega)} + \frac{2\pi N e^2}{q^2}. \tag{5.89}$$

We may pass from E_{int} to E_0 by considering the ground state energy as a function of the coupling constant, $\alpha = e^2$, which measures the strength of the electron-electron interaction. Let us formally regard α as a variable. On differentiating the ground state energy

$$E_0 = \left\langle \Psi_0 \left| \sum_i p_i^2/2m \right| \Psi_0 \right\rangle + E_{\text{int}} \tag{5.90}$$

with respect to α, we obtain

$$\begin{aligned} \frac{\partial E_0}{\partial \alpha} &= \left\langle \Psi_0 \left| \frac{\partial H}{\partial \alpha} \right| \Psi_0 \right\rangle + \left\langle \Psi_0 \left| H \right| \frac{\partial \Psi_0}{\partial \alpha} \right\rangle + \left\langle \frac{\partial \Psi_0}{\partial \alpha} \left| H \right| \Psi_0 \right\rangle \\ &= \frac{E_{\text{int}}}{\alpha} + E_0 \frac{\partial}{\partial \alpha} \langle \Psi_0 | \Psi_0 \rangle. \end{aligned} \tag{5.91}$$

Since the ground state wave function is normalized, the last term in Eq. (5.91) drops out, and we may write

$$\frac{\partial E_0}{\partial \alpha} = \frac{E_{\text{int}}}{\alpha}. \tag{5.92}$$

If now we integrate both sides of Eq. (5.92) over α, between $\alpha = 0$ and $\alpha = e^2$ (the actual coupling constant), we find

$$E_0(e^2) - E_0(0) = \int_0^{e^2} \frac{d\alpha}{\alpha} E_{\text{int}}(\alpha). \tag{5.93}$$

$E_0(e^2)$ is the actual ground state energy; $E_0(0)$ is the ground state energy in the absence of particle interaction, $\frac{3}{5}N(p_F^2/2m)$. We have, therefore, the desired theorem

$$E_0 = \frac{3}{10} \frac{N p_F^2}{m} + \int_0^{e^2} \frac{d\alpha}{\alpha} E_{\text{int}}(\alpha), \tag{5.94}$$

which permits us to pass directly from a knowledge of $\epsilon(\mathbf{q}, \omega, e^2)$ to E_0.

The expressions (5.94) and (5.89) permit us to determine independently the contribution to the ground state energy arising from each momentum transfer, that is, from $E_{\text{int}}(\mathbf{q}, \alpha)$. For example, to lowest order in q we have the *exact* result [cf. Eq. (4.57)],

$$\lim_{\mathbf{q}\to 0} S_{\mathbf{q}} = \frac{q^2}{2m\omega_p}. \tag{5.95}$$

It follows that

$$\lim_{\mathbf{q}\to 0} E_{\text{int}}(\mathbf{q}) = \frac{\omega_p}{4} - \frac{2\pi N e^2}{q^2}, \tag{5.96}$$

while the corresponding contribution to the ground state energy is (on carrying out the coupling constant integration)

$$\frac{\omega_p}{2} - \frac{2\pi N e^2}{q^2}. \tag{5.97}$$

The first term in (5.97) represents the zero-point energy of a long wavelength plasmon; the second is the self-energy of the density fluctuations the plasmons describe. The result (5.97) is just what might have been guessed once one realizes that at long wavelengths the dynamic form factor is dominated by plasmons. Under those circumstances, we would expect to find a contribution to the ground state energy from their zero-point oscillations that is equally divided between the kinetic energy, $\omega_p/4$, and the potential energy, $\omega_p/4$.

Let us compare (5.97) with the corresponding quantity calculated in the HFA. On making use of Eqs. (5.20) and (5.22), we see that the latter is given by

$$\frac{3\pi}{2}\frac{Ne^2}{qp_F} - \frac{2\pi N e^2}{q^2} \tag{5.98}$$

in the long wavelength limit. On subtracting (5.98) from (5.97), and dividing by N, we obtain $E_{\text{corr}}(\mathbf{q})$, the contribution to the *correlation energy per particle* from a momentum transfer $\mathbf{q}$,

$$E_{\text{corr}}(\mathbf{q}) = \frac{\omega_p}{2N} - \frac{3\pi e^2}{2qp_F}. \tag{5.99}$$

We see that at long wavelengths the dynamic, charge-induced correlations give rise to a substantial contribution to the correlation energy; put another way, comparison of Eq. (5.95) with (5.20) shows that the dynamic correlations lead to a substantial reduction in the mean square density fluctuation, $S_{\mathbf{q}}$, over its HFA value. Let us underline the fact that for small $\mathbf{q}$ the result (5.99) is *exact*.

CALCULATION OF E_{RPA}

On the basis of the above considerations we see that the calculation of E_{RPA}, the ground state energy evaluated in the RPA, amounts to a calculation of the interaction energy, $E_{int}^{RPA}(\mathbf{q})$, for a given momentum transfer, followed by an integration over coupling constants and a sum over momenta, $\mathbf{q}$. Let us write the interaction energy as

$$E_{int}^{RPA}(\mathbf{q}) = -\frac{2\pi e^2}{q^2}\left[N + \frac{1}{\pi}\int_0^\infty d\omega \, \mathrm{Im}\, \chi_{RPA}(\mathbf{q}, \omega)\right], \tag{5.100}$$

where

$$\chi_{RPA} = \frac{\chi^\circ}{1 - (4\pi e^2/q^2)\chi^\circ}. \tag{5.101}$$

For the explicit calculation of E_{RPA}, it is convenient to exploit the analytic behavior of $\chi(\mathbf{q}, \omega)$; by means of a suitable choice of contours* one can show that

$$\int_0^\infty d\omega \, \mathrm{Im}\, \chi_{RPA}(\mathbf{q}, \omega) = \frac{1}{2}\int_{-\infty}^\infty dw \, \chi_{RPA}(\mathbf{q}, iw), \tag{5.102}$$

where

$$\omega = iw. \tag{5.103}$$

The ground state energy may then be written:

$$E_{RPA} = \frac{E_0^{RPA}}{N} = \tfrac{3}{5}\epsilon_F{}^\circ - \sum_{\mathbf{q}} \frac{2\pi}{q^2 N}\int_0^{e^2} d\alpha \left\{N + \frac{1}{2\pi}\int_{-\infty}^\infty dw \frac{\chi^\circ(\mathbf{q}, iw)}{1 - (4\pi\alpha/q^2)\chi^\circ(\mathbf{q}, iw)}\right\}. \tag{5.104}$$

On carrying out the integration over coupling constants, we find

$$E_{RPA} = \tfrac{3}{5}\epsilon_F{}^\circ + \sum_{\mathbf{q}} \left\{-\frac{2\pi e^2}{q^2} + \frac{1}{4\pi N}\int_{-\infty}^\infty dw \ln\left[1 - \frac{4\pi e^2}{q^2}\chi^\circ(\mathbf{q}, iw)\right]\right\}. \tag{5.105}$$

If, now, we introduce

$$\begin{aligned} u &= w/qv_F{}^\circ, \\ \chi^\circ(\mathbf{q}, iw) &= -[N^\circ(0)/2\pi]Q_x(u), \\ -\frac{4\pi e^2}{q^2}\chi^\circ(\mathbf{q}, iw) &= \frac{\alpha r_s}{\pi x^2} Q_x(u), \end{aligned} \tag{5.106}$$

where x is defined in Eq. (5.69) as q/p_F, we may write Eq. (5.105) in the form

$$E_{RPA} = \left\{\frac{2.21}{r_s{}^2} + \frac{3}{4\pi\alpha^2 r_s{}^2}\int_0^\infty dx \, x^3 \int_0^\infty du \ln\left[1 + \frac{\alpha r_s}{\pi x^2} Q_x(u)\right]\right\} \text{ryd} - \sum_{\mathbf{q}} \frac{2\pi e^2}{q^2}. \tag{5.107}$$

* See, for example, Pines (1963), Appendix C.

Our result (5.107), agrees with that obtained by Gell-Mann and Brueckner (1957) by a direct summation of selected terms in the perturbation-theoretic expansion of E_o.

It is not completely trivial to obtain an explicit analytic expression for E_{RPA} from Eq. (5.107). By an examination of the structure of the corresponding perturbation theory expression, one can show that it gives rise to the exchange energy plus a series of terms of the form

$$a \ln r_s + b + cr_s + \cdots . \tag{5.108}$$

Gell-Mann and Brueckner have calculated the constants a and b, which are the only ones of importance in the high-density limit ($r_s \ll 1$) for which the RPA turns out to be a useful approximation. They find, in this limit,

$$E_{\text{RPA}} = \left(\frac{2.21}{r_s^2} - \frac{0.916}{r_s} + 0.622 \ln r_s - 0.142\right) \text{ ryd.} \tag{5.109}$$

STRUCTURE OF PERTURBATION THEORY*

It is instructive to return now to the explicit perturbation-theoretic expansion for the ground state energy. A brief survey of the various terms that appear enables us to appreciate the reasons for the appearance of the form (5.109) for the ground state energy, and to establish directly which terms are summed in the RPA. Consideration of the terms that are omitted is useful in assessing the validity of the RPA.

We have seen that the calculation to first order in V is identical to the HFA, and yields the first two terms in Eq. (5.109). According to Rayleigh–Schrödinger perturbation theory, the second-order (in V) contribution to E_o is

$$E_2 = -\sum_n \frac{\langle 0|H_{\text{int}}|n\rangle \langle n|H_{\text{int}}|0\rangle}{E_n - E_o}, \tag{5.110}$$

where H_{int} is the Coulomb interaction specified in Eq. (5.1). A typical transition introduced by H_{int} is one in which electrons in states $\mathbf{p}$ and $-\mathbf{k}$, say, are scattered respectively into states $\mathbf{p} + \mathbf{q}$ and $-\mathbf{k} - \mathbf{q}$. Because of the Pauli principle, the states $\mathbf{p}$ and $-\mathbf{k}$ must lie within the Fermi sphere, while the states $\mathbf{p} + \mathbf{q}$ and $-\mathbf{k} - \mathbf{q}$ must lie outside it. The matrix element for this transition, where allowed, is simply

$$V_{\mathbf{q}} = \frac{4\pi e^2}{q^2}.$$

The excitation energy is

$$E_{\mathbf{pk}}(\mathbf{q}) = \frac{(\mathbf{p} + \mathbf{q})^2}{2m} + \frac{(-\mathbf{k} - \mathbf{q})^2}{2m} - \frac{p^2}{2m} - \frac{k^2}{2m} = \frac{\mathbf{q} \cdot (\mathbf{p} + \mathbf{k} + \mathbf{q})}{m}. \tag{5.111}$$

* The treatment in this subsection follows closely that given in Pines (1963).

Having arrived at the excited state n, characterized by electrons in states $(\mathbf{p}+\mathbf{q})$, $(-\mathbf{k}-\mathbf{q})$, and holes in states $\mathbf{p}$, $-\mathbf{k}$, we must use H_{int} once more to return to the ground state. Here we may distinguish between two different processes, which correspond to distinct contributions to the ground state energy. In the first, the so-called "direct" process, one comes back down just the way one went up, via a matrix element $V_{\mathbf{q}}$. The direct process is depicted in Fig. 5.7a. A diagrammatic expression for such a process is shown in Fig. 5.7b, where the solid lines represent the propagating electron and hole and the

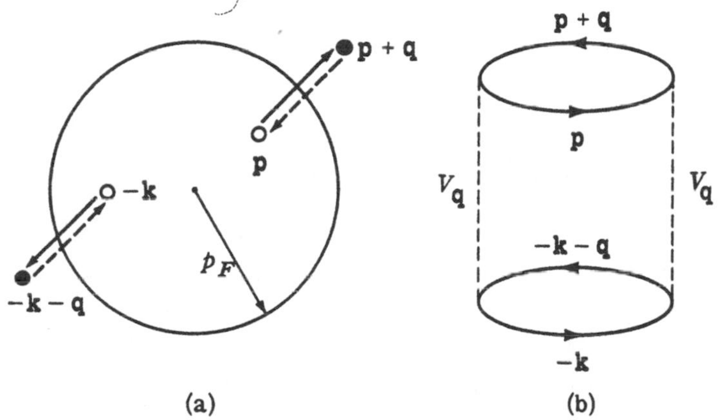

FIGURE 5.7. *Two ways of picturing the second-order direct contribution to the ground state energy, in which electrons in states* $\mathbf{p}$ *and* $-\mathbf{k}$ *are scattered to states* $\mathbf{p}+\mathbf{q}$, $-\mathbf{k}-\mathbf{q}$ *(solid lines) and back again (dashed lines); in both processes, the matrix element is* $V_{\mathbf{q}}$.

dashed lines describe the interactions with momentum transfer $\mathbf{q}$. The contribution which the direct process makes to E_2 is

$$E_2^{(a)} = -4\sum_{\mathbf{pkq}}\left(\frac{4\pi e^2}{q^2}\right)^2 \frac{m}{\mathbf{q}\cdot(\mathbf{p}+\mathbf{k}+\mathbf{q})} n_{\mathbf{p}}{}^{\circ}(1-n^{\circ}_{\mathbf{p}+\mathbf{q}})n_{\mathbf{k}}{}^{\circ}(1-n^{\circ}_{\mathbf{k}+\mathbf{q}}) \quad (5.112)$$

(the factor 4 allows for the sum over spins).

In the second, the so-called "exchange" process, the electron in the state $(\mathbf{p}+\mathbf{q})$ makes a transition to the state $-\mathbf{k}$, while that in the state $(-\mathbf{k}, -\mathbf{q})$ returns to the state $\mathbf{p}$. The matrix element for this transition is

$$-V_{\mathbf{p}+\mathbf{k}+\mathbf{q}} = -\frac{4\pi e^2}{(\mathbf{p}+\mathbf{k}+\mathbf{q})^2}.$$

The minus sign arises because a different ordering of creation and annihilation operators is involved. We note that the exchange process can only occur for states $\mathbf{p}$ and $\mathbf{k}$ which have parallel spins, while the direct process is available for states $\mathbf{p}$ and $\mathbf{k}$ of both parallel and antiparallel spins. A typical exchange process is pictured in Fig. 5.8. The contribution which exchange processes make to E_2 is given by

$$E_2^{(b)} = 2 \sum_{\mathbf{pkq}} \frac{4\pi e^2}{q^2} \frac{4\pi e^2}{(\mathbf{p}+\mathbf{k}+\mathbf{q})^2} \frac{m}{\mathbf{p}\cdot(\mathbf{p}+\mathbf{k}+\mathbf{q})} n_{\mathbf{p}}^{\circ}(1 - n_{\mathbf{p}+\mathbf{q}}^{\circ}) n_{\mathbf{k}}^{\circ}(1 - n_{\mathbf{k}+\mathbf{q}}^{\circ}). \tag{5.113}$$

The direct contribution, $E_2^{(a)}$, is logarithmically divergent in the limit of small momentum transfers. To see this, we remark that the requirement

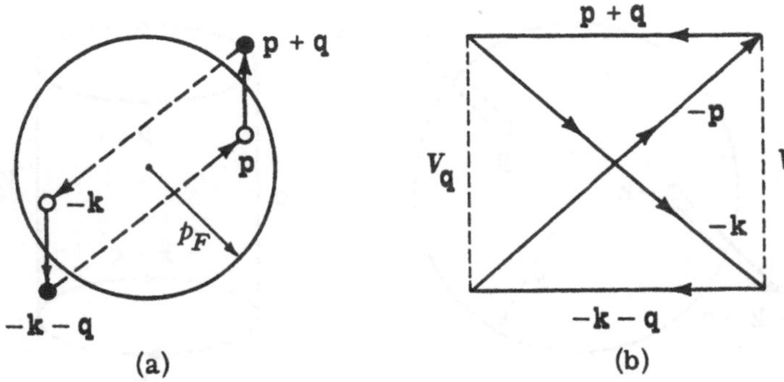

FIGURE 5.8. *Two ways of picturing the second-order exchange contribution to the ground state energy.*

$p < p_F$, $|\mathbf{p}+\mathbf{q}| > p_F$, limits the value of $\mathbf{p}$ to a thin shell in the immediate neighborhood of the Fermi surface whose width is of order q; it follows that

$$\sum_{\mathbf{p}} n_{\mathbf{p}}^{\circ}(1 - n_{\mathbf{p}+\mathbf{q}}^{\circ}) \sim (Nq/p_F) \sim \sum_{\mathbf{k}} n_{\mathbf{k}}^{\circ}(1 - n_{\mathbf{k}+\mathbf{q}}^{\circ}). \tag{5.114}$$

On changing the sum over $\mathbf{q}$ to an integral, we then find

$$\lim_{\mathbf{q}\to 0} E_2^{(a)} \sim -\int dq\, q^2 \left(\frac{4\pi e^2}{q^2}\right)^2 \frac{q^2}{p_F{}^2} \frac{m}{q p_F} \sim -\int \frac{dq}{q} \sim -\ln q. \tag{5.115}$$

This logarithmic divergence may be traced directly to the long range of the Coulomb interaction (if the interaction fell off rapidly with distance, then $V_{\mathbf{q}}$ for small q would approach a constant, and no divergence would ensue).

The higher-order terms in the perturbation expansion exhibit even stronger divergences associated with the long range of the Coulomb interaction. The structure of the terms may be made clear in the following way. In second order we have seen that for a "direct" process, in which one goes up and back with the same momentum transfer $\mathbf{q}$, one gets a divergent answer because of the "piling up" of factors $1/q^2$. On the other hand, for an exchange process there is no divergence because one has one factor of $1/q^2$ and one of $1/(\mathbf{p}+\mathbf{k}+\mathbf{q})^2$, the latter being well behaved as $\mathbf{q} \to 0$. As an example, we consider third-order contributions. The most divergent term will be that in which the electrons simply pass on the momentum transfer **q** as shown in Fig. 5.9. For this term one has initially the excitation of two electron-hole pairs, associated with a momentum transfer **q**; one of these pairs is then annihilated and created anew, again by the interaction $V_{\mathbf{q}} = 4\pi e^2/q^2$; finally the two pairs annihilate, to return the system to a state of no-pair excitation. The size of this term may be simply estimated in the limit of $\mathbf{q} \to 0$. Compared to $E_2^{(a)}$, one has an additional factor of $4\pi e^2/q^2$, an additional energy denominator which goes as

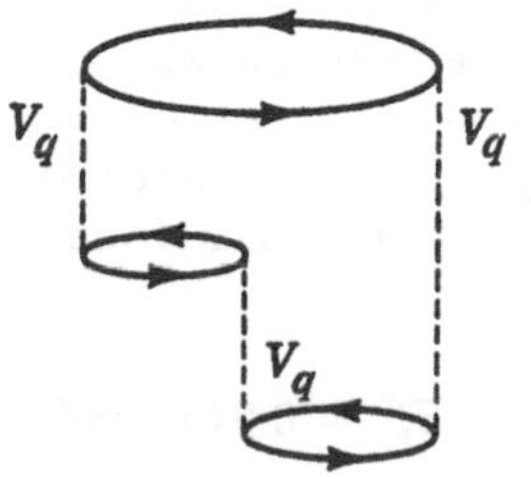

FIGURE 5.9. *The most divergent third-order contribution to the ground state energy.*

qp_F, and a reduction in the available phase space, due to Pauli principle restrictions, which goes as q/p_F. Hence the contribution from Fig. 5.9 is quadratically divergent,

$$E_3^{(a)} \propto \int_\beta \frac{dq}{q} \left(\frac{4\pi e^2}{q^2}\right) \frac{q}{p_F} \frac{1}{qp_F} \propto \frac{r_s}{\beta^2} \text{ ryd.,} \tag{5.116}$$

where we have put in a cut-off β. On the other hand, if we had permitted one of the electron-hole pairs to scatter via an exchange process, we would find

$$E_3^{(b)} \propto \int_\beta \frac{dq}{q} \frac{4\pi e^2}{(\mathbf{k} + \mathbf{p} + \mathbf{q})^2} \frac{q}{p_F} \frac{1}{qp_F} \propto r_s \ln \beta \text{ ryd.,} \tag{5.117}$$

a result which is still divergent, but not so strongly as is $E_3^{(a)}$. Similarly, terms which involve two exchange scattering processes yield

$$E_3^{(c)} \propto r_s \text{ ryd.} \tag{5.118}$$

The structure of the perturbation series is therefore clear. When analyzed from the point of view of the low-momentum behavior of the matrix elements, one finds a series

$$\begin{aligned} E_{\text{corr}} &= a + b \ln \beta + c \frac{r_s}{\beta^2} + d \frac{r_s^2}{\beta^4} + e \frac{r_s^3}{\beta^6} + \cdots \\ &\quad + a_1 + b_1 r_s \ln \beta + c_1 \frac{r_s^2}{\beta^2} + \cdots \\ &\quad + a^2 r_s + b_2 r_s^2 \ln \beta + c_2 r_s^2 + \cdots . \end{aligned} \tag{5.119}$$

If now, following the results of the previous section, we argue that β, the screening wave vector, will be proportional to $r_s^{1/2}$, we see that the first set of terms, which involve only the momentum transfer $\mathbf{q}$, gives rise to a term proportional to $\ln r_s$ and to a constant term in the correlation energy. The second set gives rise to a constant and to terms of order r_s, the third to terms of order $r_s \ln r_s$, r_s^2, etc. Thus if one could sum the first group of most divergent terms in the perturbation series, one would find both a $\ln r_s$ and a constant contribution to the correlation energy (measured in rydbergs); the next most divergent terms yield corrections of order r_s and $r_s \ln r_s$, etc. (except for the term $E_2^{(b)}$).

What Gell-Mann and Brueckner did was just that; they used a Feynman propagator approach to sum the most divergent terms, and obtained the results (5.107) and (5.109). They calculated as well a further constant term in the correlation energy, which comes from the second-order "exchange" process (5.113):

$$E_2^{(b)} = 0.046 \quad \text{ryd.} \tag{5.120}$$

These results, together with the arguments presented above, led them to write the ground state energy of the electron gas in the form

$$E_0 = \left\{\frac{2.21}{r_s^2} - \frac{0.916}{r_s} + 0.062 \ln r_s - 0.096 + ar_s + br_s \ln r_s + cr_s^2 + \cdots\right\} \text{ ryd.} \tag{5.121}$$

Thus they showed that at high densities ($r_s \ll 1$) the RPA yields a rigorous result for the correlation energy.

We note that at high densities ($r_s \ll 1$) the kinetic energy of the electrons is the dominant term in determining their behavior; the Hartree–Fock exchange energy is already a small correction, and the correlation energy terms yield smaller corrections yet. This result is perhaps not surprising—once one has developed a method of dealing with the long range of Coulomb forces, at sufficiently high densities the interaction turns out to be comparatively weak. The form of the series expansion, which contains both powers of r_s and $\ln r_s$, is perhaps surprising (as might be the fact that such a series exists at all); as we have seen, the appearance of the logarithmic terms is a simple consequence of the long range of the Coulomb force. We return later to the question of the range of r_s over which such a series is meaningful; here we remark that for $r_s \ll 1$ it seems all right; one could perhaps use it up to values of r_s of the order of unity.

We conclude this discussion by noting that one can establish directly the connection between the RPA and the perturbation-theoretic expansion, by assuming that, in Eq. (5.101), $(4\pi e^2/q^2)\chi^0$ represents a small parameter: one can then carry out a power-series expansion of the denominator [compare Eq. (5.35)]. It is straightforward to show, using Eqs. (5.100) and (5.93), that the term proportional to $\text{Im}\,\chi^0$ yields the exchange energy, while that proportional to $\text{Im}\,(\chi^0)^2$ yields the second-order direct term, (5.112), etc. We leave the calculation as a problem to the reader (see Problem 5.4).

PAIR DISTRIBUTION FUNCTION

Calculation of the pair distribution function in the RPA has been carried out by Glick and Ferrell (1959) and by Ueda (1961). Ueda's results are shown in Fig. 5.10. We see there the appearance of not-inconsiderable correlations between electrons of antiparallel spin, the correlations being more pronounced as one goes to higher densities. Thus at $r_s \cong 0.1$, one has results which differ little from the HFA, while for $r_s \cong 1$, substantial deviations from the HFA values for $g(r)$ are found. We note too that in the Ueda calculation it is found that electrons of

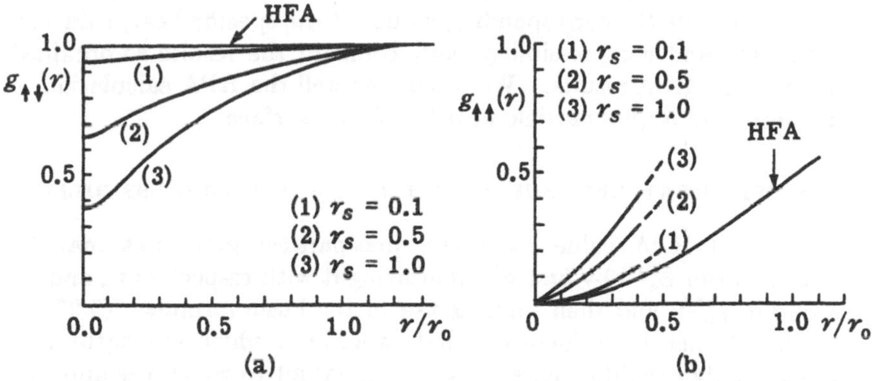

FIGURE 5.10. *Pair distribution function, as calculated in the RPA [from Ueda (1961)]: (a) $g_{\uparrow\downarrow}(r)$; (b) $g_{\uparrow\uparrow}(r)$.*

parallel spin avoid one another somewhat less in the RPA than in the HFA; nevertheless, for all these values electrons of parallel spin are considerably more adept at avoiding one another than are those of antiparallel spin. In other words, the kinetic, or spin-induced, correlations are more important than the dynamic, charge-induced ones, a result to be expected from comparison of the exchange and correlation energies at these densities.

In conclusion, we note that Glick and Ferrell found that at $r_s \cong 2$ $g_{\text{RPA}}(r)$ is negative for small r. Since $g(r)$ must be greater than 0 for all r, this result is a definite indication of the inadequacy of the RPA in this density region.

5.4. QUASIPARTICLE PROPERTIES IN THE RPA

The quasiparticles in the RPA consist of electrons accompanied by dynamic polarization clouds. As we have just seen, the screening cloud essentially acts to reduce the charge on the electron by a factor of

$1/|\epsilon(\mathbf{q}, \omega_e)|$, where ω_e is the frequency that characterizes the electron motion. Because of the screening cloud (which one might regard as a "correlation" hole surrounding the electron), the various quasiparticle properties (specific heat, spin susceptibility, etc.) are now well behaved. Moreover, in the high-density limit for which the RPA is valid, the corrections to quasiparticle properties are rather small, as might be expected from the fact that in this region both exchange and correlation contributions to the ground state energy are small.

The actual calculation of quasiparticle properties is straightforward, but often lengthy. In this section we first discuss an intuitive and simple expression for the screened interaction between quasiparticles, $f_{\mathbf{pp}'}$. We derive the corresponding values of the specific heat, spin susceptibility, and compressibility, and compare the results so obtained with the full RPA results. We discuss as well the RPA calculation of the lifetime of a quasiparticle near the Fermi surface.

SPECIFIC HEAT, SPIN SUSCEPTIBILITY, AND COMPRESSIBILITY

The correct RPA value for these quasiparticle properties may be obtained from E_0^{RPA} by first differentiating it with respect to $n_{\mathbf{p}}$ and $n_{\mathbf{p}'}$ to obtain $f_{\mathbf{pp}'}^{\text{RPA}}$, and then making use of the basic formulas, (5.27) to (5.29). Rather than following that procedure, which is lengthy and not especially enlightening, we discuss the extent to which the approximate expression

$$f_{\mathbf{p}\sigma\cdot\mathbf{p}'\sigma'} = -\frac{4\pi e^2}{|\mathbf{p} - \mathbf{p}'|^2 + q_{FT}^2}\,\delta_{\sigma\sigma'} \tag{5.122}$$

offers a consistent account of quasiparticle properties. The form (5.122) was proposed by Quinn and Ferrell (1961); it has great intuitive appeal, corresponding as it does to screening the HFA quasiparticle interaction, (5.25), with the RPA long wavelength static dielectric constant, (5.65).

According to (5.122), one has

$$f^s(x) = f^a(x) = -\frac{\pi e^2}{p_F{}^2}\frac{1}{1 - x + 2\alpha r_s/\pi}. \tag{5.123}$$

On substituting Eq. (5.123) into (5.29), we find (to order r_s):

$$\frac{C}{C_0} = \frac{m^*}{m} = \frac{1}{1 + (\alpha r_s/2\pi)[\ln(\pi/\alpha r_s) - 2]}. \tag{5.124}$$

The result (5.124) is just that calculated to the same order by Gell-Mann (1957). We note, as was to be expected in view of the dielectric screening contained in Eq. (5.122), that the specific heat is now well behaved, as are of course F_0^a and F_0^s. The appearance of corrections to the free-particle value of the order of r_s and $r_s \ln r_s$ was to be expected, in analogy with the corresponding terms in the correlation energy. (It may be noted that $(\pi e^2/p_F^2)N^\circ(0)$ is of order r_s, so that all quasiparticle corrections are at least of order r_s.)

If we substitute Eq. (5.123) into Eqs. (5.27) and (5.28), we find the following result for the compressibility (or sound velocity) and spin susceptibility:

$$\frac{\kappa_0}{\kappa} = \frac{s^2}{s_0^2} = \frac{\chi_P^\circ}{\chi_P} = 1 - \frac{\alpha r_s}{\pi} - \frac{\alpha^2 r_s^2}{\pi^2} \ln r_s + \cdots. \tag{5.125}$$

The result (5.125) is only correct to order r_s for these quantities [where, in fact, it corresponds to the HFA result, (5.31)]. We may see this directly for the compressibility, since that quantity may be determined from the ground state energy expression, (5.109). We have

$$\mu = \frac{\partial E}{\partial N} = \epsilon(\rho) + \rho \frac{\partial \epsilon(\rho)}{\partial \rho}, \tag{5.126}$$

$$\frac{ms^2}{N} = \frac{1}{\kappa} = \frac{\partial \mu}{\partial \rho} = \rho \frac{\partial^2 \epsilon(\rho)}{\partial \rho^2} + 2 \frac{\partial \epsilon(\rho)}{\partial \rho}, \tag{5.127}$$

where $\epsilon(\rho)$ is the ground state energy per particle. On making use of the relation between ρ and r_s, we obtain, after an elementary calculation,

$$\frac{\kappa_0}{\kappa} = \frac{s^2}{s_0^2} = \frac{\alpha^2 r_s^4}{6} \frac{\partial^2 \epsilon(r_s)}{\partial r_s^2} - \frac{\alpha^2 r_s^3}{3} \frac{\partial \epsilon(r_s)}{\partial r_s}. \tag{5.128}$$

On substituting (5.109) into (5.128), we find

$$\frac{\kappa_0}{\kappa_{\text{RPA}}} = \frac{s_{\text{RPA}}^2}{s_0^2} = 1 - \frac{\alpha r_s}{\pi} - \frac{\alpha^2 r_s^2}{\pi^2} (1 - \ln 2) \ln r_s. \tag{5.129}$$

It is clear then that there are corrections in the RPA to the intuitive expression (5.122) for $f_{\mathbf{pp}'}$, which are such as to change the coefficient of the $\ln r_s$ term from $-\alpha^2 r_s^2/\pi^2$ to $-(\alpha^2 r_s^2/\pi^2)(1 - \ln 2)$.

Examination of the complete expression for $f_{\mathbf{pp}'}^{\text{RPA}}$ shows that in using Eq. (5.122) one neglects screened second-order direct scattering contributions to $f_{\mathbf{pp}'}$, and that it is these which are responsible for the added term of order $r_s^2 \ln r_s$ in the compressibility. Because these give rise to an effective interaction between quasiparticles of opposite spin, one finds that the effect of electron interaction on the spin susceptibility is different from its effect on the compressibility. A calculation of χ in the RPA has been carried out by Brueckner and Sawada (1958), who found,

$$\frac{\chi_P^{\circ}}{\chi_P^{\mathrm{RPA}}} = 1 - \frac{\alpha r_s}{\pi} - \frac{\alpha^2 r_s^2}{2\pi^2} \ln r_s + \cdots \tag{5.130}$$

LIFETIME OF AN EXCITED QUASIPARTICLE

Consider a quasiparticle lying above the Fermi surface; such a particle will not remain there indefinitely; it will scatter against the particles in the Fermi sphere, and so tend to lower its energy. Such an electron state therefore possesses a finite lifetime; the same argument applies equally well to a hole

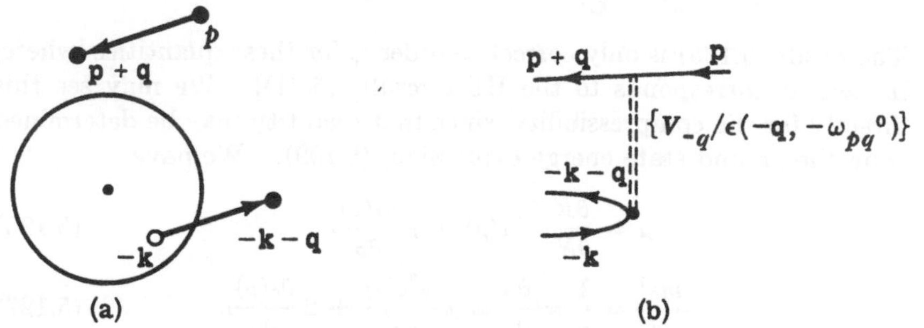

FIGURE 5.11. *The scattering of an excited quasiparticle against the ground state particles: (a) A particle of momentum* **p** *is scattered to a state* **p** + **q**, *creating a particle-hole pair* (−**k**−**q**, −**k**); *(b) a diagrammatic representation.*

inside the Fermi surface. From a field-theoretic point of view, the lifetime appears as the imaginary part of the self-energy of a quasiparticle, and it was in this way that Quinn and Ferrell (1958) first calculated the lifetime in the RPA. We may, however, derive their result from more elementary considerations in the following way [(Ritchie (1959)].

As we have seen, the physical content of the RPA is that a given electron is surrounded by an appropriate polarization cloud. Hence we may regard the process by which an electron is scattered from a state $\mathbf{p}(p > p_F)$ to some other state, $\mathbf{p} + \mathbf{q}$, which differs in energy by

$$-\omega_{\mathbf{pq}}^{\circ} = \epsilon_{\mathbf{p}}{}^{\circ} - \epsilon_{\mathbf{p+q}}^{\circ},$$

as screened by the dynamic screening factor $\epsilon(-\mathbf{q}, -\omega_{\mathbf{pq}}^{\circ})$, as shown in Fig. 5.11. We may proceed to calculate the lifetime by applying the usual golden rule of second-order perturbation theory. We have, then,

$$\frac{1}{\tau} = 2\pi \sum_{\mathbf{q}\mathbf{k}\sigma}{}' \frac{|V_{\mathbf{q}}|^2}{|\epsilon(-\mathbf{q}, -\omega_{\mathbf{pq}}^{\circ})|^2} n_{\mathbf{k}\sigma}^{\circ}(1 - n_{\mathbf{k+q}\sigma}^{\circ})\,\delta[\omega_{\mathbf{pq}}^{\circ} + \omega_{\mathbf{kq}}^{\circ}], \tag{5.131}$$

where the prime on the summation indicates that the momentum transfer **q** is subject to the condition

$$p_F^2 < |\mathbf{p} + \mathbf{q}|^2 < p^2 \tag{5.132}$$

[since the final state, $\mathbf{p} + \mathbf{q}$, must lie outside the Fermi sphere, while the electron lowers its energy in scattering against the ground state particles ($\omega^\circ_{\mathbf{pq}} < 0$)].

With the aid of Eq. (5.33) and an interchange of indices, the lifetime may be written as

$$\frac{1}{\tau} = \sum_{\mathbf{q}}{}' \frac{8\pi e^2}{q^2} \frac{\operatorname{Im} \epsilon(-\mathbf{q}, -\omega^\circ_{\mathbf{pq}})}{|\epsilon(-\mathbf{q}, -\omega^\circ_{\mathbf{pq}})|^2} = -\sum_{\mathbf{q}}{}' \frac{8\pi e^2}{q^2} \operatorname{Im} \frac{1}{\epsilon(\mathbf{q}, \omega^\circ_{\mathbf{pq}})}, \tag{5.133}$$

which is the result obtained by Quinn and Ferrell.

If we consider a quasiparticle located near the Fermi surface, Eq. (5.133) may be simplified considerably. Under these circumstances only low excitation frequencies enter the expression; on making use of Eq. (5.63) we may write

$$\begin{aligned} \frac{1}{\tau} &= \sum_{\mathbf{q}}{}' \left(\frac{4\pi^2 e^2}{q^4}\right) q_{FT}^2 \left(\frac{\omega^\circ_{\mathbf{pq}}}{q v_F{}^\circ}\right) \frac{1}{|\epsilon(\mathbf{q}, 0)|^2} \\ &= e^2 q_{FT}^2 \int_0^{2p_F} \frac{dq}{q^3 v_F{}^\circ |\epsilon(\mathbf{q}, 0)|^2} \,⨍\, dx\, \omega^\circ_{\mathbf{pq}}(x), \end{aligned} \tag{5.134a}$$

on changing the sum to an integral and introducing polar coordinates based on **p**. The restriction (5.132) manifests itself in a limitation of the maximum momentum transfer to $2p_F$. It likewise limits the range of the polar angle, $x = \cos\theta$; the latter restriction, which is schematically denoted by a bar through the integral sign, is most easily expressed if one transforms from x to the new variable,

$$\omega = \omega^\circ_{\mathbf{pq}}(x) = \frac{pqx}{m} + \frac{q^2}{2m}.$$

One finds readily that

$$⨍\, dx\, \omega^\circ_{\mathbf{pq}}(x) = \int_0^{\epsilon_p{}^\circ - \mu^\circ} \frac{d\omega}{q v_F{}^\circ}\, \omega = \frac{(\epsilon_p{}^\circ - \mu^\circ)^2}{2 q v_F{}^\circ}.$$

On substituting this result into Eq. (5.134a), we find

$$\frac{1}{\tau} = \frac{e^2 q_{FT}^2 (\epsilon_p{}^\circ - \mu^\circ)^2}{2(v_F{}^\circ)^2} \int_0^{2p_F} \frac{dq}{|\epsilon(\mathbf{q}, 0)|^2 q^4}. \tag{5.134b}$$

The last integral, as Quinn and Ferrell argue, would be severely divergent without the screening effect. Because of the screening, the integral converges, and one finds, in the high-density limit ($r_s \ll 1$),

$$\frac{1}{\tau} = \frac{\pi^2 \sqrt{3}}{128}\, \omega_p \left(\frac{\epsilon_p{}^\circ - \mu^\circ}{\mu^\circ}\right)^2 \tag{5.134c}$$

for the lifetime of a quasiparticle in the immediate vicinity of the Fermi surface. We remark that the dependence of the lifetime on $(\epsilon_p{}^o - \mu^o)^2$ is just what we have argued would appear as a consequence of the Pauli principle restrictions on the number of available scattering states. The RPA result, (5.134c), represents a special case of the more general formula (1.163) that we obtained earlier for the lifetime of a quasiparticle near the Fermi surface.

5.5. EQUATIONS OF MOTION AND THE GENERALIZED RPA

Before going on to consider approximate theories of the electron liquid at metallic densities, we pause in an attempt to place the RPA in a somewhat broader perspective. As a microscopic approximation, the RPA is clearly but one of a whole hierarchy of approximations. There are, moreover, a number of methods of arriving at this hierarchy: two of the most popular are the use of Feynman diagrams and the approximate solution of a chain of coupled equations for the many-body Green's functions. In the present section we shall consider yet another way, closely related to the latter approach, which involves the approximate solution of the coupled equations of motion of the relevant operators in the Heisenberg representation. The method, which we shall call the *equation-of-motion* method, has the principal advantage that it leads one directly to a large number of the simple microscopic approximations (HFA, RPA, BCS theory of superconductivity), with a minimum of additional mathematical definitions, rules, and complications.

In the equation-of-motion method one attempts to determine the elementary excitation spectrum and ground state energy of a many-particle system by finding an approximate self-consistent solution of various coupled Heisenberg operator equations of motion [Bohm and Pines (1953), Sawada *et al.* (1957), Anderson (1958), Suhl and Werthamer (1961)]. The steps by which one proceeds may be outlined as follows [(Pines (1962)]:

(1) Assume the ground state wave function Ψ_o to be known.

(2) Find the creation and destruction operators for the elementary excitations of the system, $O_{\mathbf{q}}{}^+$ and $O_{\mathbf{q}}$, by requiring that they satisfy oscillatory equations of motion:

$$(H, O_{\mathbf{q}}{}^+) = \omega_{\mathbf{q}} O_{\mathbf{q}}{}^+, \qquad (H, O_{\mathbf{q}}) = -\omega_{\mathbf{q}} O_{\mathbf{q}}. \tag{5.135}$$

(3) Applied to Ψ_o, $O_{\mathbf{q}}{}^+$ creates an excitation of energy $\omega_{\mathbf{q}}$. Likewise, one must have

$$O_{\mathbf{q}}\Psi_o = 0$$

if Ψ_0 is to be the ground state wave function. The latter condition is used to determine Ψ_0.

(4) Once ω_q and Ψ_0 are determined, the other properties of the system (ground state energy, etc.) can be determined more or less directly in a variety of ways.

In the present section we shall not make complete use of the possibilities inherent in the equation-of-motion method for determining the ground state energy. Rather, we shall content ourselves with the determination of allowed excitation frequencies, and the general nature of the solutions to which these correspond. As a first example of the method, we consider single-particle excitations, and rederive the Hartree and Hartree–Fock approximations. We next consider particle-hole excitations, using the method to obtain anew the density fluctuation excitation spectrum in the HFA and RPA. We then discuss a simple generalization of the RPA solution for particle-hole pairs. We use this generalized RPA as a basis for a discussion of possible microscopic instabilities in fermion systems, with special attention to their relationship to those considered in the Landau theory. We conclude the section with a discussion of the domain of validity of the RPA for the electron liquid.

SINGLE-PARTICLE EXCITATIONS: HARTREE AND HARTREE–FOCK APPROXIMATIONS

Let us consider the motion of a single particle, characterized by the creation operator $c_{p\sigma}^{+}$. On using the Hamiltonian (5.1), it is easily found that

$$[H, c_{p\sigma}^{+}] = \epsilon_p^{\circ} c_{p\sigma}^{+} + \sum_{p'\sigma' q} V_q c_{p+q,\sigma}^{+} c_{p'-q,\sigma'}^{+} c_{p'\sigma'}. \tag{5.136}$$

According to Eq. (5.136), the motion of the single particle $(\mathbf{p}, \sigma)$ is coupled to that of two particles and a hole. In general, we would next write down the equation of motion for the product $c^{+}c^{+}c$, and find it coupled to configurations with three particles and two holes, etc. This series of coupled equations of motion represents simply another way of writing the Schrödinger equation, and is, of course, too complicated to be solved exactly. The approximations enter in the way one obtains a solution by cutting the chain of equations off at a given order, and searching for a self-consistent solution.

Clearly the simplest approximation is to consider only the single equation (5.136). Since we wish $c_{p\sigma}^{+}$ to satisfy an oscillatory equation of motion of the form (5.135), we must then simplify the trilinear term on the right-hand side of Eq. (5.136). An obvious simplification is to keep only the term $\mathbf{q} = 0$ in the summation, and further, to replace the product $c^{+}_{\mathbf{p}'\sigma'}c_{\mathbf{p}'\sigma'}$ by its expectation value in the ground state. Equation (5.136) then takes the *linear* form

$$[H, c^{+}_{\mathbf{p}\sigma}] = \epsilon_{\mathbf{p}} c^{+}_{\mathbf{p}\sigma}, \tag{5.137}$$

where $\epsilon_{\mathbf{p}}$ is given by

$$\epsilon_{\mathbf{p}} = \epsilon_{\mathbf{p}}{}^{\circ} + V_{\circ} \sum_{\mathbf{p}'\sigma'} n^{\circ}_{\mathbf{p}'\sigma'}. \tag{5.138}$$

According to Eq. (5.138), $c^{+}_{\mathbf{p}\sigma}$ creates an elementary excitation (here a quasiparticle), with energy $\epsilon_{\mathbf{p}}$; Eq. (5.138) is clearly equivalent to the Hartree approximation for the energy of the quasiparticle.

We may go further along those lines, and try to linearize Eq. (5.136) *systematically*. We are thus led to make the following replacement:

$$c_1{}^{+}c_2{}^{+}c_3 \rightarrow c_1{}^{+}\langle c_2{}^{+}c_3\rangle - c_2{}^{+}\langle c_1{}^{+}c_3\rangle.$$

With this approximation, the equation of motion, (5.136), again takes the form (5.137); the energy $\epsilon_{\mathbf{p}}$ is now given by

$$\epsilon_{\mathbf{p}} = \epsilon_{\mathbf{p}}{}^{\circ} + \sum_{\mathbf{p}'\sigma'} n_{\mathbf{p}'\sigma'}[V_{\circ} - V_{\mathbf{p}-\mathbf{p}'}\delta_{\sigma\sigma'}]. \tag{5.139}$$

On comparing Eq. (5.139) with (5.23), we see that the systematic linearization of Eq. (5.136) leads directly to the Hartree–Fock approximation of the quasiparticle excitation spectrum for a uniform system.

PARTICLE-HOLE EXCITATIONS: HFA AND RPA

We next apply the equation-of-motion method to the calculation of the density fluctuation excitation spectrum for an arbitrary fermion system. Let us consider the equation of motion for a particle-hole pair, of momentum $\mathbf{q}$:

$$\rho^{+}_{\mathbf{pq}\sigma} = c^{+}_{\mathbf{p}+\mathbf{q},\sigma}c_{\mathbf{p},\sigma}.$$

It is determined by the system Hamiltonian, (5.1), and is

$$[H, \rho^{+}_{\mathbf{pq}\sigma}] = \omega^{\circ}_{\mathbf{pq}}\rho^{+}_{\mathbf{pq}\sigma} - \sum_{\mathbf{k}} (V_{\mathbf{k}}/2)\{(c^{+}_{\mathbf{p}+\mathbf{q},\sigma}c_{\mathbf{p}+\mathbf{k},\sigma} - c^{+}_{\mathbf{p}+\mathbf{q}-\mathbf{k},\sigma}c_{\mathbf{p},\sigma})\rho_{\mathbf{k}}{}^{+}$$
$$+ \rho_{\mathbf{k}}{}^{+}(c^{+}_{\mathbf{p}+\mathbf{q},\sigma}c_{\mathbf{p}+\mathbf{k},\sigma} - c^{+}_{\mathbf{p}+\mathbf{q}-\mathbf{k},\sigma}c_{\mathbf{p},\sigma})\}. \tag{5.140}$$

We see that through Eq. (5.140) a single-pair excitation is connected to a two-pair excitation; if we write down the equation of motion of the two-pair excitation, we find it coupled to both the one-pair and a three-pair excitation, etc. Once again, this series of equations cannot be solved exactly.

The general form of the solutions is, however, clear. We wish to find an operator, $\xi_{\mathbf{q}}^{+}$ say, which describes an elementary excitation contributing to the density-fluctuation excitation spectrum. $\xi_{\mathbf{q}}^{+}$ must satisfy an oscillatory equation of motion:

$$(H, \xi_{\mathbf{q}}^{+}) = \omega \xi_{\mathbf{q}}^{+}. \tag{5.141}$$

In general, if such a $\xi_{\mathbf{q}}^{+}$ exists, it will consist in linear combinations of single-pair excitations, two-pair excitations, etc. (each of net momentum $\mathbf{q}$). Thus one has

$$\xi_{\mathbf{q}}^{+} = \sum_{\mathbf{p}} A(\mathbf{p}, \mathbf{q}, \omega) c_{\mathbf{p}+\mathbf{q}}^{+} c_{\mathbf{p}} + \sum_{\mathbf{p}\mathbf{p}'} B(\mathbf{p}, \mathbf{p}', \mathbf{p}'', \mathbf{q}, \omega) c_{\mathbf{p}+\mathbf{p}'+\mathbf{q}-\mathbf{p}''}^{+} c_{\mathbf{p}''}^{+} c_{\mathbf{p}'} c_{\mathbf{p}} + \cdots, \tag{5.142}$$

where for convenience we have dropped spin indices (they may be lumped with the momentum into the single notation $\mathbf{p}$). The coefficients A, B, etc. and the excitation energy ω are to be obtained by solving the eigenvalue equation (5.141) to a given order of approximation.

Clearly the lowest-order solution is obtained by neglecting the particle interaction altogether. In that case on combining Eqs. (5.142) and (5.140), we have

$$\sum_{\mathbf{p}} A(\mathbf{p}, \mathbf{q}, \omega)(\omega - \omega_{\mathbf{pq}}^{0}) c_{\mathbf{p}+\mathbf{q}}^{+} c_{\mathbf{p}} = 0. \tag{5.143}$$

The coefficient of each pair operator, $c_{\mathbf{p}+\mathbf{q}}^{+} c_{\mathbf{p}}$, must vanish. We thus obtain the following equation for the unknown coefficient $A(\mathbf{p}, \mathbf{q}, \omega)$:

$$(\omega - \omega_{\mathbf{pq}}^{0}) A(\mathbf{p}, \mathbf{q}, \omega) = 0,$$

which possesses the solutions

$$\omega = \omega_{\mathbf{pq}}^{0}, \tag{5.144a}$$

$$A(\mathbf{p}, \mathbf{q}, \omega) = c(\mathbf{q}, \omega)\delta(\omega - \omega_{\mathbf{pq}}^{0}). \tag{5.144b}$$

The allowed frequencies are those of free particle–hole propagation; the solution (5.144) represents a simple "scattering" solution of the coupled equations, as signaled by the δ function appearing in Eq. (5.144b); we have derived the HFA pair excitation spectrum in yet another way.

We next consider the solution of Eqs. (5.141) and (5.142) which corresponds to the RPA. In that approximation we keep only the term with $\mathbf{k} = \mathbf{q}$ on the right-hand side of Eq. (5.140), and, further, replace the operators in parentheses by their expectation values in the ground state for a noninteracting particle system. Equation (5.140) thereby becomes

$$[H, c^{+}_{\mathbf{p}+\mathbf{q},\sigma}c_{\mathbf{p},\sigma}]_{\mathrm{RPA}} - \omega^{\circ}_{\mathbf{pq}}c^{+}_{\mathbf{p}+\mathbf{q},\sigma}c_{\mathbf{p},\sigma} = -V_{\mathbf{q}}(n^{\circ}_{\mathbf{p}+\mathbf{q}} - n_{\mathbf{p}}{}^{\circ})\rho_{\mathbf{q}}{}^{+}. \quad (5.145)$$

We see that the part of the interaction we have kept represents a forcing term for the motion of particle-hole pairs. We can regard this term as an average force field produced by all the particle-hole pairs in the system. Such a field, which in turn gives rise to pair excitation, must be determined in self-consistent manner. In this context, the RPA may be viewed as a *time-dependent Hartree approximation*, in which particle interaction acts only to produce an average, self-consistent, time-dependent field, with wave vector **q** and frequency ω. The terms we have neglected give rise to fluctuations about that average field, and are assumed small.

We now use Eq. (5.145) to determine the form of the operator $\xi_{\mathbf{q}}$; Since Eq. (5.145) is linear in the particle-hole operators, it is clear that we need retain only the coefficient $A(\mathbf{p}, \mathbf{q}, \omega)$ in Eq. (5.142). Let us substitute Eq. (5.142) into (5.141) and make use of (5.145). On setting the coefficient for each pair operator, $c^{+}_{\mathbf{p}+\mathbf{q}}c_{\mathbf{p}}$, equal to zero, we find the following equation for $A(\mathbf{p}, \mathbf{q}, \omega)$:

$$[\omega - \omega^{\circ}_{\mathbf{pq}}]A(\mathbf{p}, \mathbf{q}, \omega) + V_{\mathbf{q}}\sum_{\mathbf{p}'}(n^{\circ}_{\mathbf{p}'+\mathbf{q}} - n^{\circ}_{\mathbf{p}'})A(\mathbf{p}', \mathbf{q}, \omega) = 0. \quad (5.146)$$

The linear equation (5.146) possesses two kinds of eigenvectors:

(i) "Scattering" solutions, in which $A(\mathbf{p}, \mathbf{q}, \omega)$ is peaked at a particular value **p**. Such solutions correspond to incoherent single-pair excitation. To order $1/N$, they agree with the HFA.

(ii) "Collective" solutions, with frequency $\omega_{\mathbf{q}}$, in which $A(\mathbf{p}, \mathbf{q}, \omega)$ is a smooth function of **p**. To obtain the collective solutions, we note that the second term on the left-hand side of Eq. (5.146) is independent of **p**. Thus, $A(\mathbf{p}, \mathbf{q}, \omega)$ must have the form

$$A(\mathbf{p}, \mathbf{q}, \omega) = \frac{A'(\mathbf{q}, \omega)}{\omega_{\mathbf{q}} - \omega^{\circ}_{\mathbf{pq}}}. \quad (5.147)$$

On making use of this result, we see that Eq. (5.146) possesses a solution only if

$$1 = V_{\mathbf{q}}\sum_{\mathbf{p}}\frac{n_{\mathbf{p}}{}^{\circ} - n^{\circ}_{\mathbf{p}+\mathbf{q}}}{\omega_{\mathbf{q}} - \omega^{\circ}_{\mathbf{pq}}}, \quad (5.148)$$

which is the well-known RPA dispersion relation for the collective modes of an interacting fermion system.

For a repulsive interaction ($V_{\mathbf{q}} > 0$), the collective modes will be well defined as long as the condition

$$\omega_{\mathbf{q}} > \omega^{\circ}_{\mathbf{pq}} \tag{5.149}$$

is satisfied. (Their energies will be distinct from those of the continuum of particle-hole pairs.) Where this condition is not satisfied, the dispersion relation (5.148) is not well defined. We may treat the singularities at

$$\omega_{\mathbf{q}} = \omega^{\circ}_{\mathbf{pq}}$$

if we introduce an external forcing term into the equation of motion (5.145) and require that the system's response be causal [compare Eq. (5.41)]. The causality condition is satisfied if we everywhere make the replacement

$$\omega \rightarrow \omega + i\eta,$$

where η is a small positive infinitesimal number. The dispersion relation, (5.148), then becomes

$$1 = V_{\mathbf{q}} \sum_{\mathbf{p}} \frac{n_{\mathbf{p}}^{\circ} - n^{\circ}_{\mathbf{p}+\mathbf{q}}}{\omega_{\mathbf{q}} - \omega^{\circ}_{\mathbf{pq}} + i\eta}. \tag{5.150}$$

The solutions of Eq. (5.150) correspond to a complex $\omega_{\mathbf{q}}$, and thus to a damped collective mode.

One may ask: In the present context what is the coupling between particle-hole excitations that is taken into account by the RPA, and that gives rise to a collective mode? To answer this question, we consider the scattering of a particle and hole of total momentum $\mathbf{q}$, as a consequence of the particle interaction,

$$V = \sum_{\mathbf{pp'q},\sigma\sigma'} \frac{V_{\mathbf{q}}}{2} c^{+}_{\mathbf{p},\sigma} c^{+}_{\mathbf{p'}+\mathbf{q'},\sigma'} c_{\mathbf{p'},\sigma'} c_{\mathbf{p}+\mathbf{q},\sigma}. \tag{5.151}$$

Let us suppose that initially we have a particle with momentum $\mathbf{p} + \mathbf{q}$, spin σ, and a hole in the state $\mathbf{p}\sigma$. The interaction (5.151) may give rise to the three different processes shown schematically in Fig. 5.12.

(i) The particle may simply fall back into the hole in the Fermi sphere. The corresponding matrix element is $V_{\mathbf{q}}$; this "pair annihilation" process is accompanied by the creation of another particle-hole pair, $(\mathbf{p'} + \mathbf{q}, \sigma')$ and $(\mathbf{p'}\sigma')$. There is no restriction on the new spin σ'.

(ii) A second possibility is that the particle is scattered from the state $(\mathbf{p} + \mathbf{q}, \sigma)$ to a state $(\mathbf{p'} + \mathbf{q}, \sigma)$, while the hole is scattered from the state $(\mathbf{p}\sigma)$ to a state $(\mathbf{p'}\sigma)$. The matrix element for this "scattering" process is

$-V_{p'-p}$; the new particle-hole pair must have the same spin, σ, as that of the initial pair.

(iii) Finally, either the particle or the hole may scatter against the ground state particles, thereby exciting a second particle-hole pair out of the Fermi sea.

Only the annihilation or direct scattering process, (i), is taken into account in the RPA. The second process, which corresponds to an exchange scattering of the pair, involves a momentum transfer different from **q**, and is thus not included. Finally, by linearizing the equation of motion (5.140) we automatically neglect all processes of type (iii).

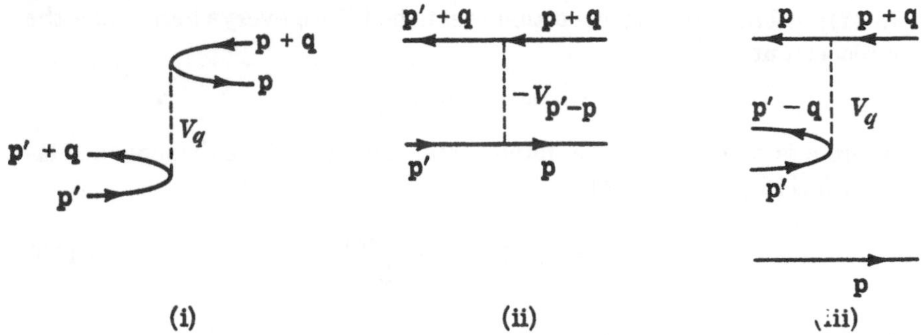

FIGURE 5.12. *Scattering processes for an excited particle-hole pair. The full lines with arrows pointing to the left represent particles, those with arrows to the right represent holes; the dotted line denotes the scattering matrix element. (i) Pair annihilation and creation (direct scattering); (ii) exchange scattering of the pair; (iii) particle scattering, accompanied by excitation of a ground state pair.*

On the other hand, in the RPA *all* processes of type (i) which go with the momentum transfer **q** are taken into account. [The summation of the series is performed automatically when one obtains the self-consistent solution to Eq. (5.145).] Put another way, the RPA provides a description of *multiple scattering* of a single particle-hole pair, via the "pair annihilation and creation process." Note that, as shown in Fig. 5.13, such scattering runs "backward" as well as forward in time.

In the present context, the continuum of individual excitations corresponds to "scattering" solutions for the problem of multiple scattering of a particle-hole pair, while the collective modes represent "bound state" solutions. As usual, the spectrum of the bound states is discrete. We shall return to this interpretation of collective modes in Chapter 7.

It should be emphasized that the RPA is essentially an "operator" approximation, which can be carried out as easily at finite temperatures as at $T = 0$. At finite temperatures, the only change in the modus operandi is that in Eq. (5.140) one replaces the operators in parentheses according to

$$c_{\mathbf{p}+\mathbf{q}}^{+}c_{\mathbf{p}+\mathbf{k}} - c_{\mathbf{p}+\mathbf{q}-\mathbf{k}}^{+}c_{\mathbf{p}} \rightarrow [n_{\mathbf{p}+\mathbf{q}}^{\circ}(T) - n_{\mathbf{p}}^{\circ}(T)]\delta_{\mathbf{q},\mathbf{k}}, \tag{5.152}$$

where $n_{\mathbf{p}}^{\circ}(T)$ is the unperturbed distribution function for the system at temperature T. For an electron gas, the finite temperature analog of Eq. (5.150) in the classical limit provides the dispersion relation for the longitudinal plasma oscillations in a classical plasma.*

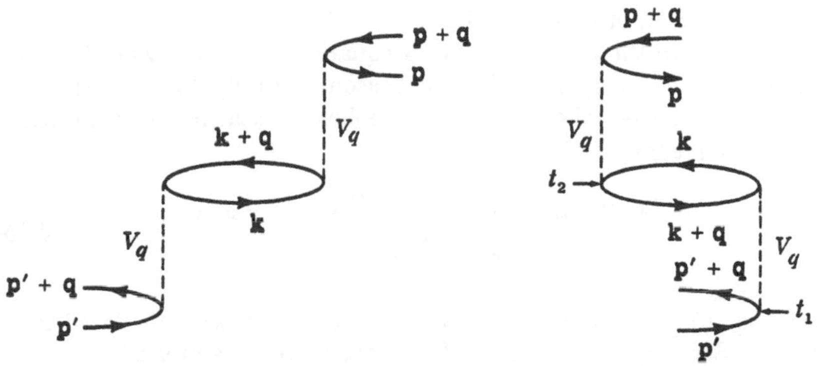

FIGURE 5.13. *Multiple scattering processes included in the RPA. Time runs from right to left; (a) corresponds to a scattering which is forward in time, (b) to a process which runs backward in time [at t_2 the pair of total momentum* **q** *is annihilated by the pair of total momentum* $-$**q***, which had been created (along with a pair* **q***) at an earlier time,* t_1*].*

GENERALIZED RPA

We have seen in the preceding subsection that the RPA corresponds to the following two approximations:

(i) *Linearization* of the equation of motion (5.140); as a result only the single-pair coefficient $A(\mathbf{p}, \mathbf{q}, \omega)$ in the expansion (5.142) for $\xi_{\mathbf{q}}^{+}$ need be retained.

(ii) Neglect of *exchange* scattering of an excited particle-hole pair (the neglected terms involve momentum transfers other than **q**).

The major simplification comes from the linearization. Indeed, if we look for the *most general* self-consistent linear solutions of Eqs. (5.141)

* For a discussion of finite temperature plasmas along the lines developed here see, for example, Pines (1960), and the references cited therein.

and (5.142), we automatically take into account the exchange scattering of an excited pair. It is left as a problem to the reader to show that if we collect systematically *all* bilinear terms on the right-hand side of Eq. (5.140), we find

$$[H, \rho^+_{\mathbf{pq}\sigma}] \cong \left\{\omega^0_{\mathbf{pq}} + \sum_{\mathbf{k}} n^0_{\mathbf{k}\sigma}(V_{\mathbf{k}-\mathbf{p}} - V_{\mathbf{k}-\mathbf{p}-\mathbf{q}})\right\} \rho^+_{\mathbf{pq}\sigma} - \sum_{\mathbf{k}\sigma'} (V_{\mathbf{q}} - V_{\mathbf{k}-\mathbf{p}}\delta_{\sigma\sigma'})(n^0_{\mathbf{p}+\mathbf{q},\sigma} - n^0_{\mathbf{p},\sigma})c^+_{\mathbf{k}+\mathbf{q},\sigma'}c_{\mathbf{k},\sigma'}. \tag{5.153}$$

The more general approximation, based on Eq. (5.153), is known as the generalized, or extended, RPA.

Solutions to the generalized RPA equation of motion will differ from those obtained in the RPA for two reasons. First, the scattering solutions are changed; the "free" particle–hole excitations now propagate at the frequency

$$\omega = \omega_{\mathbf{pq}} = \omega^0_{\mathbf{pq}} + \sum_{\mathbf{k}} n^0_{\mathbf{k}\sigma}(V_{\mathbf{k}-\mathbf{p}} - V_{\mathbf{k}-\mathbf{p}-\mathbf{q}}) = \epsilon^{\mathrm{HFA}}_{\mathbf{p}+\mathbf{q},\sigma} - \epsilon^{\mathrm{HFA}}_{\mathbf{p},\sigma} \tag{5.154}$$

[see Eq. (5.23)]: the particle-hole energies are those calculated in the HFA. Second, the character of the forcing term is altered.

We can appreciate the nature of the generalized RPA most easily by assuming that there exists an oscillatory solution, at frequency ω, of Eq. (5.153). If we take the expectation value of both sides of the equation, and pass to the long wavelength limit, we see that it becomes

$$\left\{\omega - \mathbf{q}\cdot\mathbf{v}_{\mathbf{p}}{}^{\circ} + \mathbf{q}\cdot\nabla_{\mathbf{p}}\sum_{\mathbf{k}} n^0_{\mathbf{k}\sigma}V_{\mathbf{k}-\mathbf{p}}\right\}\delta n_{\mathbf{p}\sigma}(\mathbf{q},\omega) + \mathbf{q}\cdot\nabla_{\mathbf{p}}n_{\mathbf{p}}{}^{\circ}\sum_{\mathbf{k}\sigma'}(V_{\circ} - V_{\mathbf{k}-\mathbf{p}}\delta_{\sigma\sigma'})\delta n_{\mathbf{k}\sigma'}(\mathbf{q},\omega) = 0 \tag{5.155}$$

when we make the identifications (5.48), (5.50a), and (5.50b). Comparison with the basic Landau transport equation, (1.103),

$$(\omega - \mathbf{q}\cdot\mathbf{v}_{\mathbf{p}})\delta n_{\mathbf{p}\sigma} + \mathbf{q}\cdot\nabla_{\mathbf{p}}n_{\mathbf{p}}{}^{\circ}\sum_{\mathbf{p}'\sigma'} f_{\mathbf{p}\sigma,\mathbf{p}'\sigma'}\delta n_{\mathbf{p}'\sigma'} = 0,$$

shows that for a neutral fermion system the generalized RPA reduces to a special case of the Landau transport equation, in which the quasiparticle energies and interaction are those specified in the HFA:

$$\epsilon_{\mathbf{p}\sigma} = \epsilon^0_{\mathbf{p}\sigma} - \sum_{\mathbf{p}'} n^0_{\mathbf{p}'\sigma}V_{\mathbf{p}'-\mathbf{p}} + NV_{\circ},$$
$$f_{\mathbf{p}\sigma,\mathbf{p}'\sigma'} = V_{\circ} - V_{\mathbf{p}'-\mathbf{p}}\delta_{\sigma\sigma'}.$$

Just as the RPA corresponds to a time-dependent Hartree approximation, so then does the generalized RPA correspond to a time-dependent HFA.

For the neutral system, the generalized RPA offers an account of the density fluctuation spectrum which we would expect to be valid in the weak coupling limit, since at long wavelengths it yields the Landau transport equation with the lowest order (HFA) results for $\epsilon_{\mathbf{p}}$ and $f_{\mathbf{pp}'}$. From Eq. (5.24) we see that

$$F_l^s = N^\circ(0)\{2V_o\delta_{l,o} - \langle V_{\mathbf{p}'-\mathbf{p}}\rangle_l\}, \tag{5.156a}$$

$$F_l^a = -N^\circ(0)\langle V_{\mathbf{p}'-\mathbf{p}}\rangle_l. \tag{5.156b}$$

Starting with these equations, we can apply directly the results of Chapter 1 to work out all of the various long wavelength collective modes, stability criteria, etc. in the generalized RPA. Moreover, where it is valid, the generalized RPA permits us to develop the appropriate microscopic generalization of the Landau theory, through the solution of the basic equation, (5.153), for large values of $\mathbf{q}$ and ω. In this fashion, one can discuss the Landau damping of collective modes at finite $\mathbf{q}$, microscopic instabilities, and the like.

For example, let us consider zero sound for a system where $V(\mathbf{r})$ is a contact interaction,

$$V(\mathbf{r}) = V\delta(\mathbf{r}), \tag{5.157a}$$

in which case one has

$$F_o^s = N^\circ(0)V. \tag{5.157b}$$

We introduce $\rho^s_{\mathbf{pq}} = (\rho_{\mathbf{pq}\sigma} + \rho_{\mathbf{pq}-\sigma})/2$ and note that the spin symmetric part of the general equation, (5.153), may be written in the form

$$[H, (\rho^+_{\mathbf{pq}})^s] = \omega^\circ_{\mathbf{pq}}(\rho^+_{\mathbf{pq}})^s - \frac{V}{2}(n^\circ_{\mathbf{p+q},\sigma} - n^\circ_{\mathbf{p},\sigma})\rho_{\mathbf{q}}{}^+. \tag{5.158}$$

On comparing Eq. (5.158) with (5.145), we see that in this special case the generalized RPA reduces to the RPA with the effective interaction $V_{\mathbf{q}} = V/2$. The dispersion relation for zero sound therefore takes the form [see Eq. (5.150)]

$$1 = \frac{V}{2}\sum_{\mathbf{p}} \frac{n_{\mathbf{p}}{}^\circ - n^\circ_{\mathbf{p+q}}}{\omega_{\mathbf{q}} - \omega^\circ_{\mathbf{pq}} + i\eta}. \tag{5.159}$$

On carrying out the sum over $\mathbf{p}$, we find that

$$1 = \frac{N^\circ(0)V}{2}\left\{-1 + \frac{p_F}{2q}\left(\left[\frac{(\omega_q - q^2/2m)^2}{(qv_F{}^\circ)^2} - 1\right]\ln\left|\frac{\omega_q - qv_F{}^\circ - q^2/2m}{\omega_q + qv_F{}^\circ - q^2/2m}\right| + \left[\frac{(\omega_q + q^2/2m)^2}{(qv_F{}^\circ)^2} - 1\right]\ln\left|\frac{\omega_q + qv_F{}^\circ + q^2/2m}{\omega_q - qv_F{}^\circ + q^2/2m}\right|\right)\right\}, \tag{5.160}$$

provided the collective mode is well defined ($\omega_q > qv_F{}^\circ + q^2/2m$). The reader may verify that in the long wavelength limit, Eq. (5.160) reduces to the Landau dispersion relation, (1.128b):

$$\frac{\omega_q}{qv_F{}^\circ} \ln \left| \frac{\omega_q + qv_F{}^\circ}{\omega_q - qv_F{}^\circ} \right| = 2 + \frac{2}{N^\circ(0)V}. \tag{5.161}$$

In the weak coupling limit

$$N^\circ(0)V \ll 1, \tag{5.162}$$

Eq. (5.161) possesses the solution

$$\omega_q = qv_F{}^\circ \left\{ 1 + \frac{2}{e^2} \exp\left(- \frac{2}{N^\circ(0)V} \right) \right\} \tag{5.163}$$

(here e is the basis of Napierian logarithms).

Landau damping of zero sound occurs when the collective mode spectrum overlaps the pair excitation spectrum. This happens at a wavevector, q_c, such that

$$\omega_{q_c} = q_c v_F{}^\circ + q_c{}^2/2m. \tag{5.164}$$

It is not difficult to solve Eq. (5.164) directly for q_c; in the weak coupling limit, one finds

$$q_c = \frac{2}{e} p_F \exp\left[- \frac{2}{N^\circ(0)V} \right] \tag{5.165}$$

We see from Eqs. (5.163) and (5.165) that for weak coupling, zero sound sits just above the pair excitation spectrum in the long wavelength limit, and that it is damped at a wave vector q_c which is still quite small compared to p_F. The situation is thus very different from that one encounters for Coulomb interactions; the long range of the latter is, of course, responsible for this difference.

INSTABILITIES AND PHASE TRANSITIONS

In the long wavelength limit, the stability of the ground state against spontaneous growth of a collective mode is guaranteed by the conditions (1.141). We may use Eq. (5.156) to write these stability criteria in a form applicable to the generalized RPA,

$$\begin{aligned} 1 + N^\circ(0)[2V_\circ - \langle V_{p'-p} \rangle_\circ] &> 0, \\ 1 - N^\circ(0) \frac{\langle V_{p'-p} \rangle_\ell}{2\ell + 1} &> 0. \end{aligned} \tag{5.166}$$

If one of these conditions is violated, the corresponding collective mode grows exponentially. The state $|\Phi_\circ\rangle$ which was assumed to be the ground state spontaneously evolves toward another state, of lower energy, which contains permanent fluctuations of the quantity associated with the unstable collective mode. For example, an unstable spin wave leads to a permanent spin density fluctuation, i.e., to a *ferromagnetic* state.

The generalized RPA is thus interesting, in that it provides a specific approximation to the exact Landau theory of macroscopic instabilities. However, its main virtue is that it may also be used to discuss *microscopic* instabilities, characterized by a collective mode which starts to grow at some finite value of $\mathbf{q}$. As in the long wavelength limit, the instability threshold is reached when the dispersion relation for the collective mode under consideration possesses a zero frequency solution. The stability conditions now depend on the wave vector $\mathbf{q}$; when the interaction strength is varied, the instability first shows up at a specific value of $\mathbf{q}$; the latter thus influences the nature of the new equilibrium state. An example is the "giant spin wave" considered by Overhauser (1960, 1962), in which the instability signals a transition to an ordered state in which there exists a permanent spiral spin density fluctuation.

The generalized RPA may be used as well to study instabilities at a finite temperature T, according to the prescription (5.152). It is found that raising the temperature always decreases the tendency toward instability. At some critical temperature, T_c, the instability disappears altogether. Above T_c, the "normal" state characterized by the distribution $n_{\mathbf{p}}{}^{\circ}$ is stable; at T_c one has a *second-order phase transition* from the "normal" state to the low-temperature distorted state. Below T_c the latter state becomes stable. The generalized RPA provides an approximate method to calculate the transition temperature T_c (e.g., the Curie point in the ferromagnetic transition).

We shall not attempt to describe in detail the general results of such investigations,* but only consider as an example the simple case of the long wavelength $(\mathbf{q} \to 0)$, spin wave instability, in a system with the contact interaction (5.157a). We introduce $\rho_{\mathbf{pq}}^{a} = (\rho_{\mathbf{pq}\sigma} - \rho_{\mathbf{pq}-\sigma})/$and note that the spin antisymmetric part of the general equation (5.153) is simply

$$[H, (\rho_{\mathbf{pq}}^{+})^{a}] = \omega_{\mathbf{pq}}^{\circ}(\rho_{\mathbf{pq}}^{+})^{a} + \frac{V}{2}[n_{\mathbf{p}+\mathbf{q}}^{\circ} - n_{\mathbf{p}}{}^{\circ}]\,(\rho_{\mathbf{q}\sigma}^{+} - \rho_{\mathbf{q},-\sigma}^{+}), \qquad (5.167)$$

where $\rho_{\mathbf{q}\sigma}^{+}$ is the net density fluctuation for particles with spin σ. Equation (5.167) is formally similar to the RPA result (5.145), with the choice $V_{\mathbf{q}} = -V/2$. On making use of Eq. (5.150) and of the prescription (5.152), we may write the spin wave dispersion relation at temperature T in the form

$$1 = -\frac{V}{2}\sum_{\mathbf{p}} \frac{n_{\mathbf{p}}{}^{\circ}(T) - n_{\mathbf{p}+\mathbf{q}}^{\circ}(T)}{\omega_{\mathbf{q}} - \omega_{\mathbf{pq}}^{\circ} + i\eta}. \qquad (5.168a)$$

* The interested reader is referred to the recent monograph on phase transitions by Brout (1965) for a number of interesting applications.

The threshold of the instability occurs at that temperature T_c for which $\omega_{\mathbf{q}} = 0$. On passing to the limit $\mathbf{q} \to 0$, we find

$$1 = -\frac{V}{2} \sum_{\mathbf{p}} \frac{\partial n_{\mathbf{p}}^{\circ}(T_c)}{\partial \epsilon_{\mathbf{p}}} \tag{5.168b}$$

as the equation determining T_c. Equation (5.168b) possesses a solution only if

$$N^{\circ}(0)V > 1,$$

a condition which is equivalent to the instability criterion at zero temperature, (5.160).

When a collective mode is unstable, the pair operator $\rho_{\mathbf{pq}\sigma}$ acquires a permanent fluctuation

$$\langle c^{+}_{\mathbf{p}+\mathbf{q},\sigma} c_{\mathbf{p}\sigma} \rangle,$$

which for $\mathbf{q} \to 0$ reduces to a shift $\delta n_{\mathbf{p}\sigma}$ in the particle distribution function. In the case of spin waves, $\delta n_{\mathbf{p}\sigma}$ is isotropic and spin antisymmetric: the Fermi surfaces for the two spin directions split spontaneously, and so give rise to a net magnetization. In order to find the new equilibrium state at $T = 0$, we return to the Hartree–Fock approximation, and allow for a spin-dependent distribution $\delta n_{\mathbf{p}\sigma}$. The quasiparticle energy (5.139) is also spin dependent; in the case of a contact interaction, (5.157a), it is equal to

$$\epsilon_{\mathbf{p}\sigma} = \epsilon_{\mathbf{p}}^{\circ} + VN_{-\sigma},$$

where N_σ is the total number of particles with spin σ. Let $p_{F\pm}$ be the Fermi momenta for spin up and down, $\epsilon_{F\pm}$ the corresponding quasiparticle energies. At equilibrium, the chemical potentials for both spin directions must be equal, hence

$$\epsilon_{F+} = \epsilon_{F-}.$$

This equation, together with the condition

$$N_+ + N_- = N,$$

serves to fix the difference $(p_{F+} - p_{F-})$, i.e., the magnetization. Beside the trivial solution $p_{F+} = p_{F-}$, which corresponds to the nonmagnetic (unstable) state, it possesses another "magnetic" solution if the instability criterion, $N^{\circ}(0)V > 1$ is met. The latter solution provides the structure of the real "magnetic" ground state.

A similar calculation may be carried out when the unstable collective mode has a finite wave vector $\mathbf{q}$. In that case, the new equilibrium state is obtained by a *self-consistent* solution of the Hartree–Fock equations, in which the one-particle eigenstates are adjusted to the periodic deformation of the system. Clearly, such a procedure is much more difficult to work out.

The generalized RPA may be used to find the collective modes appropriate to the new magnetic ground state. Detailed calculations are left as a problem to the reader (see Problem 5.6). It is found that the spin wave which is unstable in the nonmagnetic state becomes stable in the real ground state.

The foregoing conclusions are general; they apply to any instability, whether macroscopic or microscopic. In each case, the analysis should proceed along the following lines:

(i) Search for the possible instabilities of the ideal ground state distribution n_p°. Calculate the corresponding transition temperature.

(ii) Determine the new distorted ground state by looking for a self-consistent solution to the linearized Hartree–Fock equations of motion. The nature of the distortion is dictated by that of the unstable collective mode.

(iii) Calculate afresh the collective mode spectrum in the new ground state, by using the generalized RPA. Verify that the instabilities have been washed out by the phase transition.

The above approach will be used extensively in Chapter 7 to discuss the transition to the superfluid state.

VALIDITY OF THE RPA AND THE GENERALIZED RPA

In considering the validity of the above approximations, we must take care to distinguish between neutral and charged fermion systems. For a neutral system, only the *generalized RPA* furnishes a consistent account of the system properties; it is, moreover, valid only if the interaction between the particles is weak. For the special case of a contact interaction, one may state more precisely the range of validity of the generalized RPA; it is valid as long as

$$N^\circ(0)V \ll 1.$$

That the generalized RPA is a *weak coupling theory* is evident from the fact that both the quasiparticle energy and interaction, (5.23) and (5.24), are those specified in the HFA; the latter is itself well known to correspond to a first-order (in V) calculation in the case of a uniform system.*

Where the coupling is not weak, it is necessary to take into account

* We note that since the generalized RPA is a weak coupling approximation, its use to predict instabilities which appear only in the strong coupling limit [$N^\circ(0)|V| > 1$; cf. Eq. (5.166)], must be viewed with considerable caution.

the quadrilinear terms in the exact equation of motion, (5.140). One thus writes down the equation of motion for a two-pair operator, finds it is coupled to the motion of one-pair excitations and three-pair excitations, and then searches for a systematic method of finding an approximate solution for the full set of coupled equations. This general method resembles closely that of solving for the chain of coupled equations for the Green's functions of the system, and the latter procedure is often easier to follow. We do not enter upon the details of any more general calculations here [for an account of the Green's function approach, see Baym and Kadanoff (1962)]; we merely wish to remark that in the long wavelength, low-frequency limit, the resulting equations may be substantially simplified, since only single-pair excitations are of importance. One recovers, then, the Landau transport equation, (1.103), where the $f_{\mathbf{pp}'}$ corresponds to the full interaction between quasiparticles.

For the electron liquid, the hierarchy of systematic approximations is quite different from that which obtains for the neutral system. The reason is simple; the "direct" term, $V_{\mathbf{q}}$, in Eq. (5.153) plays a dominant role, since it diverges in the long wavelength limit. As a result, one obtains a consistent account in lowest order of approximation by making the "simple" RPA instead of the generalized RPA, that is, by neglecting the exchange terms in Eq. (5.153). If, now, one wants to go beyond the RPA (and thereby take exchange scattering of pairs into account), it is necessary to go beyond the equations of the generalized RPA, (5.153); specifically, one must include those terms in the chain of coupled equations which serve to screen the exchange scattering contribution. Systematic procedures for doing this have been worked out by Suhl and Werthamer (1961, 1962) and by Watabe (1962). One finds, in the next order of approximation, a transport equation which involves the correct RPA screened quasiparticle interaction, $f_{\mathbf{pp}'}^{\mathrm{RPA}}$ (that which gives rise to the various RPA quasiparticle properties considered in Section 5.4); one also finds all the corrections of next order in r_s to the RPA results for the plasmon dispersion relation, ground state energy, etc. We shall not enter upon these corrections here, because, as we shall now demonstrate, they are of little physical interest: they represent small corrections for the domain of high densities ($r_s \ll 1$) for which the entire perturbation-theoretic approach is appropriate.

That the RPA is a small r_s expansion is evident from the general form of the series expansion for the ground state energy, quasiparticle properties, etc. The breakdown of the RPA at densities of the order of $r_s = 1$ may be seen in a number of ways. For example, the second-order "exchange" correction to the ground state energy is found to be 0.046 ryd. At a density of $r_s = 1$, such a correction amounts to some 30% of the RPA calculation of the correlation energy (−0.142 ryd).

Even when this exchange correction is included, the resulting expression is only valid for values of $r_s \lesssim 1$; at lower densities, it fails to satisfy the requirement that the second derivative of the ground state energy with respect to the coupling constant be negative [Ferrell (1959)]. One might hope to improve the RPA by calculating the next terms in the perturbation series expansion for the polarizability, the ground state energy, etc. Such calculations have, for example, been carried by DuBois (1959) for the specific heat, and (apart from some enormously complicated integrals) for the ground state energy. They are interesting as a study of the structure of the series expansion; however, one cannot in this way arrive at an accurate account of an electron gas at metallic densities ($1.8 < r_s < 5.5$), in which the kinetic and potential energies are comparable; this fact renders it equally doubtful that any extension of the strong coupling calculations (which are valid for $r_s \gg 1$) can offer a reliable guide to metallic behavior.

5.6. EQUILIBRIUM PROPERTIES OF SIMPLE METALS

We consider now the thorny question of whether microscopic theory can tell us anything useful about the behavior of electrons in metals. We may distinguish between the role played by electron interaction in macroscopic and microscopic phenomena. For macroscopic phenomena, the Landau theory is applicable. The aim of microscopic theory must then be the calculation, from first principles, of the relevant Landau parameters. For microscopic phenomena, on the other hand, no phenomenological theory exists; only a microscopic theory will suffice.

A microscopic theory of metallic behavior must take into account at the outset the combined effects of (i) the periodic potential of the ions, a "solid state" effect; (ii) electron-electron interactions, a "many-body" effect. For many phenomena, one must also consider the influence of (iii) electron-phonon interactions. The challenge to the theorists is a formidable one; at the very least, it is necessary to develop a theory of electron-electron interactions which is valid for metallic electron densities. Moreover, solid state and many-body effects are intertwined. In allowing for the influence of the periodic potential, it is generally necessary to include the screening action of the electrons, a characteristic many-body effect; one would like as well to consider changes in a theory of electron interaction which comes about because the electrons move in a periodic potential. In view of all the difficulties involved in the construction of a theory, what is perhaps surprising is that it has been attempted for *any* metallic property.

Historically, the first microscopic calculations in which both electron-ion and electron-electron interactions were taken into account were those of Wigner and Seitz (1933) for the cohesive energy of metals. Such calculations continue to be of great interest; they are typical, in that comparison of theory with experiment furnishes a test of the combined accuracy of

(i) a "solid state" calculation (of the binding of single electrons in the ionic potential, and the choice of such potentials);
(ii) a "many-body" calculation (of the correlation energy for electrons at metallic densities);
(iii) the separation which has been made of solid state and many-body effects.

Essentially the same situation arises when one compares theory and experiment for the electronic specific heat or spin susceptibility; one is testing not simply a many-body theory or a solid state theory, but a combined many-body, solid state theory. It is evident that one cannot expect precise agreement between such a theory and experiment; so many approximations go into constructing the theory that a 15% discrepancy between theory and experiment may be regarded as a theoretical triumph—or a happy accident, depending on one's point of view.

Accordingly we shall not attempt in this section an exhaustive survey of microscopic calculations of metallic behavior; rather, we shall present only a few examples, which are intended to illustrate the "state of the art" of carrying out microscopic calculations of many-body effects at metallic densities. Our primary interest is in such calculations, and the extent to which they agree with experiment. Therefore we do not develop a firm theoretical base for the solid state effects we consider, but rather refer the interested reader to the relevant scientific literature.

We begin with a discussion of the approximation procedures which have been proposed for electron liquids at metallic densities. We then apply these microscopic calculations to the determination of the cohesive energy of alkali metals, and to the quasiparticle properties of simple, nearly isotropic, metals, such as the alkalis, Al, and Pb.

APPROXIMATIONS FOR METALLIC DENSITIES

In attempting to develop an approximation procedure which is valid at metallic electron densities it is useful to analyze in some detail why the RPA does not apply in this density regime. To do so, we consider the contributions to the correlation energy from different momentum transfers. We have seen in, for example, Eqs. (5.89) and (5.94) how it is always possible to isolate the contribution from a given momentum transfer by calculating, say, $E_{\text{int}}(\mathbf{q})$, and then carrying out the integration over the coupling constant. It is not difficult to calculate the contributions from small momentum transfers in the RPA: the leading terms come from the zero-point energy of the plasmons and agree with the *exact* result, (5.99). Moreover, the authors [Nozieres and Pines (1958a)] have shown that for larger values of $\mathbf{q}$, so long as $q \lesssim q_c$, the "exchange" corrections to the RPA calculation of $E_{corr}(\mathbf{q})$ are small, even at metallic densities. It follows that errors in the RPA come from large values of the momentum transfer—in fact from $q \gtrsim p_F$.

It is clear why the RPA runs into difficulties at large momentum transfers. As Hubbard (1957) and the authors [Nozières and Pines (1958a)] have argued, there is no distinction in the RPA between the contribution to correlation effects from electrons of parallel spin and antiparallel spin. On the other hand, physically, one would expect that electrons of parallel spin simply don't feel the short-range part of the interaction, inasmuch as they are kept apart by the Pauli principle. Mathematically, such an effect appears because for large momentum transfers the "exchange" parts of the perturbation-theoretic expansion (which occur only for electrons of parallel spin, and which are neglected in the RPA) cancel that one half of the "direct" interaction that may be attributed to electrons of parallel spin. The result: only electrons of antiparallel spin interact via the large momentum transfer part of the Coulomb interaction. The origin of such a cancellation is simple: to a given direct process for electrons of parallel spin, going by a matrix element $V_{\mathbf{q}}$, there exists an exchange "conjugate" process, going by a matrix element $-V_{\mathbf{k}-\mathbf{p}}$. (For electrons of antiparallel spin, no such exchange processes are possible.) For small momentum transfers, the exchange terms are not important; on the other hand, for large momentum transfers the exchange process will cancel that part of the direct process that is attributed to interactions between particles of parallel spin.*

These two remarks are at the basis of the approximation procedures proposed for metallic densities by Hubbard (1957), and by the authors (1958a). Hubbard's approximation is equivalent to taking the screened density-density response function to be

$$\chi_{sc}^{H}(\mathbf{q},\omega) \cong \frac{\chi^{0}(\mathbf{q},\omega)}{1 + f(q)(4\pi e^2/q^2)\chi^{0}(\mathbf{q},\omega)}, \tag{5.169}$$

* At this point one may ask: Why then does the RPA work at any density? The answer is that at high densities the important contributions to the correlation energy come from low momentum transfers, for which the RPA result is valid. It is only as one goes toward metallic densities that the high momentum transfer contributions become appreciable; where they are important one must go beyond the RPA to calculate the system properties.

where

$$f(q) = \frac{q^2}{2(q^2 + q_F^2)}. \tag{5.170}$$

With this choice of χ_{sc}, the basic quantity which determines the correlation energy, $\chi(\mathbf{q}, \omega)$, is given by

$$\chi^H(\mathbf{q}, \omega) = \frac{\chi^\circ(\mathbf{q}, \omega)}{1 - (4\pi e^2/q^2)(1 - f)\chi^\circ(\mathbf{q}, \omega)}. \tag{5.171}$$

We see that for $q \ll q_F$, χ is unchanged from its RPA value, (5.101). On the other hand, for $q \gg q_F$ one has $f \cong \frac{1}{2}$, and one can further expand the denominator in Eq. (5.171); one finds then

$$\chi^H(\mathbf{q}, \omega) \cong \chi^\circ + \frac{4\pi e^2}{q^2}\frac{(\chi^\circ)^2}{2} + \cdots. \tag{5.172}$$

The first term in the expansion, that responsible for the exchange energy, is unchanged; the second term, which gives rise to the second-order "direct" term in the perturbation-theoretic expansion for the ground state energy, is reduced by a factor of 2 for large q, in accord with the above physical arguments. The expression (5.171) offers a smooth interpolation between these two limiting regions. The results which Hubbard obtained for the correlation energy are shown in Table 5.1.

The approximation proposed by the authors (hereafter referred to as the NP approximation) represents another way of interpolating between the low-q and high-q contributions to the correlation energy. For low momentum transfers, the RPA result, $E_0^{\text{RPA}}(q)$, may be expanded in powers of q. On summing $E_0^{\text{RPA}}(q)$ up to a sufficiently small cut-off, say $\beta_1 q_F$, one finds the following contribution to the correlation energy [Nozières and Pines (1958a)]:

$$E_{\text{corr}}^{\text{l.r.}}(\beta_1) = \left\{-0.46\frac{\beta_1^2}{r_s} + 0.87\frac{\beta_1^3}{r_s^{3/2}} - 0.98\frac{\beta_1^4}{r_s^2} + 0.71\frac{\beta_1^5}{r_s^{5/2}} + \cdots\right\} \text{ ryd.} \tag{5.173a}$$

The first two terms in Eq. (5.173a) arise from the electron self-energy and plasmon zero-point energy [see Eqs. (5.90) and (5.97)]: they are *exact*. The following terms are subject to "exchange" corrections which are small at metallic densities.

TABLE 5.1. *Comparison of Different Calculations for the Correlation Energy*

r_s	$E_{\text{corr}}^{\text{Hubbard}}$	$E_{\text{corr}}^{\text{NP}}$	$E_{\text{corr}}^{\text{Wigner}}$
2	−0.099	−0.094	−0.090
3	−0.086	−0.081	−0.082
4	−0.074	−0.072	−0.075
5	−0.067	−0.065	−0.069

At large momentum transfers, it was argued that the perturbation-theoretic expansion converges sufficiently rapidly that it suffices to consider only the second-order term. In this term the exchange part effectively cancels one-half the direct part; in other words, for the large momentum transfers under consideration, one need take into account only the contribution from the interaction between electrons of antiparallel spin. Thus, the short-range contribution to the correlation energy (for momentum transfers $q > \beta_2 p_F$) may be approximated as

$$E_{\text{corr}}^{\text{s.r.}} = -\{0.025 - 0.063 \ln \beta_2 + 0.0064\beta_2^2\} \quad \text{ryd}, \qquad (5.173\text{b})$$

provided β_2 is sufficiently large. [The result (5.173b) is obtained by carrying out the summations for $E_2^{(a)}$ in Eq. (5.112), and dividing the result by a factor of 2, to allow for the fact that only the direct interactions between particles of antiparallel spin are to be included.]

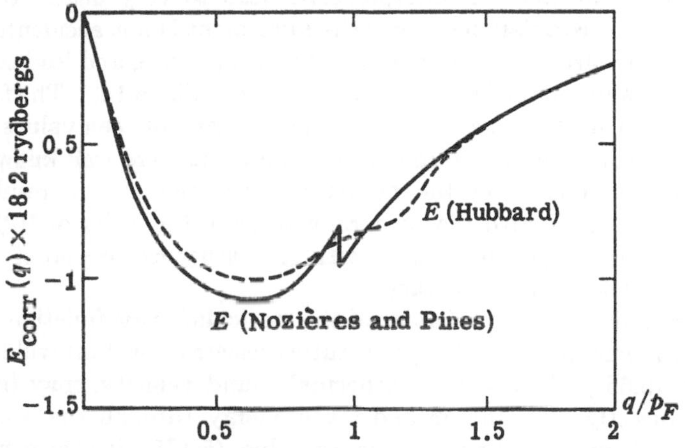

FIGURE 5.14. *Contributions to the correlation energy from different momentum transfers [Nozières and Pines (1958a)].*

What are the limits of validity for the results (5.173a) and (5.173b)? The largest value of β_1 for which Eq. (5.173a) would seem appropriate is the plasmon cut-off, $q_c/p_F = 0.47r_s^{1/2}$. The smallest value of β_2 for which Eq. (5.173b) would seem appropriate is of the order of 1.5; for smaller values of momentum transfer, we may expect that the higher-order terms in the perturbation-theoretic expansion will play an important role. In Fig. 5.14, we give a plot of $E_c(q)$, the contribution to the correlation energy from momentum transfer q, for a density of $r_s = 4$ (sodium); for $q/q_F \leq 0.47r_s^{1/2}$ we show the RPA expression, while for $q \geq 0.47r_s^{1/2}$ we show that derived by considering only the interaction between electrons of antiparallel spin, calculated in second-order perturbation theory. We see by inspection that the two curves very nearly meet at $q/q_F \cong 0.47r_s^{1/2} \cong 1$. If now we assume that $E_c(q)$ is a smoothly varying function of q, it is easy to interpolate between, say, $q/q_F \cong 0.8$ and $q/q_F \cong 1.5$ to obtain an estimate of the contribution to the correlation energy from those important *intermediate* momentum transfers for which neither the RPA nor perturbation-theory applies. In fact, since the discontinuity in $E_c(q)$ at $q/q_F = 0.47r_s^{1/2}$ is rather symmetrically located, one can obtain an explicit expression for E_{corr} by simply substituting $\beta = 0.47r_s^{1/2}$ into the expressions (5.173a) and (5.173b) for the long-range and short-range correlation energy, respectively. One then finds

$$\begin{aligned} E_{\text{corr}}^{\text{l.r.}}(\beta) &= -0.043 \quad \text{ryd}, \\ E_{\text{corr}}^{\text{s.r.}}(\beta) &= -0.072 + 0.031 \ln r_s \quad \text{ryd}, \end{aligned} \tag{5.174}$$

and, on adding,

$$E_{\text{corr}} \cong -0.115 + 0.031 \ln r_s \quad \text{ryd}. \tag{5.175}$$

The resulting numerical values for the correlation energy are shown in Table 5.1.

The Hubbard and NP interpolations lead to very nearly the same results for the correlation energy; the agreement is not accidental, since both schemes are based on the same physical picture, and lead to nearly the same values for $E_c(q)$, as may be seen from Fig. 5.14. The fact that it is not difficult to make a smooth interpolation between values of q for which the contributions to the correlation energy are well known gives one some confidence in the numerical accuracy of the results; the authors have estimated that accuracy to be of the order of 10%. We note that both expressions agree with the earlier expression of Wigner, (5.88), to this order of accuracy.

We consider next the behavior of the static dielectric function, $\epsilon(\mathbf{q}, 0)$. For low momentum transfers, the static dielectric constant will be given by Eq. (5.66), where s, the isothermal sound velocity, may be determined directly from the ground state energy through the use of Eq. (5.128). If we take the approximate value, (5.175), for the correlation energy, we find

$$\frac{\kappa_0}{\kappa} = \frac{s^2}{s_0^2} \cong 1 - \frac{\alpha r_s}{\pi} - 0.016\alpha^2 r_s^2. \tag{5.176}$$

The term in $\alpha^2 r_s^2$, which comes from the correlation energy, has only a slight influence on the compressibility compared to that arising from the exchange energy—essentially because the correlation energy is a slowly varying function of r_s. [Values of the compressibility calculated from the Wigner or Hubbard expression differ little from Eq. (5.176).] We see that electron interaction has a pronounced effect on the compressibility, and hence on the long wavelength screening action of the electron liquid: at $r_s = 4$, the compressibility is nearly four times its Fermi–Thomas value.

The Hubbard expression, (5.169), does not provide an accurate interpolation between the small q and large q limits of $\epsilon(\mathbf{q}, 0)$, since it predicts Fermi–Thomas screening in the long wavelength limit. On the other hand, it may be expected to provide a better account of short wavelength screening than does the RPA, since it allows for the role of exchange corrections in this limit. We note that insofar as the very long-range oscillations in the screening density are concerned, there is little to choose between a result based on Eq. (5.169) and that calculated in the RPA; at $q \cong 2p_F$, the two expressions are nearly the same (see Fig. 5.6). Langer and Vosko (1960) have carried out studies of the dielectric screening of a fixed impurity, using both the RPA and the Hubbard expression for $\epsilon(\mathbf{q}, 0)$; for detailed comparison with experiment it would be desirable to have a more accurate calculation than is provided by either of these expressions.

Calculations of the specific heat and spin susceptibility for an electron liquid at metallic densities have been carried out by Silverstein (1962), with the help of field-theoretic methods. The NP approximation procedure is taken as the basis for his computations. Recently Rice (1965) has used the Hubbard approximation to calculate both these quantities and the quasiparticle interaction energy $f_{\mathbf{pp}'}$. Rice has taken as his starting point the Hubbard approximation to the ground state energy E_H. To obtain E_H, one substitutes χ_H, [Eq. (5.171)] for χ_{RPA} in (5.100); one finds [compare (5.105)].

$$E_H - \tfrac{3}{5}\epsilon_F{}^0 = \sum_{\mathbf{q}} \left\{ \frac{1}{4\pi N} \int_{-\infty}^{\infty} dw \ln\left[1 - \frac{4\pi e^2}{q^2}[1 - f(q)]\chi^0(\mathbf{q}, iw)\right] - \frac{2\pi e^2}{q^2} \right\}. \quad (5.177)$$

Rice has then considered the various functional derivatives of E_H, and calculated numerically m^*, χ_P, and $f_{\mathbf{pp}'}$. His choice of f differs somewhat from Eq. (5.170), being $f = q^2/2(q^2 + q_F{}^2 + q_{FT}^2)$. The results of Silverstein (as corrected by Rice) and Rice for the spin susceptibility and effective mass for several values of r_s are given in Table 5-2. By

TABLE 5.2. *Calculations of χ_P and m^* in the Hubbard and NP Approximations*

	$\chi_P/\chi_P{}^0$		m^*/m	
r_s	NP	H	NP	H
2	1.26	1.26	1.02	0.99
3	1.28	1.40	1.05	1.02
4	1.29	1.48	1.10	1.06

TABLE 5.3 *Values of Landau Parameters, as Calculated in the Hubbard and NP Approximations*

	F_0^s		F_0^a		F_1^s	
r_s	NP	H	NP	H	NP	H
2	−0.34	−0.35	−0.19	−0.21	0.06	−0.03
3	−0.52	−0.51	−0.18	−0.27	0.15	0.06
4	−0.70	−0.69	−0.15	−0.29	0.30	0.18

combining the above calculations, and including that for the compressibility, we may arrive at values for F_0^s, F_0^a, and F_1^s. The Landau parameters calculated in the Hubbard and NP approximations are given in Table 5.3.

The interaction between quasiparticles at metallic densities is found not to be especially strong, being most effective on the compressibility. We note that the differences between the Hubbard and NP interpolation procedures are more pronounced for quasiparticle properties than for the ground state energy. In Fig. 5.15, we reproduce Rice's calculation of the quasiparticle interaction energy at two different densities. The magnitude of the interaction between quasiparticles of parallel spin is seen to be larger than that between those of antiparallel spin, an expected result in view of the fact that the former interactions

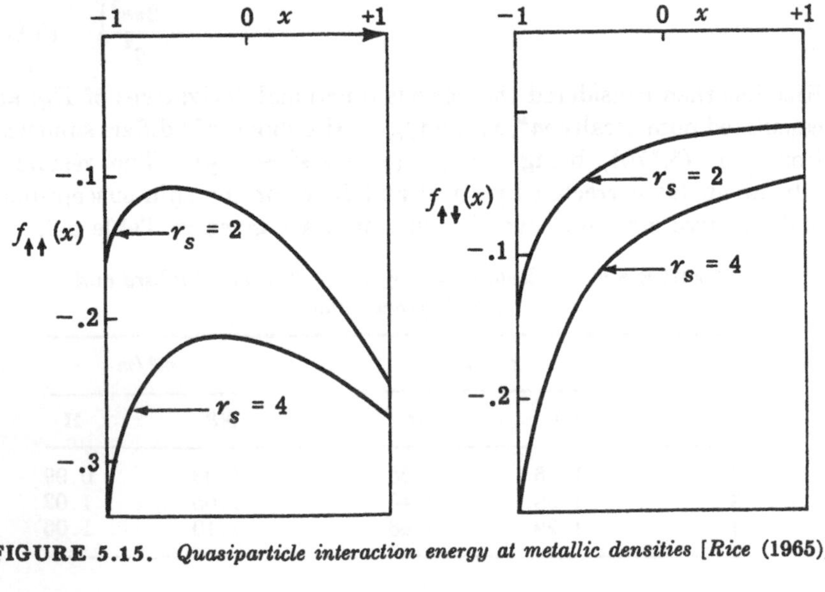

FIGURE 5.15. *Quasiparticle interaction energy at metallic densities* [*Rice* (1965)].

dominate completely in the high-density limit. (Recall that in the HFA, $f_{\uparrow\downarrow} = 0$.) We note that the interactions are more nearly equal for $r_s = 4$, an indication that at this density charge-induced correlations are beginning to play a large role. We remark also that while both $f_{\uparrow\uparrow}$ and $f_{\uparrow\downarrow}$ are negative, $f_{\uparrow\uparrow} < f_{\uparrow\downarrow}$; this last effect arises from the exchange repulsion of particles of parallel spin, and is responsible for the negative value of F_0^a.

It is difficult to estimate the accuracy of these approximate calculations of quasiparticle properties at metallic densities; it is likely that they are not as accurate as the corresponding calculations of the correlation energy. We may make one general remark—that as one goes to the very low end of metallic densities ($4 \lesssim r_s \lesssim 5.56$), there is a qualitative change in the physical behavior of the electron liquid. This may be seen by comparing, for example, the size of the zero-point energy of plasmons with the Fermi energy, $2.21/r_s^2$. With $\beta = 0.47 r_s^{1/2}$, one finds the zero-point energy of plasmons is constant, $\cong 0.10$ ryd. At $r_s \cong 4.5$, it is comparable to the "unperturbed" kinetic energy of the single particles. One would expect that such a change in the physical behavior of the system would require a corresponding modification in the approximations used to describe it. Since the above approximations are reached from the high-density side, and since they do not appreciably change character with r_s, it is likely that they are more reliable for $r_s \lesssim 3$ than for the low electron densities characteristic of the heavier alkali metals.

A further indication of difficulties with the theory at low electron densities comes from consideration of the compressibility. From Eq. (5.176) one finds that the compressibility becomes infinite at $r_s \cong 5.5$ (roughly the density of cesium). According to Section 4.1, this would mean that an electron liquid of this density is unstable. It is likely that the instability is not real (the electron solid is thought to occur at very much lower densities), and that its appearance here should be regarded as indicating a failure of the expression (5.176) in this density region.

COHESIVE ENERGY

As a first application of the microscopic calculations of the preceding subsection, we consider the cohesive energy of the alkali metals. The cohesive energy of a metal is defined as the difference in energy between a collection of free atoms and those atoms brought together in the metal.* For a monovalent metal, with an isotropic Fermi surface, the contributions to the cohesive energy may be divided into three parts:

* For a review of cohesive energy calculations in simple metals, see Brooks (1958); for recent calculations of the band structure of the alkali metals, see Ham (1962).

(1) E_{ion}, the difference between the binding energy of the most tightly bound conduction electron (that at the bottom of the conduction band) and the ionization energy of the free atom.

(2) E_{kin}, the average kinetic energy of the conduction electrons, calculated by assuming that the electrons do not interact.

(3) E_{pot}, the average contribution to the energy arising from the interaction between conduction electrons.

With this division, the quantities E_{ion} and E_{kin} are determined by a "one-electron" calculation in which the effect of the periodic ionic potential on the motion of a single electron is taken into account. E_{kin} may be written in the form

$$E_{\text{kin}} = \langle \epsilon_{\mathbf{p}}^{\circ} \rangle = \tfrac{3}{5}(p_F^{\,2}/2m_{\text{kin}}), \qquad (5.178)$$

where $\epsilon_{\mathbf{p}}^{\circ}$ is the energy of a Bloch wave of momentum $\mathbf{p}$, and the *average is carried out over the entire Fermi sphere.* Equation (5.178) serves to define the *kinetic mass,* m_{kin}. We note that m_{kin} will in general differ from the *crystalline mass,* m_c; the latter is determined by the density of states for electrons *on* the Fermi surface, according to Eq. (1.69a). We may write

$$E_{\text{kin}} = \frac{2.21}{r_s^{\,2}}\left(\frac{m}{m_{\text{kin}}}\right) \text{ ryd.} \qquad (5.179)$$

All major effects of electron interaction reside in E_{pot}, the ground state energy for the many-electron system, where, however, the contributions from the electronic kinetic energy (E_{kin}) and the binding of the electrons in the field of the positive core (E_{ion}) are not included. E_{pot} differs from the corresponding terms in the ground state energy of an electron liquid of the same density in the following respects:

(a) Electrical neutrality is brought about by discrete positive ions rather than a uniform background of positive charge. Since we have allowed for the electron-positive ion interaction (through E_{ion}), it is necessary that we include the Coulomb self-energy of the electronic charge distribution (in the electron *liquid*, this contribution is canceled by the uniform distribution of positive charge). The electronic self-energy is easily shown to be $1.2/r_s$ ryd per electron.

(b) In computing the exchange and correlation energy for the many-electron system, it is in principle necessary to correct for the fact that the electrons move in the ionic potential, rather than in a uniform distribution of positive charge.

We now make the following important approximation: we *assume* that the exchange and correlation energy are the same as for the corresponding electron liquid. In that case, the cohesive energy takes the form

$$E_{coh} = \left(E_{ion} + \frac{2.21}{r_s^2} (m/m_{kin}) + 0.28/r_s + E_{corr} \right) \text{ ryd}, \quad (5.180)$$

where we have used the value (5.11) for the exchange energy per electron, and E_{corr} is the correlation energy per electron for the electron liquid.

The last approximation we have made is a plausible one, in that the greater part of the exchange and correlation energies comes from microscopic values of the momentum transfer, that is, from comparatively short-range interactions between the electrons. Two electrons which are close together do not feel appreciably the influence of the ion cores; they see only one another, so to speak. As a result, the effects of periodicity on the exchange and correlation energies should be comparatively slight. We do not know how to calculate such solid state effects; we cannot therefore justify their neglect. We can only make the assumption that they are negligible, and then compare our calculation with experiment to see whether any obvious inconsistencies appear. Thus, as we have mentioned already, by comparing the theoretical calculations of the cohesive energy with experiment, one obtains a sense of the combined accuracy of the one-electron "solid state" calculations, the microscopic many-body calculations of E_{corr} at metallic electron densities, and the separation which has been made of many-body and solid state effects.

The most accurate cohesive energy calculations which have thus far been carried out are for the alkali metals. A comparison between theory and experiment is given in Table 5.4; the band structure calculations are those of Brooks (1953); the correlation energies used are those obtained with the NP interpolation approximation. We comment on several features of interest in this table.

TABLE 5.4. *Contributions to the Cohesive Energy of the Alkali Metals*[a]

Metal	r_s	E_{ion}	m_{kin}/m	E_{kin}	E_{corr}	E_{pot}	E_{coh}	E_{expt}
Li	3.22	−87.2	1.45	47.3	−25.6	2.4	−37.5	−36.5
Na	3.96	−71.3	0.98	46.5	−23.2	−0.4	−25.2	−26.0
K	4.87	−51.6	0.93	32.4	−21.2	−2.6	−21.8	−22.6
Rb	5.11	−47.6	0.89	29.9	−20.7	−3.2	−20.9	−18.9
Cs	5.57	−43.9	0.83	27.8	−20.0	−3.8	−19.9	−18.8

[a] After Pines (1955). Energies are given in kg cal/mole (1 ryd = 311 kg cal/mole).

(1) The total contribution from the potential energy of electron interaction is relatively small. Electrons tend to avoid one another to such an extent that if one electron is found in a given unit cell, it is not likely that another electron will be found there. It is not a bad approximation to take $E_{pot} = 0$, thus neglecting *all* effects of electron interaction. [This was recognized long ago by Wigner and Seitz (1933).]

(2) E_{pot} decreases as one goes from Li to Cs; thus, as the electron density is decreased, electrons become more adept at avoiding each other. As we have noted, as a result, one finds in the very low-density limit an electron solid.

(3) Given the number of approximations which have been made in obtaining E_{coh}, the agreement with experiment is little short of startling. This measure of agreement cannot, in and of itself, be taken as indicating that all approximations which have gone into E_{coh} are that accurate; still, the good agreement with experiment should not be held against the methods of approximation that have been employed.

QUASIPARTICLE PROPERTIES

We now consider the microscopic calculation of quasiparticle properties in metals. In dealing with such properties, we must at the outset confine our attention to metals which are nearly isotropic. We have remarked in Chapter 3 that, in the case of a highly anisotropic metal, although the Landau theory may apply it will involve so many independent parameters as to be virtually useless; the outlook for a microscopic description of such metals is scarcely better. Thus we shall consider in this section relatively simple metals, such as the alkalis, Al, and Pb. We shall furthermore confine our attention to the spin susceptibility and specific heat of these metals.

We have seen in Section 3.8 that for a metal with an isotropic Fermi surface, application of the Landau theory shows that [cf. Eqs. (3.160) and (3.164)]

$$\frac{C}{C_0} = \frac{m^*}{m} = \left(\frac{m_c}{m}\right)(1 + F_1{}^s/3), \tag{5.181}$$

$$\frac{\chi_P}{\chi_P{}^0} = \left(\frac{m_c}{m}\right)\frac{(1 + F_1{}^s/3)}{1 + F_0{}^a}, \tag{5.182}$$

where C_0 and $\chi_P{}^0$ are the specific heat and spin susceptibility for a noninteracting free electron gas, and m^* is the *thermal* effective mass, defined through the specific heat ratio, (5.181). In these equations, m_c is the crystalline mass, defined through Eq. (3.157) in terms of the current carried by a quasiparticle,

$$\mathbf{j}_\mathbf{p} = \mathbf{p}/m_c. \tag{5.183}$$

In principle, as we have seen in Section 3.8, m_c may be determined from a measurement of the optical conductivity in the collisionless regime, at frequencies such that interband transitions play no role. In practice, such data are not yet available for the metals we consider; m_c must therefore be determined from band theory. It is for this reason that we have not attempted in Chapter 3 to construct a table of experimentally determined values for the various parameters, $F_0{}^a$, $F_1{}^s$, etc.

In the calculation of m_c, as in the calculation of E_{ion}, many-body and solid state effects are intertwined, since the effective ionic periodic potential one assumes in a band-theory calculation is itself determined in part by electron screening effects. Moreover, assuming one knows m_c, how does one calculate $F_0{}^a$, etc? First, one needs a microscopic calculation of the quasiparticle interaction which is valid for an electron liquid of metallic density. Next, one must decide whether that calculation is, in turn, influenced by periodic solid state effects, since the electrons are moving in the periodic field of the ions rather than in a uniform background of positive charge. Finally, one must take into account the possible influence of electron-phonon interactions on the property under study.

We have given, in the earlier part of this section, the results of approximate calculations of $F_0{}^a$, etc. for electron liquids at metallic densities. We now *assume* that the interaction between quasiparticles in a metal is unaffected by the periodic ionic potential so that we may write

$$f_{\mathbf{pp'}}^{\text{metal}} \cong f_{\mathbf{pp'}}^{\text{liquid}}. \tag{5.184}$$

Our justification for this assumption is identical with that produced for the metallic correlation energy: we argue that the major contributions to $f_{\mathbf{pp'}}$ come from microscopic values of the momentum transfer, and that the short-range part of the interaction between quasiparticles is not appreciably influenced by the ions. Again, we have no way of estimating the accuracy of Eq. (5.184), other than checking to see whether there exist otherwise inexplicable inconsistencies between theory and experiment.

Field-theoretic calculations are required to see whether a given quasiparticle property is influenced by the electron-phonon interaction. Simkin (1963) has shown that the spin susceptibility and compressibility are not altered, while it has been known for some time that the specific heat is changed by this interaction. Recently Kadanoff and Prange (1964) have studied a number of transport properties, and have concluded that in almost all cases (apart from the specific heat) electron-phonon interaction plays no role. We may therefore apply Eq. (5.182) directly to a consideration of the spin susceptibility, while Eq. (5.181) must be modified. A straightforward generalization of the results of Kadanoff and Prange shows that the specific heat (and thermal effective mass) now take the form

$$\frac{C}{C_0} = \frac{m^*}{m} = \left(\frac{m^*_{\rm ph}}{m}\right)\left(\frac{m^*_{\rm el}}{m}\right), \tag{5.185}$$

where m^* is the thermal mass measured in a specific heat experiment, $m^*_{\rm el}$ is the thermal mass determined by taking only electron-electron interactions and crystal structure into account [it is thus given by Eq. (5.181)], and $m^*_{\rm ph}$ is the thermal mass calculated taking only electron-phonon interactions into account. Detailed calculations of $m^*_{\rm ph}$ have recently been carried out for Na, Al, and Pb by Ashcroft and Wilkins (1965), and we shall refer to them shortly.

In comparing theory with experiment for the spin susceptibility and specific heat we shall begin with the alkali metals, primarily because it is for these metals that solid state effects have been most thoroughly investigated. Ham (1962) has carried out a detailed theoretical study of the Fermi surface of all the alkali metals. He found that sodium is to all intents and purposes isotropic, with a density of states at the Fermi surface equal to its free electron value. Potassium is only slightly anisotropic, Rb more so, while Li and Cs display decided anisotropy; the extent of the anisotropy is made evident by comparing $(m_c/m)_{\rm sph}$, the crystalline mass as calculated assuming a spherical Fermi surface, and m_c/m, the crystalline mass calculated by averaging the state density over the actual distorted Fermi surface. Ham's results are shown in Table 5.5.

The degree of anisotropy for Rb, Li, and Cs is such as to cast serious doubt on any attempt at taking into account the influence of quasi-particle interaction on χ_P and m^*; for these metals it would seem desirable to carry out some more realistic averaging over angles in $f_{pp'}$ than that implied by the assumption of isotropy. Assuming this could be

TABLE 5.5. *Crystalline Mass for the Alkali Metals*[a]

	Li	Na	K	Rb	Cs
$(m_c/m)_{\rm sph}$	1.32	1.00	1.02	0.99	1.06
m_c/m	1.66	1.00	1.09	1.21	1.71

[a] As calculated by Ham (1962).

done, one must next overcome the following difficulty: all microscopic calculations of the influence of electron interaction on quasiparticle properties have assumed isotropy. Thus in passing from the intermediate density approximations to the metallic case, one necessarily winds up with a value for $f_{pp'}$ which depends only on the angle between **p** and **p**′ and is therefore applicable only to the isotropic case. This difficulty is not a serious one for the correlation energy or cohesive energy calculations, where all effects associated with anisotropy tend to be averaged out; it appears well nigh insurmountable for the precise calculation of quasiparticle properties. We shall, nevertheless, carry out a comparison of theory with experiment for all the alkali metals, keeping clearly in mind the deficiencies of the theoretical calculations.

We consider first the spin susceptibility, since that is unaffected by electron-phonon interactions. A direct experimental measurement of the spin susceptibility for conduction electrons in metals was first carried out by Schumacher, Carver, and Slichter (1954), who obtained χ_P from a comparison of the size of the electron spin resonance signal with that associated with the nuclear magnetic resonance in the same sample. Experimental data is now available for Li and Na. We therefore use Eq. (5.182) to combine the results of Ham with the calculations of Silverstein and Rice for the spin susceptibility of an electron liquid, to arrive at a theoretical expression which may be compared with experiment. The results of such a comparison are given in Table 5.6. We see that the influence of electron-electron interactions is appreciable for both metals. The close agreement for Li between the Silverstein calculation of χ_P and that measured experimentally would

TABLE 5.6. *Comparison of Theoretical and Experimental Values of the Spin Susceptibility.* $\chi_P{}^H$ *Denotes the Theoretical Value Obtained with the Hubbard Approximation,* χ_P^{PN} *That Using the NP Approximation*

Metal	(m_c/m)[a]	$(\chi_P{}^H/\chi_P{}^o)$[b]	$(\chi_P^{NP}/\chi_P{}^o)$[c]	$\chi_P^{exp}/\chi_P{}^o$
Li	1.66	3.26	2.60	2.57 ± 0.14[d]
Na	1.00	1.48	1.29	1.58 ± 0.25[e]

[a] Ham (1962).
[b] Rice (1965).
[c] Silverstein (1963), as corrected by Rice (1965).
[d] Schumacher and Slichter (1956).
[e] An average of the following results for χ_P^{exp} (in c.g.s. $\times 10^6$): 0.95 ± 0.10 [Schumacher and Slichter (1956)]; 0.89 ± 0.04 [Schumacher and Vehse (1960)]; 1.13 ± 0.05 [Schumacher and Vehse (1963)]. The substantial variation in results is due to the difficulty in choosing a base line for integrating over a Lorentzian line shape.

TABLE 5.7. *Comparison of Theory with Experiments for the Thermal Effective Mass. The Symbols Are Defined in the Text*

Metal	$\frac{m_c}{m}$	$\left(\frac{m^*_{el}}{m}\right)_H$	$\left(\frac{m^*_{el}}{m}\right)_{NP}$	$\frac{m^*_{ph}}{m}$	$\left(\frac{m^*}{m}\right)_H$	$\left(\frac{m^*}{m}\right)_{NP}$	$\left(\frac{m^*}{m}\right)_{exp}$
Li	1.66[a]	1.74	1.80				2.19[c]
Na	1.00[a]	1.06	1.10	1.18[b]	1.25	1.30	1.27[c]
K	1.09[a]		1.30				1.25[d]
Rb	1.21[a]		1.52				1.26[d]
Cs	1.76[a]		2.61				2.82[e]
Al	1.06[b]	1.05	1.09	1.49[b]	1.57	1.62	1.45[f]
Pb	1.12[b]	1.12	1.16	2.05[b]	2.30	2.38	2.00[g]

[a] Ham (1962).
[b] Ashcroft and Wilkins (1965).
[c] Martin (1961).
[d] Lien and Phillips (1964).
[e] Martin and Heer (1964).
[f] Otter and Mapother (1962).
[g] Decker, Mapother, and Shaw (1958); Mapother (private communication).

seem fortuitous in view of the disagreement for Na, which is so nearly isotropic.

We turn next to the specific heat of simple metals. The results of calculations based on Eq. (5.185) are given in Table 5.7, where they are compared with experiment; the effects of electron-phonon interaction have been included for Na, Al, and Pb. The subscripts H and NP refer to the Fermi liquid calculations using either the Hubbard or NP approximation procedures [Silverstein (1962), (1963); Rice (1965)].

The agreement with experiment is surprisingly good, considering the number of different calculations, each of an approximate nature, required in order to carry out the comparison between theory and experiment. It is clear that electron-phonon interactions give the dominant correction to the one-electron theory, particularly in the case of a "strong coupling" metal such as Pb. The thermal mass values obtained using the Hubbard approximation are sufficiently close to those using the NP approximation that present calculations do not indicate that one set is to be preferred to the other.

CONCLUSION

With this presentation of microscopic theories of the electron liquid, we conclude our study of normal Fermi liquids. The approximate theories of the behavior of an electron liquid at metallic densities, and the approximations used to relate these calculations to the observed properties of simple metals, represent a striking departure from the earlier parts of this book (and this chapter), in which the primary emphasis had been placed on deriving exact results, and exploring well-defined models. They represent a necessary departure if one hopes to pursue the consequences of electron interaction in real metals.

It is to be expected that with the development of more accurate approximation methods will come quantitative changes in some of the results we have quoted. We do not, however, expect future results to be qualitatively different: thus the strength of the quasiparticle interaction could turn out to differ by 30% from the results given here; we do not expect it to be changed by a factor of 3. We conclude that electron-electron interactions influence the spin susceptibility of metals to a significant and measurable extent, but that their influence on the specific heat is difficult to pin down experimentally, as it is small compared to the large measurable effect of the electron-phonon interactions. Finally, there is an appreciable change in the electron compressibility associated with electron-electron interactions; the effect is, however, difficult to put into evidence experimentally.

PROBLEMS

5.1. Derive the exact expressions (5.27)–(5.29) for the compressibility, spin susceptibility, and specific heat of a Fermi liquid.

5.2. Obtain an explicit analytic expression for the threshold wave vector, q_c, at which (within the RPA) it first becomes possible for a plasmon to decay into a particle-hole pair. Show that at metallic electron densities, Eq. (5.58) furnishes a satisfactory approximation.

5.3. Calculate in the Fermi–Thomas approximation the total electronic charge displaced by a negatively charged impurity. Show that at small distances from the impurity more charge is apparently pushed out than existed there in the absence of the impurity.

5.4. Use Eq. (5.100) to show that when one carries out a formal perturbation-theoretic expansion,

$$\chi_{\mathrm{RPA}} = \frac{1}{1 - (4\pi e^2/q^2)\chi^{\circ}} = 1 + \frac{4\pi e^2}{q^2}\chi^{\circ} + \left(\frac{4\pi e^2}{q^2}\right)^2 (\chi^{\circ})^2 + \cdots,$$

the term proportional to $\mathrm{Im}\,\chi^{\circ}$ yields the exchange energy, while that proportional to $\mathrm{Im}\,(\chi^{\circ})^2$ yields the second-order direct contribution to the ground state energy.

5.5. Derive the equation of motion, (5.153), for $\rho_{\mathbf{p}\mathbf{q}\sigma}$ in the generalized RPA.

5.6. Calculate the dispersion relation in the long wavelength limit for the transverse spin waves appropriate to a Fermi liquid which possesses a stable, magnetized ground state.

5.7. (i) Write down the interband and intraband oscillator strengths, in the long wavelength limit, for a system of noninteracting electrons in a solid.

Hint: Make use of the exact results of Problem 4.2.

(ii) Obtain an explicit expression for the density-density response function in this approximation, $\chi^c(\mathbf{q}, \omega)$.

(iii) The RPA for electrons in a solid is obtained by setting $\chi_{sc}(\mathbf{q}, \omega) = \chi^c(\mathbf{q}, \omega)$. Discuss the behavior of

$$\lim_{\mathbf{q}\to 0} \epsilon^c_{\mathrm{RPA}}(\mathbf{q}, 0) = 1 - \frac{4\pi e^2}{q^2}\chi^c(\mathbf{q}, 0)$$

for the following systems:

(a) Insulator.

(b) Semiconductor, for which conduction electrons form a classical system (the presence of the valence electrons must be taken into account).

(c) Conduction electrons in a metal.

(iv) Give a qualitative discussion of the behavior of $\epsilon^c_{\mathrm{RPA}}(0, \omega)$ for the above cases, as ω varies from a frequency less than any interband excitation frequency to one greater than any interband excitation frequency.

5.8. Calculate the value of $f_{\mathbf{pp}'}$ obtained in the approximation in which only the exchange energy and the second-order direct term are kept in the ground state energy.

5.9. Calculate the value of $f_{\mathbf{pp}'}$ obtained in the RPA, by differentiating directly the RPA expression for the ground state energy. Compare the result you obtain with Eq. (5.122), and with the result of Problem 5.8, and discuss the significance of your result.

5.10. (a) Show that the effective *magnetic* interaction between electrons is given by

$$H_{\mathrm{magn}} = \sum_{\substack{\mathbf{pp}' \\ \mathbf{q}\mu}} \frac{4\pi e^2(\mathbf{J_{pq}} \cdot \mathbf{n}_{\mathbf{q}\mu})(\mathbf{J_{p'q}} \cdot \mathbf{n}_{\mathbf{q}\mu})}{\omega^2 - \omega_p{}^2 - c^2q^2 - 4\pi e^2\chi_\perp^{\mathrm{RPA}}(\mathbf{q}, \omega)},$$

where

$$\mathbf{J_{pq}} = \frac{\mathbf{p}}{m} c^+_{\mathbf{p}} c_{\mathbf{p+q}}$$

is the current associated with a particle-hole pair of momentum $\mathbf{p}$.

Hint: Start with the Hamiltonian, (4.209), for electrons interacting with the transverse electromagnetic field, and follow the procedure analogous to that used to obtain the phonon-induced effective interaction between electrons.

(b) Show that for small q, the *static magnetic* interaction resembles the Coulomb interaction, but is attractive, and of order v^2/c^2 compared to the latter.

REFERENCES

Anderson, P. W. (1958), *Phys. Rev.* **112,** 1900.
Ashcroft, N. and Wilkins, J. (1965), *Phys. Letters* **14,** 285.
Bardeen, J. (1936), *Phys. Rev.* **50,** 1098.
Baym, G. and Kadanoff, L. P. (1962), *Quantum Statistical Mechanics*, W. A. Benjamin, New York.
Bohm, D. and Pines, D. (1953), *Phys. Rev.* **92,** 609.
Brooks, H. (1953), *Phys. Rev.* **91,** 1027.
Brooks, H. (1958), *Nuovo Cimento Suppl.* **7,** 165.
Brout, R. (1965), *Phase Transitions*, W. A. Benjamin, New York.
Brueckner, K. A. and Sawada, K. (1958), *Phys. Rev.* **112,** 328.
Decker, P. L., Mapother, D. E., and Shaw, R. W. (1958), *Phys. Rev.* **112,** 1888.
DuBois, D. F. (1959), *Ann. Phys.* **7,** 174.
Ehrenreich, H. and Cohen, M. (1959), *Phys. Rev.* **115,** 786.
Ferrell, R. A. (1957), *Phys. Rev.* **107,** 450.
Ferrell, R. A. (1959), *Phys. Rev. Letters* **1,** 443.
Fock, V. (1930), *Z. Physik* **61,** 126.
Friedel, J. (1958), *Nuovo Cimento Suppl.* **7,** 287.
Gell-Mann, M. (1957), *Phys. Rev.* **106,** 369.
Gell-Mann, M. and Brueckner, K. A. (1957), *Phys. Rev.* **106,** 364.
Glick, A. and Ferrell, R. A. (1959), *Ann. Phys.* **11,** 359.
Goldstone, J. and Gottfried, K. (1959), *Nuovo Cimento* [X] **13,** 849.
Ham, F. (1962), *Phys. Rev.* **128,** 2524.
Hartree, D. R. (1928), *Proc. Cambridge Phil. Soc.* **24,** 89.
Hubbard, J. (1957), *Proc. Roy. Soc. (London)* **A243,** 336.
Kadanoff, L. P. and Prange, R. (1964), *Phys. Rev.* **134,** A566.
Klimontovitch, Y. and Silin, V. P. (1952), *Zh. Exp. Teor. Fiz.* **23,** 151.
Kohn, W. and Vosko, S. (1960), *Phys. Rev.* **119,** 912.
Langer, J. and Vosko, S. (1960), *J. Phys. Chem. of Solids* **12,** 196.
Lien, W. H. and Phillips, N. E. (1964), *Phys. Rev.* **133,** A1370.
Lindhard, J. (1954), *Kgl. Danske Videnskab. Selskab., Mat-fys. Medd.* **28,** 8.
Macke, W. (1950), *Z. Naturforsch.* **5a,** 192.
Martin, B. D. and Heer, C. V. (1964), *Bull. Am. Phys. Soc.* **9,** 230.
Martin, D. L. (1961), *Proc. Roy. Soc.* **A263,** 378; *Phys. Rev.* **124,** 438.
Nozières, P. and Pines, D. (1958a), *Phys. Rev.* **111,** 442.
Nozières, P. and Pines, D. (1958b), *Nuovo Cimento* [X], **9,** 470.
Otter, F. A. and Mapother, D. E. (1962), *Phys. Rev.* **125,** 1171.
Overhauser, A. W. (1960), *Phys. Rev. Letters* **4,** 462.
Overhauser, A. W. (1962), *Phys. Rev.* **128,** 1437.
Pines, D. (1950), Ph.D. Thesis, Princeton Univ., unpublished.
Pines, D. (1953), *Phys. Rev.* **92,** 636.
Pines, D. (1955), *Advan. Solid State Phys.* **1,** 368.
Pines, D. (1960), *Physica* **26,** S103.
Pines, D. (1962), *The Many Body Problem*, W. A. Benjamin, New York.
Pines, D. (1963), *Elementary Excitations in Solids*, W. A. Benjamin, New York.
Pines, D. and Bohm, D. (1952), *Phys. Rev.* **85,** 338.
Quinn, J. J. and Ferrell, R. A. (1958), *Phys. Rev.* **112,** 812.
Quinn, J. J. and Ferrell, R. A. (1961), *J. Nucl. Energy, Pt. C,* **2,** 18.
Rice, M. (1965), *Ann. Phys.* **31,** 100.
Ritchie, R. N. (1959), *Phys. Rev.* **114,** 644.

Rowland, T. (1960), *Phys. Rev.* **119**, 900.
Sawada, K., Brueckner, K. A., Fukuda, N., and Brout, R. (1957), *Phys. Rev.* **108**, 507.
Schumacher, R. T., Carver, T., and Slichter, C. P. (1954), *Phys. Rev.* **95**, 1089.
Schumacher, R. T. and Slichter, C. P. (1956), *Phys. Rev.* **101**, 58.
Schumacher, R. T. and Vehse, W. E. (1960), *Bull. Am. Phys. Soc.* **4**, 296.
Schumacher, R. T. and Vehse, W. E. (1963), *J. Chem. Phys. Solids* **24**, 297.
Seitz, F. (1940), *Modern Theory of Solids*, McGraw-Hill, New York.
Silverstein, S. D. (1962), *Phys. Rev.* **128**, 631.
Silverstein, S. D. (1963), *Phys. Rev.* **130**, 912, 1703.
Simkin, D. (1963), Ph.D. Thesis, Univ. Illinois, unpublished.
Suhl, H. and Werthamer, R. N. (1961), *Phys. Rev.* **122**, 359.
Suhl, H. and Werthamer, R. N. (1962), *Phys. Rev.* **125**, 1402.
Ueda, S. (1961), *Prog. Theoret. Phys.* **26**, 45.
Watabe, M. (1962), *Prog. Theoret. Phys.* **28**, 265.
Wigner, E. P. (1934), *Phys. Rev.* **46**, 1002.
Wigner, E. P. (1938), *Trans. Farad. Soc.* **34**, 678.
Wigner, E. P. and Seitz, F. (1933), *Phys. Rev.* **43**, 804.
Wigner, E. P. and Seitz, F. (1934), *Phys. Rev.* **46**, 509.

APPENDIX*

SECOND QUANTIZATION

Let us consider a collection of N particles, contained in a box of volume Ω, on which we impose periodic boundary conditions. To describe the state of such a system, we ordinarily use the wave function $\psi(\mathbf{r}_1 \cdots \mathbf{r}_N)$, depending on the $3N$ coordinates of the particles in "configuration" space. Such a representation allows a natural generalization of methods applied to a single particle. It is not the only possible representation; there are others, much more practical.

Let us consider a complete set of *single-particle states*—for example, plane waves of momentum $\mathbf{p}$ (we ignore spin for the moment). One can specify the state of a system by indicating the number $n_\mathbf{p}$ of particles found in the "box" $\mathbf{p}$. We are thus led to choose as a basis for the space of *states of the total system* the vectors

$$|n_1, \ldots, n_\mathbf{p}, \ldots\rangle,$$

where each of the $n_\mathbf{p}$ can take on any positive integral value. (Note that the total number of particles is no longer conserved.) Such a state is simply a product (eventually symmetrized or antisymmetrized) of plane waves, one for each particle.

We can define a "creation" operator $a_\mathbf{p}^+$ by the relation

$$a_\mathbf{p}^+|\cdots, n_\mathbf{p}, \cdots\rangle = \sqrt{n_\mathbf{p}+1}\,|\cdots, n_\mathbf{p}+1, \cdots\rangle. \quad \text{(A.1)}$$

In the present representation, known as the "occupation-number" representation, $a_\mathbf{p}^+$ has a single nonzero matrix element. Its complex

* This appendix follows closely Appendix B of P. Nozières, *Theory of Interacting Fermi Systems*, W. A. Benjamin, New York (1963).

conjugate, or "destruction" operator, is defined by

$$a_{\mathbf{p}}|\cdots, n_{\mathbf{p}}, \cdots\rangle = \sqrt{n_{\mathbf{p}}}\,|\cdots, n_{\mathbf{p}}-1, \cdots\rangle. \quad \text{(A.2)}$$

Our objective is to express all the properties of the system in terms of these creation and destruction operators.

Let us first take a boson gas, and let us consider the product

$$a_{\mathbf{p}}^{+}a_{\mathbf{p}'}^{+}|\,\rangle.$$

We have created a particle $\mathbf{p}'$, then another, $\mathbf{p}$. The same result would be obtained by inverting the order, as the total wave function is invariant with respect to permutation of two particles. Hence we have the relation

$$[a_{\mathbf{p}}^{+}, a_{\mathbf{p}'}^{+}] = 0. \quad \text{(A.3)}$$

By the same method it is easily shown that

$$[a_{\mathbf{p}}, a_{\mathbf{p}'}] = 0, \qquad [a_{\mathbf{p}}, a_{\mathbf{p}'}^{+}] = \delta_{\mathbf{pp}'}. \quad \text{(A.4)}$$

The operator

$$N_{\mathbf{p}} = a_{\mathbf{p}}^{+}a_{\mathbf{p}} \quad \text{(A.5)}$$

measures the number of particles in the state $\mathbf{p}$ (since $N_{\mathbf{p}}|\cdots, n_{\mathbf{p}}, \cdots\rangle = n_{\mathbf{p}}|\cdots, n_{\mathbf{p}}, \cdots\rangle$).

We now consider the fermion gas; we denote the creation operator for a particle of momentum $\mathbf{p}$ by $c_{\mathbf{p}}^{+}$, the destruction operator by $c_{\mathbf{p}}$, and proceed to build up a set of states in a fashion directly analogous to the boson case. The difference between the two systems is manifested in the symmetry properties of the many-body wave function. That for fermions must be antisymmetric on the interchange of any two particles; the operators $c_{\mathbf{p}}$ must therefore satisfy *anticommutation relations*,

$$[c_{\mathbf{p}}^{+}, c_{\mathbf{p}'}^{+}]_{+} = [c_{\mathbf{p}}, c_{\mathbf{p}'}]_{+} = 0, \qquad [c_{\mathbf{p}}^{+}, c_{\mathbf{p}'}]_{+} = \delta_{\mathbf{pp}'}. \quad \text{(A.6)}$$

It can be seen from Eq. (A.6) that $c_{\mathbf{p}}^{2} = 0$; a state can accommodate only a single particle. Moreover, the occupation number operator

$$N_{\mathbf{p}} = c_{\mathbf{p}}^{+}c_{\mathbf{p}}$$

is such that $N_{\mathbf{p}}^{2} = N_{\mathbf{p}}$ [as can be verified with the help of Eqs. (A.6)]. Therefore, $N_{\mathbf{p}}$ can equal only 0 or 1.

We note that for fermions, the equations analogous to (A.1) and (A.2) read

$$c_{\mathbf{p}}|\cdots, n_{\mathbf{p}}, \cdots\rangle = (-1)^{\sum_{j<p} n_j}\, n_{\mathbf{p}}|\cdots, n_{\mathbf{p}}-1, \cdots\rangle, \quad \text{(A.1a)}$$

$$c_{\mathbf{p}}^{+}|\cdots, n_{\mathbf{p}}, \cdots\rangle = (-1)^{\sum_{j<p} n_j}\, [1-n_{\mathbf{p}}]|\cdots, n_{\mathbf{p}}+1, \cdots\rangle. \quad \text{(A.2a)}$$

The factor $(-1)^{\sum_{j<p} n_j}$ is needed to fix the sign of the many-fermion wave function.

Actually, fermions have a nonzero spin. Each state is characterized by a wave vector **p** and a spin index σ. The creation and destruction operators thus become $c^+_{\mathbf{p}\sigma}$ and $c_{\mathbf{p}\sigma}$. They satisfy anticommutation relations similar to Eq. (A.6):

$$\begin{aligned} [c_{\mathbf{p}\sigma}, c_{\mathbf{p}'\sigma'}]_+ &= [c^+_{\mathbf{p}\sigma}, c^+_{\mathbf{p}'\sigma'}]_+ = 0, \\ [c^+_{\mathbf{p}\sigma}, c_{\mathbf{p}'\sigma'}]_+ &= \delta_{\mathbf{p}\mathbf{p}'}\delta_{\sigma\sigma'}. \end{aligned} \tag{A.6a}$$

The normalized wave function for a plane wave of momentum **p** is

$$\frac{1}{\sqrt{\Omega}} e^{i\mathbf{p}\cdot\mathbf{r}}.$$

The probability amplitude for the destruction of a boson of momentum **p** taking place at the point **r** is thus represented by the operator

$$\frac{1}{\sqrt{\Omega}} e^{i\mathbf{p}\cdot\mathbf{r}} a_{\mathbf{p}}.$$

(with a corresponding expression for the case of a fermion). This leads us to introduce the destruction operator at the point **r**, $\psi(\mathbf{r})$, defined by

$$\begin{aligned} \psi(\mathbf{r}) &= \frac{1}{\sqrt{\Omega}} \sum_{\mathbf{p}} a_{\mathbf{p}} e^{i\mathbf{p}\cdot\mathbf{r}}, \\ a_{\mathbf{p}} &= \frac{1}{\sqrt{\Omega}} \int d^3\mathbf{r}\, \psi(\mathbf{r}) e^{-i\mathbf{p}\cdot\mathbf{r}}. \end{aligned} \tag{A.7}$$

The operator $\psi(\mathbf{r})$ describes the destruction at **r** of a particle of any momentum whatever. Similarly, the creation operator $\psi^+(\mathbf{r})$ is defined by

$$\psi^+(\mathbf{r}) = \frac{1}{\sqrt{\Omega}} \sum_{\mathbf{p}} a_{\mathbf{p}}^+ e^{-i\mathbf{p}\cdot\mathbf{r}}.$$

The commutation relations which ψ and ψ^+ satisfy are easily deduced from Eqs. (A.3) and (A.4) or from (A.6). They can be written as follows:

$$\begin{aligned} &\text{Bosons,} && [\psi(\mathbf{r}), \psi(\mathbf{r}')] = [\psi^+(\mathbf{r}), \psi^+(\mathbf{r}')] = 0, \\ & && [\psi(\mathbf{r}), \psi^+(\mathbf{r}')] = \delta(\mathbf{r}-\mathbf{r}'); \\ &\text{Fermions:} && [\psi_\sigma(\mathbf{r}), \psi_{\sigma'}(\mathbf{r}')]_+ = [\psi_\sigma^+(\mathbf{r}), \psi_{\sigma'}^+(\mathbf{r}')]_+ = 0, \\ & && [\psi_\sigma(\mathbf{r}), \psi_{\sigma'}^+(\mathbf{r}')]_+ = \delta(\mathbf{r}-\mathbf{r}')\delta_{\sigma,\sigma'}. \end{aligned} \tag{A.8}$$

In the case of fermions we have explicitly introduced spin coordinates. It is important not to confuse the *operators* ψ, ψ^+ with the state *vectors* $|\varphi\rangle$.

For the sake of simplicity, we shall consider only Fermi gases. Just as the operator $N_{\mathbf{p}} = c_{\mathbf{p}}^+ c_{\mathbf{p}}$ represents the number of particles in the state **p**, the operator

$$\rho(\mathbf{r}, \sigma) = \psi_\sigma^+(\mathbf{r})\psi_\sigma(\mathbf{r}) \tag{A.9}$$

represents the density of particles at the point **r**. The total number of particles in the system is given by

$$N = \sum_\sigma \int d^3\mathbf{r}\, \psi_\sigma^+(\mathbf{r})\psi_\sigma(\mathbf{r}) = \sum_{\mathbf{p}\sigma} c_{\mathbf{p}\sigma}^+ c_{\mathbf{p}\sigma}. \tag{A.10}$$

These results are extended without difficulty to the study of other physical properties of the system. For example, the kinetic energy

$$T = \sum_i \frac{p_i^2}{2m}$$

can be written in the form

$$T = \sum_{\mathbf{p},\sigma} \frac{p^2}{2m} N_{\mathbf{p}\sigma} = \sum_{\mathbf{p},\sigma} \frac{p^2}{2m} c_{\mathbf{p}\sigma}^+ c_{\mathbf{p}\sigma}. \tag{A.11}$$

By a Fourier transformation, we can go back to ψ, ψ^+. Equation (A.11) then becomes

$$T = -\frac{\hbar^2}{2m} \sum_\sigma \int \psi_\sigma^+(\mathbf{r})\nabla^2\psi_\sigma(\mathbf{r})\, d^3\mathbf{r} = \frac{\hbar^2}{2m} \sum_\sigma \int \nabla\psi_\sigma^+(\mathbf{r}) \cdot \nabla\psi_\sigma(\mathbf{r})\, d^3\mathbf{r}. \tag{A.12}$$

Let us turn now to the study of the potential energy V, which we assume results from a binary interaction, depending only on position,

$$V = \frac{1}{2} \sum_{i \neq j} V(\mathbf{r}_i - \mathbf{r}_j).$$

On using the expansion for a gas of point particles,

$$\rho(\mathbf{r}) = \sum_i \delta(\mathbf{r} - \mathbf{r}_i),$$

we can rewrite the interaction in the form

$$\begin{aligned} V &= \tfrac{1}{2} \iint d^3\mathbf{r}\, d^3\mathbf{r}'\, V(\mathbf{r} - \mathbf{r}')[\rho(\mathbf{r})\rho(\mathbf{r}') - \rho(\mathbf{r})\delta(\mathbf{r} - \mathbf{r}')], \\ \rho(\mathbf{r}) &= \sum_\sigma \rho(\mathbf{r}, \sigma), \end{aligned} \tag{A.13}$$

where the second term eliminates the interaction of a charge with itself. By using (A.9) and the commutation relations, we can transform (A.13) into

$$V = \frac{1}{2} \sum_{\sigma,\sigma'} \iint d^3\mathbf{r}\, d^3\mathbf{r}'\, V(\mathbf{r} - \mathbf{r}')\psi_\sigma^+(\mathbf{r})\psi_{\sigma'}^+(\mathbf{r}')\psi_{\sigma'}(\mathbf{r}')\psi_\sigma(\mathbf{r}). \quad \text{(A.14)}$$

In general we shall prefer an expression using the operators $c_\mathbf{p}$ rather than the operators $\psi(\mathbf{r})$. Let us introduce the Fourier transform of the binary potential,

$$V_\mathbf{q} = \int d^3\mathbf{r}\, e^{-i\mathbf{q}\cdot\mathbf{r}} V(\mathbf{r}). \quad \text{(A.15)}$$

Equation (A.14) can then be written as

$$V = \frac{1}{2\Omega} \sum_{\mathbf{pkq},\sigma,\sigma'} V_\mathbf{q} c^+_{\mathbf{p}+\mathbf{q},\sigma} c^+_{\mathbf{k}-\mathbf{q},\sigma'} c_{\mathbf{k},\sigma'} c_{\mathbf{p},\sigma}. \quad \text{(A.16)}$$

From Eq. (A.16) and the commutation relations, (A.6a), it follows that the matrix element for scattering a pair of fermions $(\mathbf{p}, \sigma; \mathbf{k}, \sigma')$ to new states $(\mathbf{p} + \mathbf{q}, \sigma; \mathbf{k} - \mathbf{q}, \sigma')$ is given by

$$\langle \cdots ; \mathbf{p} + \mathbf{q}, \sigma; \mathbf{k} - \mathbf{q}, \sigma'; \cdots |V| \cdots ; \mathbf{p}, \sigma; \mathbf{k}; \sigma'; \cdots \rangle = V_\mathbf{q} - V_{\mathbf{k}-\mathbf{q}-\mathbf{p}} \delta_{\sigma\sigma'}, \quad \text{(A.17)}$$

provided the states $\mathbf{p}, \mathbf{k}$ are occupied, and those $\mathbf{p} + \mathbf{q}$, $\mathbf{k} - \mathbf{q}$ are empty. The first term is that for direct scattering, the second corresponds to an exchange scattering of the pair, and occurs only if the particles have the same spin. The analogous expression for spin zero bosons is

$$V_\mathbf{q} + V_{\mathbf{k}-\mathbf{p}-\mathbf{q}}; \quad \text{(A.18)}$$

the difference is again a matter of statistics, that is, of the symmetry of the many-particle wave function.

It is easy to calculate in the same way the expression for any physical quantity in this representation (current, etc.). For example, let us consider the Fourier component of the density,

$$\rho_\mathbf{q} = \int d^3\mathbf{r}\, \rho(\mathbf{r}) e^{-i\mathbf{q}\cdot\mathbf{r}}. \quad \text{(A.19)}$$

Because of Eqs. (A.7) and (A.9), this can be written as

$$\rho_\mathbf{q} = \sum_{\mathbf{p},\sigma} c^+_{\mathbf{p}\sigma} c_{\mathbf{p}+\mathbf{q},\sigma}. \quad \text{(A.20)}$$

In the same way, the current density, $\mathbf{J_q}$, becomes

$$\mathbf{J_q} = \sum_{\mathbf{p},\sigma} \frac{\mathbf{p}}{m} c^+_{\mathbf{p}\sigma} c_{\mathbf{p}+\mathbf{q},\sigma}. \tag{A.21}$$

In conclusion, let us remind the reader of the rules for quantizing lattice waves (phonons). The basic Hamiltonian for the lattice waves is

$$H = \sum_{\mathbf{q}} \frac{P_{\mathbf{q}}{}^+ P_{\mathbf{q}}}{2} + \omega_{\mathbf{q}}{}^2 \frac{Q_{\mathbf{q}}{}^+ Q_{\mathbf{q}}}{2}, \tag{A.22}$$

where $P_{\mathbf{q}}$ and $Q_{\mathbf{q}}$ are the momentum and coordinate, respectively, of the longitudinal normal modes of the lattice field. They satisfy the harmonic oscillator commutation rules,

$$\begin{aligned} [P_{\mathbf{q}}, P_{\mathbf{q}'}] &= [Q_{\mathbf{q}}, Q_{\mathbf{q}'}] = 0, \\ [P_{\mathbf{q}}, Q_{\mathbf{q}'}] &= -i\delta_{\mathbf{q}\mathbf{q}'}. \end{aligned} \tag{A.23}$$

The transformation to a *phonon* representation, in which the operators which act to create or destroy a phonon, $a_{\mathbf{q}}{}^+$ and $a_{\mathbf{q}}$, are introduced explicitly, is given by

$$\begin{aligned} P_{\mathbf{q}} &= i\left(\frac{\omega_{\mathbf{q}}}{2}\right)^{1/2} (a_{\mathbf{q}}{}^* - a_{-\mathbf{q}}), \\ Q_{\mathbf{q}} &= \frac{1}{(2\omega_{\mathbf{q}})^{1/2}} (a_{\mathbf{q}} + a^+_{-\mathbf{q}}). \end{aligned} \tag{A.24}$$

The reader may easily verify that the operators $a_{\mathbf{q}}$ and $a_{\mathbf{q}}{}^+$ satisfy the boson commutation relations, (A.1), and that the Hamiltonian (A.22) takes the form

$$H = \sum_{\mathbf{q}} \omega_{\mathbf{q}}(a_{\mathbf{q}}{}^+ a_{\mathbf{q}} + \tfrac{1}{2}). \tag{A.25}$$

The interaction between electrons and lattice vibrations thus takes the form (in the second quantized representation)

$$H_{\text{int}} = \sum_{\mathbf{q}} v_{\mathbf{q}}{}^i \rho_{\mathbf{q}}{}^+ Q_{\mathbf{q}} = \sum_{\mathbf{q}\mathbf{p}\sigma} \frac{v_{\mathbf{q}}{}^i}{\sqrt{2\omega_{\mathbf{q}}}} c^+_{\mathbf{p}+\mathbf{q},\sigma} c_{\mathbf{p}\sigma} (a_{\mathbf{q}} + a^+_{-\mathbf{q}}). \tag{A.26}$$

Let us emphasize that we have discovered nothing new in this appendix: we have just chosen a particularly convenient representation for treating the problems of interest to us.

INDEX